Applied Factor Analysis in the Natural Sciences

Applied Factor Analysis in the Natural Sciences

RICHARD A. REYMENT
K. G. JÖRESKOG
University of Uppsala
Appendix by LESLIE F. MARCUS
American Museum of Natural History
Queens College, CUNY

CAMBRIDGE
UNIVERSITY PRESS

PUBLISHED BY THE PRESS SYNDICATE OF THE UNIVERSITY OF CAMBRIDGE
The Pitt Building, Trumpington Street, Cambridge CB2 IRP

CAMBRIDGE UNIVERSITY PRESS
The Edinburgh Building, Cambridge CB2 2RU, United Kingdom
40 West 20th Street, New York, NY 10011-4211, USA
10 Stamford Road, Oakleigh, Melbourne 3166, Australia

Original edition published in 1976 by Elsevier Scientific Publishing Company, Amsterdam
Second edition published by Cambridge University Press 1993

First published 1993
First paperback edition 1996

Printed in the United States of America

Library of Congress Cataloging-in-Publication Data is available.

A catalog record for this book is available from the British Library.

ISBN 0-521-41242-0 hardback
ISBN 0-521-57556-7 paperback

Contents

Preface

Sixteen years have elapsed since the appearance of *Geological Factor Analysis*. The book has been out of print for several years now but it is still frequently cited in the literature. A Russian translation was issued nine years ago. In response to repeated requests from around the world, we have decided to make our work available again, but this time with a wider scope so as to cater for the needs and interests of a greater range of natural scientists.

Although the information presented in the First Edition is far from outmoded, the past few years have witnessed an increasing awareness of several topics of fundamental importance in applied multivariate statistics such as, for example, the stability of eigenvectors, the identification of atypical and influential observations, the analysis of compositional data, the rise of tensor biometry as a valuable tool in evolutionary biology, and canonical correspondence analysis.

In the Preface to the First Edition, we made it quite clear that we used the term *factor analysis* in the vernacular mode. Despite all caution, however, we were completely misunderstood by *one* geological reviewer. We therefore repeat most emphatically that "factor analysis" is used here in an informal manner to signify the statistical application of the algebra of eigenvalues and eigenvectors to the analysis of single multivariate samples. To a certain degree, this usage agrees with *l'analyse factorielle des correspondances* of the francophone literature.

The rise of the personal computer has brought the possibilities of computing to anybody who so wishes. It is becoming increasingly common to supply a diskette of programs with texts in applied statistics. For this reason, Professor Leslie F. Marcus (New York) has joined us.

Aspects of the revised work have been commented upon by several colleagues. Among these we wish to mention Professor H. J. B. Birks, Botany Department, University of Bergen, Norway, Professor F. James Rohlf, SUNY, New York, and Dr. Fred L. Bookstein, Center for Human Growth, University of Michigan. We wish to thank Linda Pratt of Science Typographers, New York, for sterling assistance.

R. A. Reyment and K. G. Jöreskog
Uppsala

Glossary of the most commonly used symbols

N = the number of objects (specimens, observations) in a sample; it denotes the size of the sample

p = the number of variables (characters, attributes)

k = the number of factors

$\mathbf{X}$ = the data matrix (the order of which is $N \times p$)

$\mathbf{Z}$ = the standardized data matrix

$\mathbf{W}$ = the row-normalized data matrix of Imbrie Q-mode factor analysis

$\mathbf{R}$ = the sample correlation matrix

$\mathbf{S}$ = the sample covariance matrix

$\mathbf{\Sigma}$ = the population covariance matrix

$\mathbf{H}$ = the association matrix of some Q-mode methods

$\mathbf{\Lambda}$ = the population and sample diagonal matrix of eigenvalues; the diagonal elements of this matrix are λ_i

$\mathbf{U}$ = the sample matrix of eigenvectors as columns

$\mathbf{F}$ = the matrix of factor scores

$\mathbf{A}$ = the matrix of factor loadings

$\mathbf{E}$ = the matrix of residuals or error terms

Preface To First Edition

Our decision to write this book stems from the fact that the kind of analysis involved in what we group under the heading of "Geological Factor Analysis" has become on of the most frequently used sets of multivariate statistical techniques in geology and the general concepts of which many geologists have at least a vague understanding. In putting all of these techniques into the same bag, we recognize the fact that the term "factor analysis" has come to be applied by geologists to a particular kind of analytical procedures of which only a few belong to the classical factor model such as it is conceived by psychometricians.

It is the aim of our text to introduce students of geology to the powerful technique of factor analysis and to provide them with the background necessary in order to be able to undertake analyses on their own. For this reason, we have tended perhaps to be overexplicit when dealing with the introductory requirements for understanding the calculations and to have paid less attention to theoretical details. Clearly, we have definitely not written a text for statisticains.

The analysis of homogeneous multivariate populations in the earth sciences has grown into a primary research branch of almost unlimited potential; this development, largely made possible by the rise of the electronic computer, has greatly altered methodology in the petroleum industry, mining geology, geochemistry, stratigraphy, paleontology, chemical geology, environmental geology, sedimentology, and petrology.

Chapter 1 introduces the concept of multivariate data analysis by factoring methods. In Chapter 2, we present the basic concepts of multivariate algebra (linear algebra, matrix algebra) and the most commonly occurring matrix arithmetic operations of factor analysis. We also consider the rotation of coordinate systems and the role of eigenvalues and eigenvectors.

In Chapters 3 and 4, we take up theoretical concepts of factor analysis and the statistical interpretation of models, in which the presentation is made in terms of fixed mode and random mode: we give an account of the methods of principal components and "true" factor analysis. In Chapter 5, Q-mode factor analysis, principal coordinates, and correspondence analysis are presented. Chapter 6 is concerned with every day practical problems you are liable to run into when you plan and carry out factor analyses. Here, such diverse topics as selection of the most suitable method, choosing the number of meaningful factors, and data transformations, are discussed.

Chapter 7 takes you through examples of each of the major techniques. We end up this section with a set of reviews of randomly chosen applications of factor analysis from the literature.

It is hoped that this book, the result of the collaboration between a professional statistician and two geologists with experience in various fields of application of the methods presented here, will prove useful to students and research workers alike.

It is expected that most users of our text will have little knowledge of the general field of statistics and we have, therefore, chosen to develop our subject at an elementary level. We wish, nevertheless, to recommend strongly to those who lack a background in statistics to do some introductory reading in the subject.

The opportunity of testing our approach to Factor Analysis was given to us in the Spring of 1974, when we gave a postgraduate course on the subject at Uppsala University. This cooperation at the teaching level provided a valuable sequel to the year of preparation behind the book and helped to iron out difficulties concerning presentation of the text. Thanks to a grant from the Natural Research Council of Canada, Klovan was able to spend the greater part of the Academic year of 1973–1974 at Uppsala.

Several colleagues have aided in furthering the development of the book. In particular, we wish to mention Mr. Hans-Åke Ramdén, Uppsala Datacentral, Mr. John Gower, Rothamsted Experimental Station, U.K., Dr. M. David, Ecole Polytechnique, Montréal, Canada, Dr. M. Hill, Natural Environment Research Council, Bangor, U.K., Dr. A. T. Miesch, U.S. Geological Survey, Denver, U.S.A., and Mr. Colin Banfield, Rothamsted, U.K.

K. G. Jöreskog, J. E. Klovan, R. A. Reyment

1 Introduction

1.1 STRUCTURE IN MULTIVARIATE DATA

Commonly, almost all natural scientists make a great number of measurements in their daily activities, for example, on the orientation of strata, geochemical determinations, mineral compositions, rock analyses, measurements on fossil specimens, properties of sediments, ecological factors, genetics, and many other kinds. You need only reflect on the routine work of a geological survey department in order that the truth of this statement may become apparent.

Scientific data are often multivariate. For example, in a rock analysis, determinations of several chemical elements are made on each rock specimen of a collection. You will all be familiar with the tables of chemical analyses that issue from studies in igneous petrology and analytical chemistry. Petrologists have devised many kinds of diagrams in their endeavor to identify significant groupings in these data lists. The triangular diagrams of petrology permit the relationships between three variables at a time to be displayed. Attempts at illustrating more highly multivariate relationships have led to the use of ratios of elements and plots on polygonal diagrams (cf. Aitchison, 1986).

Obviously, one can only go so far with the graphical analysis of a data table. The logical next step is to use some type of quantitative method for summarizing and analyzing the information hidden in a multivariate table. It is natural to enquire how the variables measured for a homogeneous sample are connected to each other and whether they occur in different combinations, deriving from various relationships in the population. One may, on the other hand, be interested in seeing how the specimens or objects of the sample itself are interrelated, with the thought in mind of looking for natural groupings. In both cases, we should be seeking structure in the data.

Geologists and biologists came into touch with the concept of factor analysis and the study of multivariate data structure through the contacts between paleontologists and biologists. The biologists, in their turn, learnt the techniques from psychometricians. Thus, the French zoologist Teissier studied multivariate relationships in the carapace of a species of crabs (Teissier, 1938), using a centroid first-factor solution of

1

a correlation matrix. He interpreted this "general factor" as one indicating differential growth.

Let us now look briefly at a few typical problems that may be given meaningful solutions by an appropriately chosen variety of eigenanalysis.

A geochemist has analyzed several trace elements in samples of sediment from a certain area and wishes to study the relationships between these elements in the hope of being able to draw conclusions on the origin of the sediment.

A mining geologist is interested in prospecting an area for ores and wants to use accumulated information on the chemistry and structural geology of known deposits in the region to help predict the possibilities of finding new ore bodies.

A paleontologist wishes to analyze growth and shape variation in the shell of a species of brachiopods on which he has measured a large number of characters.

A petroleum company wants to reduce the voluminous accumulations of data deriving from paleoecological and sedimentological studies of subsurface samples to a form that can be used for exploring for oil-bearing environments.

In an oceanological study, it is desired to produce graphs that will show the relationships between bottom samples and measurements made upon them on a single diagram, as a means of relating organisms to their preferences for a particular kind of sediment.

1.2 AN EXAMPLE OF PRINCIPAL COMPONENT FACTOR ANALYSIS

At this point, we think it would be helpful to you if we gave you an inkling of what is obtained in a principal component factor analysis. (The reason for making this distinction will become clear later on.) We have chosen an artificial mining example by Klovan (1968) because it not only introduces the geological element at an early stage but also provides a good practically oriented introduction to the subject.

Imagine the following situation. We wish to carry out exploration for lead and zinc in an area containing a high-grade lead–zinc ore. The area has been well explored geologically and the bedrock is made up of an altered carbonate–shale sequence. The map area and the sampling grid are displayed in Fig. 1.1.

The three controls, paleotemperature (T), strength of deformation of the bedrock (D), and the permeability of the rock (P) are considered to determine the occurrence of lead and zinc, for the purposes of our example. It is assumed that these controls are determinable from

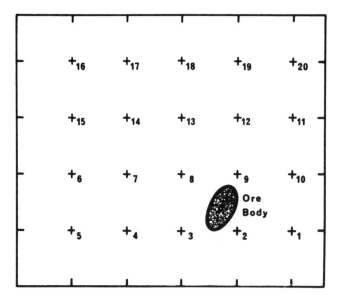

Figure 1.1. The sampling grid for the prospecting example.

observations on 10 chemical, mineralogical, and rock-deformational variables. The distribution of these causes will in reality never be known but, for this example, we shall imagine that they are distributed as shown in Fig. 1.2. You will note that the lode lies at the intersection of these causes at certain specified levels. These are, for paleotempera-ture, 80–90, for deformation, 35–45, and for permeability of the country rock, 45–50. Accepting that a geological survey of the area would have given as clear an indication as our manufactured example, it would not be unreasonable to expect that target areas for intensive prospecting would occur in localities where the intersection situation is repeated.

The three controls cannot, of course, be estimated directly. They can, however, be measured indirectly from geological properties that are a reflection of them. The arrays shown in Table 1.I list the artificial data, as well as the information used in constructing this set of observations. The left array of numbers gives the "amount" of each of the three controls at each of the localities; the upper array states precisely the degree to which each of the geological variables is related to the causes. Multiplication and summation of every row of the left array by every column of the top array yields the large array (corresponding to raw data) at the bottom. Naturally, in a real study, you would not know the left-hand and top arrays of Table 1.I. All you would have at your disposal would be the large array, or data matrix, the result of a detailed geological survey and a laboratory study of the samples col-lected.

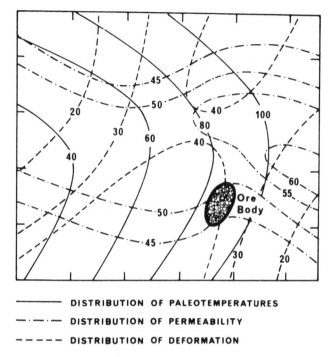

DISTRIBUTION OF PALEOTEMPERATURES
— · — · — DISTRIBUTION OF PERMEABILITY
— — — — DISTRIBUTION OF DEFORMATION

Figure 1.2. Distribution of controls imposed at the locations of the samples.

The question to be answered now is, how can we determine the existence of structure in such a large array of numbers? The technique of factor analysis turns out to be a useful way of providing plausible answers.

Simply put, factor analysis creates a minimum number of new variables, which are linear combinations of the original ones such that the new variables contain most or all of the information.

The starting point is provided by the correlations between the variables measured, 10 in all. The matrix of correlation coefficients is listed in Table 1.II. It was subjected to principal component factor analysis for which three significant factors were obtained. Thus, we began with 10 characters but can now "explain" the total variability of the sample in terms of 3 new variables or factors.

The principal-factor matrix is listed in Table 1.III; it shows the "composition" of the factors in relation to the original variables. As these factors are usually not readily interpretable, it is accustomed practice to rotate the reference axes by some appropriate method in order to bring out the important contributing loadings and to diminish the loadings on nonsignificantly contributing variables. The visual result

Table 1.1. *The raw data matrix for the lead–zinc prospecting problem*

		Causes			Geological properties									
					Mg in calcite	Fe in sphalerite	Na in muscovite	Sulfide	Crystal size of carbonates	Spacing of cleavage	Elongation of ooliths	Tightness of folds	Vein material per m²	Fractures per m²
Causes	T				0.95	0.75	0.75	0.33	−0.20	0.05	0.20	0.10	0.00	0.05
	D				0.00	0.10	0.20	0.33	0.60	0.95	0.70	0.85	0.10	0.25
	P				0.05	0.15	0.05	0.34	0.60	0.00	0.10	0.05	0.90	0.70
Locality	T	D	P				Data matrix							
1	121	21	46	117.25	99.75	97.25	62.50	16.00	26.00	43.50	32.25	43.50	43.50
2	96	35	42	93.30	81.80	81.10	57.51	27.00	38.05	47.90	41.45	41.30	42.95
3	78	54	49	76.55	71.25	71.75	60.22	46.20	55.20	58.30	56.15	49.50	51.70
4	63	51	49	62.30	59.70	59.90	54.28	47.40	51.60	53.20	52.10	49.20	50.20
5	42	44	44	42.10	42.50	42.50	43.34	44.40	43.90	43.60	43.80	44.00	43.90
6	39	26	54	39.75	39.95	37.15	39.81	40.20	26.65	31.40	28.70	51.20	46.25
7	52	36	52	52.00	50.40	48.80	46.72	42.40	36.80	40.80	38.40	50.40	48.00
8	67	46	54	66.35	62.95	62.15	55.65	46.60	47.05	51.00	48.50	53.20	52.65
9	90	37	51	88.05	78.85	77.45	59.25	34.80	39.65	49.00	43.00	49.60	49.45
10	108	27	61	105.65	92.85	89.45	65.29	31.20	31.05	46.60	36.80	57.60	54.85
11	112	33	59	109.35	96.15	93.55	67.91	32.80	36.95	51.40	42.20	56.40	55.15
12	91	38	59	89.40	80.90	78.80	62.63	40.00	40.65	50.70	44.35	56.90	55.35
13	76	39	54	74.90	69.00	67.50	56.31	40.60	40.85	47.90	43.45	52.50	51.35
14	63	30	51	62.40	57.90	55.80	48.03	36.00	31.65	38.70	34.35	48.90	46.35
15	43	19	55	43.60	42.40	38.80	39.16	35.80	20.20	27.40	23.20	51.40	45.40
16	68	16	42	66.70	58.90	56.30	42.00	21.20	18.60	29.00	22.50	39.40	36.80
17	77	27	41	75.20	66.60	65.20	48.26	25.40	29.50	38.40	32.70	39.60	39.30
18	93	37	43	90.50	79.90	79.30	57.52	29.40	39.80	48.80	42.90	42.40	44.00
19	102	47	48	99.30	88.40	88.30	65.49	36.60	49.75	58.10	52.55	47.90	50.45
20	120	36	46	116.30	100.50	99.50	67.12	25.20	40.20	53.80	44.90	45.00	47.20

Table 1.II. Correlations among the 10 geological properties

	Mg	Fe	Na	S	Crystal	Cleavage	Ooliths	Folds	Veins	Fractures
1	1.000									
2	0.998	1.000								
3	0.994	0.998	1.000							
4	0.905	0.931	0.941	1.000						
5	−0.574	−0.519	−0.493	−0.171	1.000					
6	0.125	0.178	0.231	0.479	0.628	1.000				
7	0.572	0.618	0.657	0.833	0.274	0.883	1.000			
8	0.275	0.328	0.377	0.610	0.533	0.988	0.945	1.000		
9	0.011	0.056	0.035	0.288	0.541	0.196	0.228	0.222	1.000	
10	0.256	0.312	0.313	0.593	0.556	0.540	0.616	0.588	0.907	1.000

Table *1.III. Results of the factor analysis*

Variable	Communality	Factors		
		1	2	3
Principal factors of the correlation matrix				
1	1.0000	0.7933	−0.6029	0.0850
2	1.0000	0.8302	−0.5501	0.0902
3	1.0000	0.8505	−0.5247	0.0373
4	1.0000	0.9742	−0.2073	0.0892
5	1.0000	0.0376	0.9992	−0.0111
6	1.0000	0.6609	0.5989	−0.4522
7	1.0000	0.9310	0.2365	−0.2781
8	1.0000	0.7693	0.5001	−0.3976
9	1.0000	0.3389	0.5369	0.7726
10	1.0000	0.6765	0.5364	0.5046
Variance		54.614	31.928	13.459
Cumulative variance		54.614	86.542	100.000
Varimax factor matrix				
1	1.0000	0.9974	0.0716	0.0056
2	1.0000	0.9916	0.1201	0.0479
3	1.0000	0.9839	0.1771	0.0245
4	1.0000	0.8776	0.3992	0.2656
5	1.0000	−0.6206	0.5952	0.5105
6	1.0000	0.0533	0.9882	0.1435
7	1.0000	0.5125	0.8391	0.1823
8	1.0000	0.2055	0.9637	0.1705
9	1.0000	0.0014	0.0534	0.9986
10	1.0000	0.2227	0.4059	0.8864
Variance		44.66	33.41	22.00

of the rotation will then be that some of the loadings will have been augmented while others will have become greatly lower. In our example, we used the varimax rotation technique. The varimax factor matrix displayed in Table 1.III demonstrates what we have just described, and you will see this if you compare entries in the two upper listings of the table, entry by entry. The rotated factor matrix contains 10 rows and 3 columns, each latter representing a factor. Reading down a column, the individual numbers tell us the contribution of a particular variable to the composition of the factor; in fact, each column can be thought of as a factor equation in which each loading is the coefficient of the corresponding original variable.

Table *1.III* (*cont.*)

Locality	Factors		
	1	2	3
Varimax factor score matrix			
1	1.6820	−1.1085	−0.8445
2	0.5988	0.2442	−1.3125
3	−0.2144	1.8308	0.0837
4	−0.7787	1.5001	0.0481
5	−1.5367	0.8901	−0.8567
6	−1.5277	−1.0713	0.5476
7	−1.0955	−0.0325	0.3471
8	−0.5739	0.9167	0.8009
9	0.3768	0.2012	0.1881
10	1.1764	−0.9354	1.6917
11	1.2853	−0.3045	1.4412
12	0.4317	0.1081	1.5106
13	−0.1745	0.2776	0.7082
14	−0.6279	−0.5465	0.1032
15	−1.3205	−1.7494	0.6197
16	−0.3619	−1.6486	−1.5528
17	−0.0900	−0.5506	−1.5758
18	0.4709	0.4016	−1.1222
19	0.7671	1.2609	−0.1759
20	1.5394	0.3161	−0.6498

A third chart of numbers emerges from the factor analysis, the varimax factor score matrix, shown in Table 1.III. This gives the amounts of the new variables at each of the sample localities. With this matrix, we are able to map the distributions of these new factor variables on the sample grid.

It requires sound geological reasoning in order to interpret the results of a factor analysis. From Table 1.III, you will see that the first factor is mainly concerned with the variables "Mg in calcite," "Fe in sphalerite," and "Na in muscovite," a combination indicating temperature dependence. The second factor is heavily loaded with the variables "spacing of cleavage," "elongation of ooliths," and "tightness of folds," a combination speaking for rock deformation. The third factor is dominated by the variables "vein material/m^2" and "fractures/m^3," interpretable as being a measure of permeability of the country rock.

The distribution of the three sets of factor scores is shown in Fig. 1.3. The patterns of Fig. 1.2 are almost exactly duplicated. By comparing the nature of the intersections around the known ore body, and searching the diagram for a similar pattern, you will see that at least one other

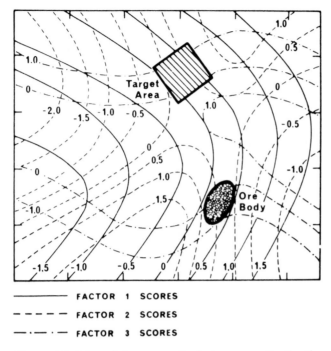

FACTOR 1 SCORES

FACTOR 2 SCORES

FACTOR 3 SCORES

Figure 1.3. Map of the composite factor scores for the three factors of the ore-prospecting example.

area on the map has the same special conditions. The marked square is thus the first-order target for further exploration. This is an artificial example, contrived to give a good result. Under actual exploration conditions, you would not expect things to fall out so nicely and your geological knowledge would be put to a greater test. This is only a brief summary of the example. If you want to work through all the steps, we refer you to Klovan (1968).

You will find a good deal of simple algebra in the ensuing pages, much of which you might think unnecessary, bearing in mind the ubiquitousness of computer programs for doing most of the things occurring in factor analysis. Obviously, nobody today is going to suggest seriously that you try inverting matrices, or extracting eigenvalues and eigenvectors, on your own, although we have devoted some space to the arithmetic of this topic. There are numerous excellent programs for doing these calculations at any computer installation and a wide selection of PC software. It is our considered opinion, nevertheless, that you should have some idea of what is done by the computer in performing these operations. Moreover, despite the fact that many varieties of factor analysis are available at most installations, not all of them are to be recommended for general use. We have therefore made a point of introducing you to the most useful and mathematically best defined procedures so that you will be able to make a satisfactory choice among the programs for factor analysis available to you.

Several recent updatings of principal component analysis have appeared. It is relevant to our revision to see what topics have been taken up in those texts.

Jolliffe (1986) covered some of the fields mapped out in the First Edition. He included the highly appropriate data-analytical topics of robust estimation procedures, determination of the "correct" number of principal components, influential observations, and the isolation of atypicalities in the data. He correctly noted that although *true factor analysis* and *principal component analysis* may, in some respects, have similar aims, they are different techniques. This was clearly enunciated in the First Edition, but we opted for a nonspecific data-analytical use of the concept of "factor analysis," which permits greater latitude in discussing methods that reduce to an eigensolution.

One of the bones of contention in principal component analysis is the question of *rotating* eigenvectors in the manner usually thought proper to true factor analysis. Jolliffe (1986, p. 118) observed that there may be circumstances in which rotation of a subset of principal components can prove advantageous – the main positive effect of this maneuver is that it tends to simplify the factor loadings, or rotated principal component coefficients, without the implication that a factor model is being assumed. The most recent treatment of this subject is Chapter 8 in Jackson (1991).

The volume by Flury (1988) provides an uncontroversial introduction to principal component analysis. Such debatable questions as the rotation of eigenvectors are not discussed at all. The main theme of that text concerns *common* principal component analysis, a variant of standard principal components, in which attention is paid to differences in covariance matrices due to "pooling" (hence the adjective "common") and the distributional theory of eigenvalues and eigenvectors. Canonical correlation is included as a peripheral technique.

The book by Preisendorfer (1988), completed posthumously by C. D. Mobley, gives an interesting insight into applications in meteorology and oceanography. Canonical correlation is also included among the techniques considered. Rather surprisingly, Preisendorfer's way of treating some problems discloses ignorance of well-established methods of multivariate statistical analysis as well as a lack of familiarity with standard terminology. This deficiency occurs despite the fact that the appropriate references appear in his bibliography. An example is the problem of testing that an eigenvector agrees with a given vector. Factor analysis is also included but without exemplifying the role of true factor analysis (perhaps weather forecasting could yield a suitable example). Rotation of eigenvectors is accepted as a matter of course by Preisendorfer as is also Procrustean superposition of "vector frames."

The volume compiled by Fornell (1982) contains several valuable pointers to future areas of development. One of the illuminating topics is that concerning the revival of interest in the method of *path analysis* of Sewall Wright. Other subjects covered are canonical correlation and the analysis of redundancy. This topic is also taken up by Jöreskog and Sörbom (1989).

Another informative volume is that of Digby and Kempton (1987). This is an ecologically oriented treatment in the hands of statisticians. It encompasses many of the methods that we gather beneath the umbrella of factor analysis and that were included in the First Edition: these include methods of ordination, principal coordinates, correspondence analysis, and the analysis of asymmetry. A very recent reference is the text on principal component analysis by Jackson (1991). As a matter of interest, the rotation of axes to simple structure is not disputed as being a useful technique.

The slim volume by Gordon (1981) is a compact reference, replete with essential information for the methods of principal components, principal coordinates, correspondence analysis, and Gabriel's biplot. Useful sources of information on factor analysis in the geosciences are to be found in the journals *Mathematical Geology* and *Computers & Geosciences*. Articles of biological significance appear regularly in *Biometrics*, *Biometrika*, and *Evolution*. The applied scientist can find much of interest in the pages of *Applied Statistics*, in which problems of biological relevance appear frequently.

1.3 OVERVIEW OF PROBLEMS AMENABLE TO TREATMENT BY FACTOR-ANALYTICAL TECHNIQUES

The present section reviews briefly a randomly chosen set of articles in which factor analysis of some variety has been used in order to solve a scientific problem. We do this in the hope that this will give you a better insight into the types of problems amenable to treatment by the factor class of techniques.

Relationships between organisms and sedimentary facies

In a study of the Pleistocene–Holocene environment of the northwestern part of the Persian Gulf, Melguen (1971) used correspondence analysis to explore relationships between ecological and sedimentological facies in the estuaries of the Rud Hilla and Rud Mund. The study material was derived from sediment cores taken from depths ranging between 8 and 60 m. Thirty-three components were determined on samples from the cores, including counts on the abundances of shell-bearing organisms, serpulid tubes, fish remains, plants, fecal pellets, argillaceous lumps, minerals, and rock fragments.

Petrology

Saxena (1970) studied a multicomponent, multiphase system of minerals by using the principal components of the correlation matrix and plotting the transformed observations. On such representations, he demonstrated that by plotting certain relative positions of all coexisting minerals as well as the components of the multiphase system, lines joining points representing pairs of coexisting minerals are significant in the same sense as in concentration diagrams (Gibb's triangle, for instance). There is a clear advantage offered by the components approach in that the lines stand for the influence of all the components of the system.

Sedimentary petrology

Osborne (1967, 1969) has employed factor analysis for grouping Ordovician limestones, on the basis of characters determinable in thin sections. He succeeded (Osborne, 1967) in attaching paleoecological significance to the factors extracted. Within much the same frame of reference, McCammon (1968) made a comparative study of factor-analytical methods of grouping facies of Recent carbonates of the Bahama Bank. A similar study for Jurassic limestones of the northern Alps has been done by Fenninger (1970).

Mineralogy

Middleton (1964) used factor analysis to elucidate a complicated mineralogical problem in scapolites. By applying principal components and factor analysis to major- and trace-element data, he could identify the marialite–meionite solid solution in scapolites and propose the possible existence of an independent end-member bearing Mg and OH. Three significant groupings of trace elements were deduced. Mineralogical analyses often require special treatment (Aitchison, 1986).

Stratigraphy

R-mode factor analysis was used by McElroy and Kaesler (1965) on well data from the Upper Cambrian Reagan Sandstone on the Central Kansas Uplift. The four factors extracted were interpreted in terms of subsidence during the time of deposition of the sandstone, regional distributional patterns, and periods of uplift or nonsubsidence.

Biofacies relationships

Cheetham (1971) made *R*- and *Q*-mode factor analyses of weight-percentage abundances of major biotic constituents in a calcareous mound in the Danian of southern Scandinavia (transforming to minimize the constant sum constraint). From the *R*-mode loadings, three kinds of influences involving bryozoans and corals could be recognized, as well as the spatial relationships of biofacies.

Intertidal environment

A question of paleoecological significance concerns the identification of communities in tidal sediments. Cassie and Michael (1968) tried several multivariate methods in a well-documented study of this problem and came to the conclusion that principal component analysis proved to be the most versatile of them in that it permits both the diagnosis of the community structure and a plausible contouring of the communities in space.

Heavy minerals

Imbrie and Van Andel (1964) studied occurrences of heavy minerals from the Gulf of California and the Orinoco–Guyana shelf by *R*- and *Q*-mode factor analysis. The two areas have quite different sedimentological histories. Factor analysis of the simple situation represented by the Californian material yielded results similar to those obtained by conventional inspection, although more meaningful detail was revealed. The more remote petrographical sources of the Orinoco–Guyana shelf produce a more complicated situation with much mixing of the minerals. The factor analysis yielded a mineral distributional pattern greatly different from the impression given by mere inspection of the data. These results could be interpreted convincingly in terms of transportation during the post-Pleistocene rise of sea level.

Vertebrate paleontology

Gould (1967), analyzing pelycosaurs by *R*- and *Q*-mode factor analysis, was able to demonstrate far-reaching agreement for his computational results with the accepted scheme of phylogeny. Mahé (1974), in what is essentially a review of a comprehensive study of Madagascan fossil lemurs, advocated the pilot application of correspondence analysis to a few specimens of a sample in order to identify the most meaningful variables for a multivariate analysis. Using this variety of factor analysis, he established the phylogenetic relationships among lemurs, using craniometrical characters, but noted that the approach works best at the generic level.

Geochemistry of magmas

Teil and Cheminée (1975) analyzed major and trace elements in Ethiopian lavas by correspondence analysis, whereby meaningful associations between elements and samples could be shown to exist. The results turned out to be in agreement with accepted chemically based considerations for fractionation of magmas.

Palynology

In a study of Flandrian pollen data, Birks (1974) made a principal component analysis of percentage data on frequencies of pollen types. The component scores for the individual samples were plotted in relation to the stratigraphical position of the sample, thus forming composite "curves" of the original pollen variables. Birks makes here a highly significant suggestion, namely, that "a pollen zone can be delimited on the basis of stratigraphically adjacent samples with similar compositional scores."

Geochemistry of Cambrian alum shale

Armands (1972) studied in detail the geochemistry of uranium, molybdenum, and vanadium in Swedish alum shale. In this treatise, principal components and factor analysis were used to determine the paragenesis of elements in alum shales. Partly with the help of the results of these analyses, Armands found that Upper and Middle Cambrian alum shales can be divided into five categories, notably, detrital, authigenic, carbonate, sulfide, and organic.

Paleoecology

Variations in the relative frequencies of different species in samples may be interpreted in terms of major environmental factors to which the organisms react. Reyment (1963) used principal components and factor analysis to unravel paleoecological relationships between environmental forces and 17 species of Paleocene ostracods. The statistical analysis succeeded in separating euryoic species from stenoöic ones (see Section 8.3). Birks and Gordon (1985) give several examples of multivariate paleoecological studies.

2 Basic mathematical and statistical concepts

2.1 SOME DEFINITIONS

The exposition of factor analysis may be approached in several ways. We shall treat it in a manner combining linear algebra, multidimensional geometry, and statistics. Many of the ideas involved will certainly be unfamiliar to some of you, and it is the purpose of this chapter to define and illustrate some of the fundamental concepts underlying the technique of factor analysis. We make no pretext of writing a treatise on linear algebra and so, if you wish to delve deeply into the algebra of vectors and matrices, you should consult a formal text on the subject. A very readable account is that of Davis (1965); another is Smith (1978).

We use a simple, geologically relevant example as a frame of reference in illustrating the examples of the various operations considered here.

The chart of numbers listed in Table 2.I is typical of data collected by geologists, apart from its smallness. It contains information on the mineralogical composition of several rock specimens and it will be used to illustrate many of the basic concepts necessary for understanding factor analysis.

Data matrix

The table of numbers of Table 2.I is an example of a data matrix. There are eight rows, one for each rock specimen, and four columns, one for each mineral. A *data matrix* is then an array of p characters measured on N specimens.

Matrix

A matrix is defined as any array of numbers with one or more rows and one or more columns. Table 2.I is a matrix with eight rows and four columns. An entry in the table is called an *element*, when we are referring to the table as a matrix. An element, in this example, gives the amount of a particular mineral in a specific rock specimen.

In this book, a matrix is symbolized by capital, boldface letters. Thus, **X** may be used to refer to the entire array of numbers in Table 2.I.

15

Table 2.I. *Example of a data matrix*

Rock specimens	Mineral species			
	Quartz	Hornblende	Biotite	Feldspar
1	4.51	2.66	0.42	4.10
2	6.07	0.58	1.77	1.54
3	6.42	1.32	4.65	4.05
4	4.46	2.16	1.41	4.47
5	8.92	2.54	4.66	4.50
6	7.60	2.39	4.14	4.49
7	5.29	1.69	3.66	3.77
8	4.73	2.65	3.29	5.08

The *order* of the matrix refers to the size of the matrix in terms of the numbers of rows and columns. The order is denoted by means of subscripts attached to the letter symbolizing the matrix; the number of rows is given first, followed by the number of columns. In our example, $X_{(8 \times 4)}$ indicates that we are referring to the matrix X, which has eight rows and four columns. A matrix denoted $A_{(N \times p)}$ has N rows and p columns. The order is given verbally as "N by p."

In referring to the elements of a matrix, one uses the lowercase equivalent of the letter denoting the matrix. The position of an element in a matrix is given by row and column subscripts. Thus, x_{43} refers to the element in the fourth row and the third column of X, which, in Table 2.I, is the quantity 1.41, the amount of biotite in specimen 4. In general, a_{nj} is the element in the nth row and the jth column of matrix A.

The beauty of the matrix notation you have just been introduced to is that it allows us to write matrix equations in shorthand form, thus getting around the need to have to put down all the elements of an array.

Transpose matrix

In the course of manipulating matrices, it is often necessary to "flip" one onto its side. This operation is termed *transposition*. The transpose of a matrix is the original matrix with all the rows and columns interchanged, whereby the first column of the original matrix becomes the first row of the transpose and the first row of the original matrix becomes the first column of the transpose, and so on for all rows and columns. The transpose of the data matrix of Table 2.I is shown in Table 2.II. The transpose of a matrix is symbolized by the same capital letter as the original matrix, but with a prime attached to it. Thus, X' is the transpose of X. The order of the transpose is the reverse of the

Table 2.II. *Transpose of the data matrix of Table 2.I*

Mineral species	Rock specimen number							
	1	2	3	4	5	6	7	8
Quartz	4.51	6.07	6.42	4.46	8.92	7.60	5.29	4.73
Hornblende	2.66	0.58	1.32	2.16	2.54	2.39	1.69	2.65
Biotite	0.42	1.77	4.65	1.41	4.66	4.14	3.66	3.29
Feldspar	4.10	1.54	4.05	4.47	4.50	4.49	3.77	5.08

order of the original. In the example we have been using, $\mathbf{X}_{(8 \times 4)}$ and $\mathbf{X}'_{(4 \times 8)}$ are the respective orders. In $\mathbf{X}'$ of Table 2.II, the rows represent mineral species and the columns the rock specimens.

Vector

A vector may be defined as a matrix with only one row or column. A *column vector* is a column of vector elements; a *row vector* is a row of vector elements.

To denote a vector, we use a letter printed in lowercase boldface italics; this is a column vector. To indicate a row vector, we write it as the transpose of a column vector, marked typographically by placing a prime after the lowercase letter. Thus, if the vector has elements 1, 2, and 3, in this order, we write

$$x = \begin{bmatrix} 1 \\ 2 \\ 3 \end{bmatrix} \quad \text{and} \quad x' = [1, 2, 3].$$

Furthermore, we shall sometimes write

$$x = (x_i)$$

to denote the column vector x with elements $x_1, x_2, \ldots, x_p$.

A matrix may, therefore, be considered as a bundle of column or row vectors. The matrix $\mathbf{X}$ of Table 2.I consists of four column vectors; it is also composed of eight row vectors. One may refer to a column or row vector of a matrix by using a subscript to denote the appropriate row or column of the matrix.

Scalar

A scalar may be thought of as a matrix with one column and one row. Scalars are the everyday numbers we all are familiar with; they have, of course, no subscripts.

Objects

In a data matrix, as occurring in this book, the rows will always represent some sort of geological or biological entity upon which observations have been made. The entity may be a rock specimen, a particular fossil, a sample of subsurface brine, an ecological factor, and so forth. As a general term used for referring to the thing measured, we employ the word "object." In the example of Table 2.I, the objects are rock specimens. Synonyms for object are specimen, entity, individual, case, and observational vector. This last term is more characteristically found in literature on the theory of statistics.

Variables

The columns of a data matrix represent characters measured on the specimens. The first column of Table 2.I lists determinations of the content of quartz in each of the eight rock specimens. We can therefore say that each column of this data matrix lists a set of observations on one of the four variables.

2.2 GEOMETRICAL INTERPRETATION OF DATA MATRICES

Variables and objects as vectors

A vector can be interpreted both algebraically and geometrically. A vector can be defined as a directed line segment.

Consider Fig. 2.1, which was drawn using the first and second columns of Table 2.I. This diagram describes the eight objects (rock specimens) in terms of their composition of the variables quartz and hornblende. The reference axes are at right angles (orthogonal), the one in terms of quartz, the other in terms of hornblende. Each rock specimen can now be located with reference to its coordinates on the two axes. Thus, the coordinates for rock specimen 1 are 4.51 and 2.66. By joining the point denoted by these coordinates to the origin, we obtain the geometrical vector representation of object 1, and similarly for the other objects.

Variable space

If, as in Fig. 2.1, the variables form the coordinate axes for the objects, then the objects are said to be in the coordinate space defined by these variables. This may be termed "variable space".

In Fig. 2.1, we used only two coordinate axes, representing quartz and hornblende, respectively. If we add a third column vector, that is, a

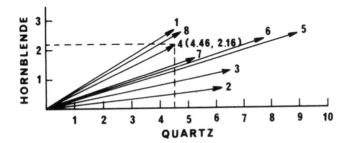

Figure 2.1. Representation of objects as vectors in two-dimensional variable space; data from Table 2.I.

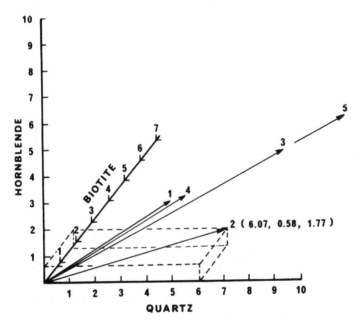

Figure 2.2. Representation of objects in three-dimensional variable space; data selected from Table 2.I.

third column of Table 2.I, the objects can be thought of as vectors located in a three-dimensional space (Fig. 2.2). To portray the objects in terms of the four variables of Table 2.I would require four mutually perpendicular axes. This is naturally impossible in three-dimensional space and so, for the quadrivariate situation, we must think in terms of abstract space. The four elements of a row vector of observations give the coordinates of the object measured in four-dimensional space.

Expanding this reasoning, we arrive at the general concept of a data matrix, which can be said to consist of N rows and p columns. The objects may be thought of as N vectors of observations on them,

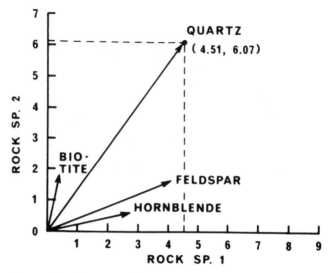

Figure 2.3. Representation of variables as vectors in two-dimensional object space; data from Table 2.I.

situated in p-dimensional space. This space may be further specified as Euclidean variable space.

Object space

A similar line of reasoning may be used to interpret the columns of a data matrix., According to this viewpoint, each object can be interpreted as representing an axis mutually perpendicular to all other object axes. The elements of a column vector are the coordinates of a variable with reference to the axes of the objects. In Fig. 2.3, we illustrate the four vectors of variables located in the two-dimensional space defined by rock specimens 1 and 2.

In the general case of an N-by-p data matrix, the variables may be thought of as p vectors of variables situated in N-dimensional object space.

2.3 ELEMENTARY VECTOR OPERATIONS

Now that the concepts of vectors of variables and objects have been introduced, we can turn to the mathematical manipulation of vectors. The conventions and rules employed are, with minor modifications, also applicable to matrices. For those who wish to read more deeply on the subject, we can recommend the text by Bellman (1960), rather old now, but still good.

Equality of vectors

Two vectors are said to be equal if each of their corresponding elements are equal. For the p-element row vectors v' and w',

$$v' = w', \quad \text{if } v_k = w_k \text{ for } k = 1, \ldots, p. \qquad [2.1]$$

Addition of vectors

The sum of two vectors is a vector containing the sums of corresponding elements. In the general case, for the p-element row vectors v' and w',

$$v' + w' = (v_k + w_k) \qquad [2.2]$$

Subtraction of vectors

Subtraction of two vectors yields a vector the elements of which contain the difference between the corresponding elements of the two vectors. As for addition, the two vectors must be of equal order.

Multiplication of vectors

Minor product. There are several ways of making vector products. First, we consider the *minor product*, a term introduced by Horst (1963) for what is also known under the names of the scalar, inner, or dot product. This product is made by the multiplication of a row vector by a column vector. The left vector (*prefactor*) is a row vector, the right vector (*postfactor*) a column vector. The minor product is a scalar that is the sum of products of corresponding elements of the two vectors; this can be summarized algebraically as

$$x' \cdot y = \sum_{k=1}^{p} x_k y_k. \qquad [2.3]$$

The minor product is defined only for vectors of equal order.

We note that

$$v'w = w'v \qquad [2.4]$$

and that

$$cv'w = c(v'w), \qquad [2.5]$$

where c is a scalar.

We shall now demonstrate how the calculation of a minor product is done. The vectors $x' = [2\ 4\ 1]$ and $y' = [3\ 1\ 5]$ will be multiplied to yield the scalar *a*.

$$a = \begin{bmatrix} 2 & 4 & 1 \end{bmatrix} \begin{bmatrix} 3 \\ 1 \\ 5 \end{bmatrix} [15],$$

where $a = 2 \times 3 + 4 \times 1 + 1 \times 5 = 15$.

Major product. The *major product* of two vectors may be described as follows. When a column vector is postmultiplied by a row vector, the result is a matrix sometimes termed the major product. The *ij*th element of the major product is the product of the *i*th element of the prefactor with the *j*th element of the postfactor. For vectors *x* and *y*, the product *xy'* can be shown diagrammatically as

$$\begin{bmatrix} y_1 & y_2 & \cdots & y_p \end{bmatrix}$$

$$\begin{bmatrix} x_1 \\ x_2 \\ \cdot \\ x_p \end{bmatrix} \begin{bmatrix} x_1 y_1 & x_1 y_2 & \cdots & x_1 y_p \\ x_2 y_1 & x_2 y_2 & \cdots & x_2 y_p \\ \cdot & \cdot & \cdots & \cdot \cdot \\ x_p y_1 & x_p y_2 & \cdots & x_p y_p \end{bmatrix} \qquad [2.6]$$

A numerical example of a major product follows.

Using the same two vectors, *x* and *y* and the diagrammatical method, we shall compute $A = xy'$:

$$\begin{bmatrix} 3 & 1 & 5 \end{bmatrix}$$

$$\begin{bmatrix} 2 \\ 4 \\ 1 \end{bmatrix} \begin{bmatrix} 2 \times 3 & 2 \times 1 & 2 \times 5 \\ 4 \times 3 & 4 \times 1 & 4 \times 5 \\ 1 \times 3 & 1 \times 1 & 1 \times 5 \end{bmatrix} = \begin{bmatrix} 6 & 2 & 10 \\ 12 & 4 & 20 \\ 3 & 1 & 5 \end{bmatrix}$$

Minor and major product moments

The minor product of a column vector premultiplied by its transpose is called a *minor product moment* and represents the sum of the squared vector elements. The *major product moment* is defined as a column vector postmultiplied by its transpose. The result is a matrix. An example is given after equation [2.11].

Unit vector and null vector

The *unit vector* is a vector all the elements of which are 1. A unit column vector is denoted by *1*. The order of a unit vector may be deduced from the context in which it is used.

The *null vector* is a vector all the elements of which are 0.

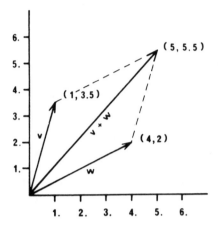

Figure 2.4. The addition of vectors $v' = [1.0\ \ 3.5]$ and $w' = [4.0\ \ 2.0]$ yields the coordinates of the resultant vector, $v' + w' = [5.0\ \ 5.5]$.

2.4 THE GEOMETRY OF VECTORS

Some vector operations can be given a geometrical interpretation, which is often useful in understanding factor analysis. We illustrate our examples for two dimensions, but the relations hold for any number of dimensions.

Resultant vector

The vector representing the sum of two other vectors gives the coordinates of the resultant vector. This is illustrated in Fig. 2.4.

Difference vector

The vector formed by the differences between the corresponding elements of two vectors is known as the difference vector. The idea is illustrated in Fig. 2.5.

Scalar multiples of vectors

The effect of multiplying a vector by a scalar is illustrated in Fig. 2.6. Scalars used in this fashion can be thought of as coefficients "stretching" or "contracting" vectors to which they are applied.

Vector length

The length of a vector can be found from the square root of the sum of its squared projections onto the coordinate axes. This is illustrated in

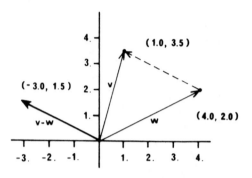

Figure 2.5. The subtraction of two vectors yields a difference vector, the elements of which supply the coordinates of the distance vector. Thus, if $v' = [1.0\ 3.5]$ and $w' = [4.0\ 2.0]$, $v' - w' = [-3.0\ 1.5]$.

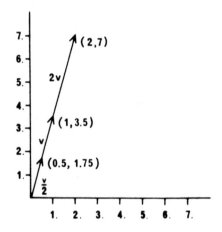

Figure 2.6. The effect of multiplying a vector by scalars:

$$v = [1.0 \quad 3.5], \qquad 2v = [2 \quad 7], \qquad v/2 = [0.5 \quad 1.75].$$

Fig. 2.7. This corresponds to the square root of the minor product moment. The length of a vector, written as $|v|$ is thus defined as

$$|v| = (v'v)^{1/2}. \qquad [2.7]$$

The angle between two vectors

The cosine of the angle between two vectors, v and w, is obtained from the inner product of the two vectors, divided by the product of their lengths (also termed magnitudes of the vectors):

$$\cos \theta = \frac{v'w}{(v'v)^{1/2}(w'w)^{1/2}}. \qquad [2.8]$$

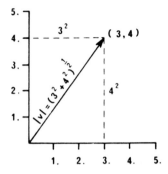

Figure 2.7. Pythagoras's theorem applied in two-dimensional space for finding the length of a vector: $v = [3 \ 4]$ and $|v| = [3^2 + 4^2]^{1/2} = 5$.

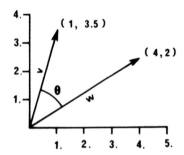

Figure 2.8. The angle between two vectors.

In our factor terminology, this is the ratio of the minor product (inner product) of the vectors to the product of the square roots of the minor product moments (lengths, magnitudes). In the example of Fig. 2.8, $v' = [1.0 \ 3.5]$ and $w' = [4 \ 2]$, $v'w = 11$, $(v'v)^{1/2} = 3.64$, and $(w'w)^{1/2} = 4.47$. Using now [2.8], we have that

$$\cos \theta = (11.0)/[(3.64)(4.47)] = 0.67,$$
$$\theta = 47.5°.$$

The cosine of the angle of separation may be used to illustrate the mutual positions of two vectors. A cosine of 1.0 indicates that the vectors are collinear, whereas a cosine of 0.0 indicates that they are at right angles to each other (orthogonal).

Distance

The Euclidean distance between two points is the length of the distance vector and is found by Pythagoras's theorem from the square root of the minor product moment of their difference vector:

$$d = |v - w| = [(v - w)'(v - w)]^{1/2} = v'v + w'w - 2w'v. \quad [2.9]$$

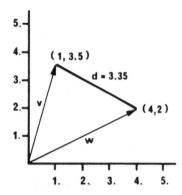

Figure 2.9. The Euclidean distance between two points.

The example illustrated in Fig. 2.9 is for the vectors $v' = [1 \; 3.5]$ and $w' = [4 \; 2]$, with the difference vector $v - w = [-3 \; 1.5]$. Applying [2.9], we have that the distance $d = (9 + 2.25)^{1/2} = 3.35$.

2.5 TYPES OF MATRICES

Rectangular matrices

Any matrix with more rows than columns, or vice versa, is termed a rectangular matrix. The data matrix of Table 2.I is a rectangular matrix.

Square matrices

A matrix with the same number of rows as columns is called a square matrix. The elements, the row and column subscripts of which are the same, are said to form the *principal diagonal* of the matrix. These are the elements extending from the upper left corner to the lower right corner. The *trace* is defined as the sum of the elements of the principal diagonal. The usually employed method of denoting this is tr A, the trace of matrix **A**.

Symmetrical matrices

A symmetrical matrix is a square matrix in which an element in the ith row and jth column is equal to the element in the jth row and the ith column. A matrix **A** is symmetrical if $a_{ij} = a_{ji}$ for all i and j.

Diagonal matrices

A diagonal matrix is a square symmetrical matrix, the off-diagonal elements of which are all zero. A special case of the diagonal matrix is the *identity matrix*, which has ones in the principal diagonal. The identity matrix serves the same function in linear algebra as the number 1 in scalar arithmetic.

Null matrix

A null matrix is a matrix all elements of which are zero. This matrix is the analog of zero in scalar algebra.

2.6 ELEMENTARY MATRIX ARITHMETIC

Equality of matrices

Two matrices are equal if all corresponding elements are equal. It follows that the matrices must be of the same order.

Addition and subtraction of matrices

As with vectors, the addition of two matrices is carried out by adding together corresponding elements of the matrices. The result is a matrix of the same order as the two matrices being summed. Subtraction follows the same rule. We shall now give an example of matrix addition in which matrices **A** and **B** are summed to yield the matrix **C**:

$$\mathbf{A} = \begin{bmatrix} 1 & 3 \\ 4 & 1 \\ 2 & 5 \end{bmatrix}, \quad \mathbf{B} = \begin{bmatrix} 5 & 2 \\ 0 & 1 \\ 4 & 4 \end{bmatrix},$$

$$\mathbf{C} = \mathbf{A} + \mathbf{B} = \begin{bmatrix} 1+5 & 3+2 \\ 4+0 & 1+1 \\ 2+4 & 5+4 \end{bmatrix} = \begin{bmatrix} 6 & 5 \\ 4 & 2 \\ 6 & 9 \end{bmatrix}.$$

Note that both of these matrices are rectangular and that the result, **C**, is also rectangular.

Multiplication of matrices

Multiplication of matrices is a somewhat more complex procedure than vector multiplication but, because many of the mechanisms of factor analysis involve the products of matrices, the technique of multiplication must be well understood.

There is one condition that must be satisfied before two matrices can be multiplied, namely, that the number of columns of the matrix in the prefactor must be equal to the number of rows of the postfactor. Thus, before the product of matrices **A** and **B** can be made, their orders must

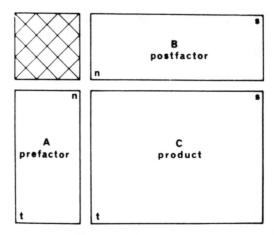

Figure 2.10. The rectangles represent the matrix multiplication $C = AB$. If drawn so that the orders are to relative scale, the cross-hatched area will make a square for conformable matrices. The product will fit into the area to the lower right of the two factors.

be known. If A is of order t by n and B is of order m by s, then n must equal m before a product can be formed.

An informative way of checking for compatibility of matrices is by drawing a diagram of the two matrices to be multiplied, as shown in Fig. 2.10. Here the two rectangles A and B are the two factors; C is the product. The orders of the matrices involved are as shown. If the dimensions of the rectangles are drawn so that they are proportional to their orders, the cross-hatched area to the upper left must make a square, thus signifying that A has the same number of columns as B has rows. Also, the product will fit into the area to the lower right. You will now appreciate that the order in which matrices are multiplied is important. The phrase "A multiplied by B" may be misleading and the more specific "A postmultiplied by B" or "B premultiplied by A" is preferable.

The actual mechanics of multiplication are readily understood in terms of vector products. To form the product $C = AB$, assuming the orders compatible, every element of C is equal to the minor product of the corresponding row vector of A and the column vector of B.

We shall now demonstrate the truth of these statements by recourse to a worked example. Consider the matrices

$$A = \begin{bmatrix} 2 & 3 & 1 \\ 1 & 2 & 0 \end{bmatrix} \quad \text{and} \quad B = \begin{bmatrix} 1 & 4 \\ 2 & 2 \\ 3 & 1 \end{bmatrix}.$$

Since A is 2 by 3 and B is 3 by 2, a product can be formed. To form

$C = AB$, the element c_{11} is made from the minor product of the first row vector of A and the first column vector of B:

$$c_{11} = 2 \times 1 + 3 \times 2 + 1 \times 3 = 11.$$

Similarly, c_{12} is found as

$$\begin{bmatrix} 4 \\ 2 \\ 1 \end{bmatrix}$$
$$[2 \quad 3 \quad 1][15].$$

That is, $c_{12} = 15$. You can run through all the multiplications on your own to see that you have got the hang of things. You will soon find the labor can be reduced for symmetric matrices as corresponding upper versus lower triangular elements are equal. In the present case, this is, however, not so, for

$$c_{21} = 1 \times 1 + 2 \times 2 + 0 \times 3 = 5.$$

Summing up for the present multiplication exercise, we have then that $C = AB$ is given by

$$\begin{bmatrix} 1 & 4 \\ 2 & 2 \\ 3 & 1 \end{bmatrix}$$
$$\begin{bmatrix} 2 & 3 & 1 \\ 1 & 2 & 0 \end{bmatrix} \begin{bmatrix} 11 & 15 \\ 5 & 8 \end{bmatrix}.$$

In the multiplication of scalars, the order of the factors is not important. Multiplication of matrices is, in general, noncommutative, that is, AB is not usually the same as BA. We shall now show that in the preceding example, changing the positions of A and B in the multiplication leads to a different result. We shall multiply out a few of the vector products for $E = BA$, but we shall no longer use the diagrammatical representation. The rest are left to you.

$$e_{11} = [1 \quad 4]\begin{bmatrix} 2 \\ 1 \end{bmatrix} = 6; \quad e_{12} = [1 \quad 4]\begin{bmatrix} 3 \\ 2 \end{bmatrix} = 11.$$

In summary, the multiplication of B followed by A is

$$E = BA = \begin{bmatrix} 1 & 4 \\ 2 & 2 \\ 3 & 1 \end{bmatrix} \begin{bmatrix} 2 & 3 & 1 \\ 1 & 2 & 0 \end{bmatrix} = \begin{bmatrix} 6 & 11 & 1 \\ 6 & 10 & 2 \\ 7 & 11 & 3 \end{bmatrix}.$$

The general equation for the product of two matrices of the same order is

$$C = AB; \quad c_{ij} = \sum_{k=1}^{n} a_{ik}b_{kj}. \qquad [2.10]$$

Major and minor product moments

Two types of products of certain matrices with themselves are important in multivariate statistics. The *major product moment* is defined as the product of a matrix, postmultiplied by its transpose. Thus, for the N-by-p matrix $\mathbf{X}$, the major product moment is

$$\mathbf{C} = \mathbf{XX'}. \qquad [2.11]$$

Matrix $\mathbf{C}$, the major product moment, is a square symmetrical matrix of order N by N. The major product moment of the data matrix of Table 2.I is now illustrated.

Multiplication of two matrices to form the major product moment $\mathbf{C} = \mathbf{XX'}$, using the data of Table 2.I, is required:

$$c_{nm} = \sum_{j=1}^{p} x_{nj} x_{mj}.$$

Thus, an element c_{nm} contains the sum of products of the nth and the mth row elements of $\mathbf{X}$. The principal diagonal of $\mathbf{C}$ contains the sum of the corresponding squared row elements of $\mathbf{X}$:

$$\mathbf{X} = \begin{bmatrix} 4.51 & 2.66 & 0.42 & 4.10 \\ 6.07 & 0.58 & 1.77 & 1.54 \\ 6.42 & 1.32 & 4.65 & 4.05 \\ 4.46 & 2.16 & 1.41 & 4.47 \\ 8.92 & 2.54 & 4.66 & 4.50 \\ 7.60 & 2.39 & 4.14 & 4.49 \\ 5.29 & 1.69 & 3.66 & 3.77 \\ 4.73 & 2.65 & 3.29 & 5.08 \end{bmatrix};$$

$$\mathbf{X'} = \begin{bmatrix} 4.51 & 6.07 & 6.42 & 4.46 & 8.92 & 7.60 & 5.29 & 4.73 \\ 2.66 & 0.58 & 1.32 & 2.16 & 2.54 & 2.39 & 1.69 & 2.65 \\ 0.42 & 1.77 & 4.65 & 1.41 & 4.66 & 4.14 & 3.66 & 3.29 \\ 4.10 & 1.54 & 4.05 & 4.47 & 4.50 & 4.49 & 3.77 & 5.08 \end{bmatrix};$$

$$\mathbf{C} = \mathbf{XX'} = \begin{bmatrix} 44.46 & 35.97 & 51.03 & 44.83 & 67.44 & 60.82 & 45.35 & 50.62 \\ 35.97 & 42.67 & 54.17 & 37.69 & 70.80 & 61.75 & 45.35 & 43.89 \\ 51.03 & 54.17 & 80.91 & 56.13 & 100.49 & 89.35 & 68.44 & 69.73 \\ 44.82 & 37.69 & 56.13 & 46.55 & 71.98 & 64.98 & 49.25 & 54.18 \\ 67.44 & 70.80 & 100.49 & 71.98 & 128.05 & 113.40 & 85.50 & 87.15 \\ 60.82 & 61.75 & 89.35 & 64.98 & 113.40 & 100.79 & 76.31 & 78.74 \\ 45.35 & 45.35 & 68.44 & 49.25 & 85.50 & 76.31 & 58.43 & 60.70 \\ 50.62 & 43.89 & 69.73 & 54.18 & 87.15 & 78.74 & 60.70 & 66.06 \end{bmatrix}.$$

The *minor product moment* is defined as the premultiplication of a matrix by its transpose. For an N-by-p matrix $\mathbf{X}$, the minor product moment $\mathbf{E}$ is

$$\mathbf{E} = \mathbf{X'X}. \qquad [2.12]$$

E is a square matrix of order p by p. An example of the minor product moment, using the data of Table 2.I, is given next. The above two terms are due to Horst (1963). The usual term is "sums of squares and cross-products matrices." Horst's expressions are more precise.

The minor product moment $\mathbf{E} = \mathbf{X'X}$, where $\mathbf{X}$ is the data matrix of Table 2.I, is now shown. Elements of $\mathbf{E}$ are obtained from $e_{ij} = \sum_{k=1}^{N} x_{ki}x_{kj}$. The element e_{ij} contains the sum of products of the ith and jth columns of $\mathbf{X}$ and elements in the principal diagonal contain the sum of squared elements of the columns.

$$\mathbf{X'} = \begin{bmatrix} 4.51 & 6.07 & 6.42 & 4.46 & 8.92 & 7.60 & 5.29 & 4.73 \\ 2.66 & 0.58 & 1.32 & 2.16 & 2.54 & 2.39 & 1.69 & 2.65 \\ 0.42 & 1.77 & 4.65 & 1.41 & 4.66 & 4.14 & 3.66 & 3.29 \\ 4.10 & 1.54 & 4.05 & 4.47 & 4.50 & 4.49 & 3.77 & 5.08 \end{bmatrix},$$

$$\mathbf{X} = \begin{bmatrix} 4.51 & 2.66 & 0.42 & 4.10 \\ 6.07 & 0.58 & 1.77 & 1.54 \\ 6.42 & 1.32 & 4.65 & 4.05 \\ 4.46 & 2.16 & 1.41 & 4.47 \\ 8.92 & 2.54 & 4.66 & 4.50 \\ 7.60 & 2.39 & 4.14 & 4.49 \\ 5.29 & 1.69 & 3.66 & 3.77 \\ 4.73 & 2.65 & 3.29 & 5.08 \end{bmatrix},$$

$$\mathbf{E} = \mathbf{X'X} = \begin{bmatrix} 306.00 & 96.00 & 156.73 & 192.00 \\ 96.00 & 35.92 & 48.00 & 68.85 \\ 156.73 & 48.00 & 90.00 & 99.63 \\ 192.00 & 68.85 & 99.63 & 136.00 \end{bmatrix}.$$

Products of matrices and vectors

A matrix premultiplied by a conformable row vector yields a row vector:

$$\begin{bmatrix} 1 & 3 & 0 \end{bmatrix} \begin{bmatrix} 3 & 2 \\ 1 & 0 \\ 2 & 1 \end{bmatrix} = \begin{bmatrix} 6 & 2 \end{bmatrix}.$$

The premultiplication of a matrix by a unit row vector yields a row vector, the elements of which are the sums of the corresponding columns of the matrix.

A matrix postmultiplied by a conformable column vector yields a column vector. If the vector is a unit vector, the product is a column vector containing the row sums of the matrix. As an example of postmultiplication consider

$$\begin{bmatrix} 3 & 1 \\ 4 & 0 \\ 1 & 2 \end{bmatrix} \begin{bmatrix} 1 \\ 3 \end{bmatrix} = \begin{bmatrix} 6 \\ 4 \\ 7 \end{bmatrix}.$$

The triple product, $I'XI$, is a scalar that is the sum of all the elements of X.

A matrix premultiplied by a diagonal matrix has the elements in its nth row multiplied by the nnth element of the diagonal matrix:

$$\begin{bmatrix} 2 & 0 & 0 \\ 0 & 3 & 0 \\ 0 & 0 & 1 \end{bmatrix} \begin{bmatrix} 3 & 2 \\ 1 & 0 \\ 2 & 1 \end{bmatrix} = \begin{bmatrix} 6 & 4 \\ 3 & 0 \\ 2 & 1 \end{bmatrix}.$$

A matrix postmultiplied by a diagonal matrix has elements in its jth column multiplied by the jjth element of the diagonal matrix:

$$\begin{bmatrix} 3 & 2 \\ 1 & 0 \\ 2 & 1 \end{bmatrix} \begin{bmatrix} 2 & 0 \\ 0 & 3 \end{bmatrix} = \begin{bmatrix} 6 & 6 \\ 2 & 0 \\ 4 & 3 \end{bmatrix}.$$

Because premultiplication or postmultiplication by diagonal matrices causes all the elements in a given row or column to be multiplied by a constant, these operations can be used for the differential scaling of columns or rows of a matrix.

Determinant

Another topic in matrix algebra concerns the theory of determinants. The subject is well treated at the elementary level in Davis (1965, Chapter 5). A determinant is a scalar derived from operations on a square matrix. The determinant of the square matrix A is denoted as $|A|$. For a 2-by-2 matrix A, the determinant is obtained as follows:

$$|A| = a_{11}a_{22} - a_{12}a_{21}.$$

For a 3-by-3 matrix A, the expression is

$$a_{11}a_{22}a_{33} - a_{11}a_{23}a_{32} - a_{12}a_{21}a_{33} + a_{12}a_{23}a_{31} + a_{13}a_{21}a_{32} - a_{13}a_{22}a_{31}.$$

A matrix whose determinant is zero is called a *singular matrix*. *Nonsingular matrices* have nonzero determinants.

Minor

A minor is a special kind of determinant. Given a square matrix A, the minor M_{ij} is defined as the determinant of the matrix formed by deleting the ith row and jth column of $\mathbf{A}$. There will therefore be one minor corresponding to each element of $\mathbf{A}$.

Of special interest here are the minors associated with the principal diagonal of square symmetric matrices. A square symmetric matrix $\mathbf{A}_{p \times p}$ is said to be *positive definite* if all the minors associated with the elements of the principal diagonal are greater than zero.

Matrix inversion

The scalar analog of division requires the calculation of the inverse of a matrix. This is a rather involved computational exercise and for details of the various methods available, we refer you to one of the standard books on practical matrix algebra, for example, Horst (1963). For the sake of continuity, we offer the following brief explanation.

In scalar arithmetic, the reciprocal of a number may be defined as a number that, when multiplied by the original number, yields 1. For example, for the scalars b and c, $bc = 1$. Here, c is the reciprocal of b, provided that b is not zero. Division of any number by b is then equivalent to multiplication by the reciprocal of b. That is, $y/b = yc$ if c is the reciprocal of b.

In matrix algebra, the reciprocal $\mathbf{C}$ of a square matrix $\mathbf{B}$ satisfies the relationships

$$\mathbf{BC} = \mathbf{CB} = \mathbf{I}.$$

The identity matrix $\mathbf{I}$ is the analog of the number 1 in scalar arithmetic. The reciprocal of $\mathbf{B}$ as just defined, $\mathbf{C}$, is called the inverse of $\mathbf{B}$; it is denoted $\mathbf{B}^{-1}$.

A *singular* matrix does not possess an inverse. Although we are not concerned with the subject in factor analysis as presented in this book, singular and rectangular matrices can be given a kind of generalized inverse that does not obey the preceding rules.

The most commonly occurring type of inverse in statistics concerns square symmetrical positive definite matrices.

Matrix inversion is used in the solution of systems of simultaneous equations. For example, if $\mathbf{A}$ contains the known coefficients, the vector x the unknowns, and y is the vector of coefficients to be yielded by the solution, the entire system of equations can be expressed as

$$\mathbf{A}x = y.$$

For example, if $\mathbf{A}$ is a 2-by-2 matrix, the equations appear as

$$a_{11}x_1 + a_{12}x_2 = y_1,$$
$$a_{21}x_1 + a_{22}x_2 = y_2.$$

In general, if **A** is nonsingular, a solution to the equations may be found by multiplying both sides of the expression by $\mathbf{A}^{-1}$:

$$\mathbf{A}^{-1}\mathbf{A}x = \mathbf{A}^{-1}y,$$

which gives $x = \mathbf{A}^{-1}y$.

2.7 NORMAL AND ORTHONORMAL VECTORS AND MATRICES

We noted earlier that a matrix may be considered as a set of column or row vectors. We now introduce concepts that unite the properties of vectors and matrices into a useful whole.

Normalized vectors

A vector is said to be normalized if it is of unit length. Any vector x can be normalized by dividing each of its elements by the length of the vector: $x/(x'x)^{1/2}$.

As an example of vector normalization, let us consider the vector $x' = [2 \ 1 \ 2]$:

$$x'x = \begin{bmatrix} 2 & 1 & 2 \end{bmatrix} \begin{bmatrix} 2 \\ 1 \\ 2 \end{bmatrix} = 9$$

$$(x'x)^{1/2} = 3$$

$$v = x/(x'x)^{1/2} = \begin{bmatrix} 2/3 \\ 1/3 \\ 2/3 \end{bmatrix}$$

The length of the normalized vector, v, is always 1, as shown here:

$$v'v = \begin{bmatrix} 2/3 & 1/3 & 2/3 \end{bmatrix} \begin{bmatrix} 2/3 \\ 1/3 \\ 2/3 \end{bmatrix} = 1$$

Orthogonal vectors

Two vectors are said to be orthogonal if their minor product is zero; if $u'v = 0$, then u and v are orthogonal to each other. For two vectors u

and v, the property of orthogonality implies that the vectors are separated by 90°, for

$$\cos \theta = u'v/(|u| \, |v|) = 0.$$

The arccosine of 0 is 90°.

Orthogonal matrices

These concepts can be applied to the column vectors of matrices. If the minor product moment of a matrix X is a diagonal matrix, then X is said to be orthogonal. That is,

$$X'X = D,$$ [2.13]

where D is a diagonal matrix, implies that all possible pairs of column vectors of X are orthogonal. This is illustrated as follows:

$$X'X = \begin{bmatrix} 1 & 1 & 1 \\ 2 & -3 & 1 \end{bmatrix} \begin{bmatrix} 1 & 2 \\ 1 & -3 \\ 1 & 1 \end{bmatrix} = D = \begin{bmatrix} 3 & 0 \\ 0 & 14 \end{bmatrix}.$$

When the minor product moment of a matrix is an identity matrix, the matrix is said to be *orthonormal*. This implies that all the column vectors of the matrix are normalized vectors and that they are orthogonal to each other. Of particular importance in factor analysis is the square orthonormal matrix, say Q, which has the properties that

$$Q'Q = I$$ [2.14]

and

$$QQ' = I$$ [2.15]

There are an infinite number of square orthonormal matrices of any given order (Horst, 1963).

2.8 DESCRIPTIVE STATISTICS IN MATRIX NOTATION

As an aid in understanding the matrix development of factor analysis, we present some of the well-known equations of simple statistics in matrix form, using, in the practical illustrations, the data matrix of Table 2.I. The rows of this matrix represent objects, the columns observations on variables.

The mean

In matrix terms, the formula for calculating a mean is for the p means of the N-by-p data matrix X:

$$\bar{x} = I'X/I'I,$$

where $\bar{x}$ is the p-component vector of mean values. Recall that I is a vector of ones.

Deviate scores

The difference between an observation on a variable and the mean of that variable is referred to as the deviation. If we refer to the original value x_{nj} as the *raw score*, then the deviate score, y_{nj}, is defined as

$$y_{nj} = x_{nj} - \bar{x}_j. \qquad [2.16]$$

The sum of deviate scores for a variable is, of course, zero. By converting variables to deviate scores, the origin of each variable is, in effect, shifted to the mean. The means of variables expressed in deviate form are all zero.

Variance

The variance of a variable is a measure of the scatter of individual values about the mean. It is defined as the average of the sum of squared deviation scores. Using [2.16] for deviate scores, for the jth variable, the variance is

$$s_j^2 = \sum_{n=1}^{N} y_{nj}^2/N.$$

Here, and sometimes elsewhere, we use division by N for reasons of expediency. That is, deviations are divided by N, the total sample size, rather than $N - 1$. The latter divisor gives a so-called unbiased estimate of the variance or covariance. You should, however, be aware of the fact that in statistical connections, division by $N - 1$ is almost always employed.

Using vector notation, the foregoing equation is written

$$s_j^2 = y_j'y_j/I'I$$

Covariance

The covariance expresses the relationship between two variables. For variables x_i and x_j, the covariance is defined as the average of the sum of the products of the deviation scores for all objects:

$$s_{ij} = \sum_{n=1}^{N} y_{ni} y_{nj} / N.$$

By comparing with the formula for the variance, you will see that it is a special case of the formula for the covariance. For an entire data matrix, the matrix of variances and covariances, termed in this book the *covariance matrix*, is obtained from

$$S = Y'Y/1'1 \qquad [2.17]$$

The covariance matrix is therefore the minor product moment of the data matrix, expressed in deviate form, each element divided by N, the number of objects in the sample. It is a square symmetric matrix of order p. The elements in the principal diagonal are the variances of the variables.

Standard deviation

The standard deviation is defined as the positive square root of the variance. You will recall that in Section 2.4, the length of a vector was defined as $|y| = (y'y)^{1/2}$; here, y is a vector of deviate scores of a variable. We see that the standard deviation of a variable is proportional to the length of the vector of observations on the variable. Thus

$$s_j = |y| N^{-1/2}. \qquad [2.18]$$

This relationship allows the interpretation of the standard deviation in terms of vector lengths of variables in deviate form.

Standardized scores

It is often useful to be able to express observations in terms of their deviations from the mean, using the standard deviation as the unit of divergence. The standardized score for a particular object for variable j is given by

$$z_{nj} = (x_{nj} - \bar{x}_j)/s_j$$

Standardized variables have a mean of zero and a variance (and standard deviation) of 1. The lengths of the vectors of variables are $N^{1/2}$.

The formula for standardizing all the variables in a data matrix is

$$\mathbf{Z} = \mathbf{YD}^{-1/2}, \qquad [2.19]$$

where $\mathbf{D}$ denotes the diagonal matrix formed by the diagonal elements of the covariance matrix $\mathbf{S}$, and $\mathbf{Y}$ is the matrix of deviate scores. The use of this formula and other pertinent calculations are shown in the following text.

Correlation

A measure of association between two variables of great importance in factor analysis is the Pearsonian product–moment correlation coefficient.

This measure is defined as the ratio of the covariances of the two variables to the product of their standard deviations:

$$r_{ij} = s_{ij}/s_i s_j. \qquad [2.20]$$

For variables in deviate form, the formula has the appearance

$$r_{ij} = \frac{\displaystyle\sum_{n=1}^{N} y_{ni} y_{nj}}{\left(\displaystyle\sum_{n=1}^{N} y_{ni}^2 \sum_{n=1}^{N} y_{nj}^2\right)^{1/2}}.$$

For a matrix of standardized data, the formula may be conveniently represented as

$$\mathbf{R} = \mathbf{Z}'\mathbf{Z}/N. \qquad [2.21]$$

The correlation matrix is a square symmetric matrix. It is the minor product moment of the standardized data matrix with each element divided by the size of the sample, N.

Steps in computing the correlation and covariance matrices from the data matrix of Table 2.I are as follows:

We start with the matrix of deviate scores, the elements of which were obtained by [2.16]. The covariance matrix $\mathbf{S}$ is found from $\mathbf{S} = \mathbf{Y}'\mathbf{Y}/\mathbf{N}$.

$$\begin{bmatrix} -1.49 & 0.07 & 0.42 & -1.54 & 2.92 & 1.60 & -0.71 & -1.27 \\ 0.66 & -1.42 & -0.68 & 0.16 & 0.54 & 0.39 & -0.31 & 0.65 \\ -2.58 & -1.23 & 1.65 & -1.59 & 1.66 & 1.14 & 0.66 & 0.29 \\ 0.10 & -2.46 & 0.05 & 0.47 & 0.50 & 0.49 & -0.23 & 1.08 \end{bmatrix} \begin{bmatrix} -1.49 & 0.66 & -2.58 & 0.10 \\ 0.07 & -1.42 & -1.23 & -2.46 \\ 0.42 & -0.68 & 1.65 & 0.05 \\ -1.54 & 0.16 & -1.59 & 0.47 \\ 2.92 & 0.54 & 1.66 & 0.50 \\ 1.60 & 0.39 & 1.14 & 0.49 \\ -0.71 & -0.31 & 0.66 & -0.23 \\ -1.27 & 0.65 & 0.29 & 1.08 \end{bmatrix} = \begin{bmatrix} 18.00 & 0.00 & 12.73 & 0.00 \\ 0.00 & 3.92 & 0.00 & 4.85 \\ 12.73 & 0.00 & 18.00 & 3.63 \\ 0.00 & 4.85 & 3.63 & 8.00 \end{bmatrix}$$

$$\mathbf{S} = \mathbf{Y'Y}/8 = \begin{bmatrix} 2.25 & 0.00 & 1.59 & 0.00 \\ 0.00 & 0.49 & 0.00 & 0.61 \\ 1.59 & 0.00 & 2.25 & 0.45 \\ 0.00 & 0.61 & 0.45 & 1.00 \end{bmatrix}.$$

The next step, the correlation matrix, **R**, requires the diagonal matrix formed by the reciprocals of the square roots of the diagonal elements of **S**.

$$\mathbf{D}^{-1/2} = (\text{diag } \mathbf{S})^{-1/2} = \begin{bmatrix} 0.67 & 0 & 0 & 0 \\ 0 & 1.43 & 0 & 0 \\ 0 & 0 & 0.67 & 0 \\ 0 & 0 & 0 & 1.00 \end{bmatrix}.$$

Using **Y**, we compute a matrix of standard scores, $\mathbf{Z} = \mathbf{Y}\mathbf{D}^{-1/2}$, which is

$$\begin{bmatrix} 0.67 & 0.00 & 0.00 & 0.00 \\ 0.00 & 1.43 & 0.00 & 0.00 \\ 0.00 & 0.00 & 0.67 & 0.00 \\ 0.00 & 0.00 & 0.00 & 1.00 \end{bmatrix}$$

$$\begin{bmatrix} -1.49 & 0.66 & -2.58 & 0.10 \\ 0.07 & -1.42 & -1.23 & -2.46 \\ 0.42 & -0.68 & 1.65 & 0.05 \\ -1.54 & 0.16 & -1.59 & 0.47 \\ 2.92 & 0.54 & 1.66 & 0.50 \\ 1.60 & 0.39 & 1.14 & 0.49 \\ -0.71 & -0.31 & 0.66 & -0.23 \\ -1.27 & 0.65 & 0.29 & 1.08 \end{bmatrix} \begin{bmatrix} -0.99 & 0.95 & -1.72 & 0.10 \\ 0.05 & -2.03 & -0.82 & -2.46 \\ 0.28 & -0.97 & 1.10 & 0.05 \\ -1.03 & 0.23 & -1.06 & 0.47 \\ 1.95 & 0.78 & 1.11 & 0.50 \\ 1.07 & 0.56 & 0.76 & 0.49 \\ -0.48 & -0.44 & 0.44 & -0.23 \\ -0.85 & 0.93 & 0.20 & 1.08 \end{bmatrix}.$$

The correlation matrix is

$$\mathbf{R} = \mathbf{Z}'\mathbf{Z}/8 = \begin{bmatrix} 1.00 & 0.00 & 0.71 & 0.00 \\ 0.00 & 1.00 & 0.00 & 0.87 \\ 0.71 & 0.00 & 1.00 & 0.30 \\ 0.00 & 0.87 & 0.30 & 1.00 \end{bmatrix}.$$

The geometrical interpretation of the correlation matrix plays an important part in factor analysis. We shall now study this subject.

From [2.22], which expresses the correlation coefficient in terms of vectors in deviate form,

$$r_{ij} = y_i'y_j / \left[(y_i'y_i)^{1/2} (y_j'y_j)^{1/2} \right], \qquad [2.22]$$

it will be seen that the correlation coefficient is the minor product moment of two vectors, divided by the product of their respective lengths. This is the same equation as we used for the angle between two vectors ([2.8]). From [2.17], the numerator of [2.22] is

$$y_i'y_j = Ns_{ij}.$$

From [2.18], $|y_i| = N^{1/2}s_i$ and $|y_j| = N^{1/2}s_j$. Thus

$$\cos \theta + Ns_{ij} / \left(N^{1/2}s_i N^{1/2}s_j \right) = s_{ij}/s_i s_j = r_{ij}.$$

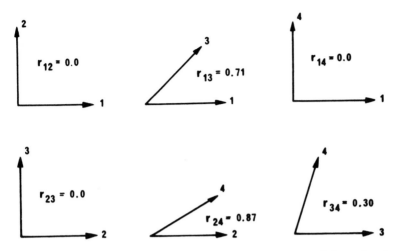

Figure 2.11. Some examples of correlations computed from the data matrix of Table 2.I.

For standardized variables, the equation is

$$\cos \theta = z'_i z_j / (|z_i| \, |z_j|).$$

This reduces to $\cos \theta = z'_i z_j / N$ because the length of a standardized vector is $N^{1/2}$. By [2.21], $z'_i z_j = N r_{ij}$. Therefore,

$$\cos \theta = z'_i z_j / N = r_{ij}.$$

These formulas lead to the interpretation of correlations between variables as the angles between vectors of variables in object space. The correlation matrix, **R**, contains the cosines of all possible angles of separation between all the variables. This concept is illustrated in Fig. 2.11.

Another way of interpreting the correlation coefficient is by reference to scatter diagrams. In Fig. 2.12, the objects of the data matrix of Table 2.I are plotted as points situated in the space of pairs of standardized variables. The correlation coefficient measures the tendency of the points to cluster along a line. Both of these interpretations are useful in factor analysis.

Regression, multiple and partial correlation

Some of the basic ideas of factor analysis rest on concepts like regression, multiple correlation, and partial correlation. These are very important concepts in multivariate statistics and for a full account of them we refer to Seber (1984). Only a brief account will be given here.

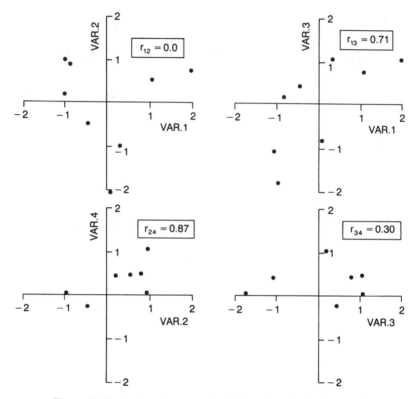

Figure 2.12. Scatter diagrams for Table 2.I with the correlation coefficients shown.

Consider p variables $y_1, y_2, \ldots, y_p$ on which we have N observations. For convenience, we assume that each variable is measured as a deviation from its mean so that the data matrix $\mathbf{Y}$ of order $N \times p$ is a deviation score matrix. A common procedure for studying the linear interrelationships among the variables is to consider the *linear regression* of each variable on all the others. Suppose $p = 4$ and consider y_1. Then this regression is written

$$y_1 = b_{12}y_2 + b_{13}y_3 + b_{14}y_4 + u_1, \qquad [2.23]$$

where u_1 denotes the residual.

The regression coefficients b_{12}, b_{13}, and b_{14} are usually determined by least squares, which means that

$$\sum_{n=1}^{N} (y_{n1} - b_{12}y_{n2} - b_{13}y_{n3} - b_{14}y_{n4})^2$$

is minimized. It can be shown that (see, e.g., Guttman, 1953; Seber, 1984)

$$b_{1j} = \frac{s^{1j}}{s^{11}}, \qquad j = 2, 3, 4,$$

where s^{1j} and s^{11} are elements in the inverse S^{-1} of the covariance matrix S. The linear combination

$$y_1^* = b_{12} y_2 + b_{13} y_3 + b_{14} y_4$$

is the best linear estimate of y_1 in the least-squares sense that can be obtained from y_2, y_3, and y_4. The *residual* u_1 in [2.23] represents that part of y_1 that cannot be explained by or predicted from y_2, y_3, and y_4. It may be shown that y_1^* and u_1 are uncorrelated, so that the variance of y_1 is the sum of the variance of y_1^* and the variance of u_1. The latter is called the *residual variance* and may be computed as $1/s^{11}$. The proportion of the variance of y_1 explained by y_2, y_3, and y_4 is

$$R_1^2 = \frac{s_{11} - 1/s^{11}}{s_{11}} = 1 - \frac{1}{s_{11} s^{11}}$$

and is called *the squared multiple correlation of* y_1. The positive square root of this is the correlation between y_1 and y_1^*.

So far we have focused on the first variable y_1. However, in a similar fashion one can proceed to consider the regression of y_2 on all the other variables (including y_1), of y_3 on all the other variables, and so on. In compact form we can now write

$$y = By + u$$

where y is the vector of all p variables, u is the vector of all p residuals, and B is a matrix whose diagonal elements are zero (since y_i is not involved on the right side of the ith equation) and the off-diagonal elements are the regression coefficients. Then

$$B = I - (\text{diag } S^{-1})^{-1} S$$

and the residual variances are given by the reciprocals of the diagonal elements of S^{-1}.

To introduce the concept of partial correlation, consider again the $p = 4$ variables and the regressions of y_1 and y_2 on y_3 and y_4, that is,

$$y_1 = b_{13} y_3 + b_{14} y_4 + u_1,$$
$$y_2 = b_{23} y_3 + b_{24} y_4 + u_2. \qquad [2.24]$$

The correlation between u_1 and u_2 is called the *partial correlation*

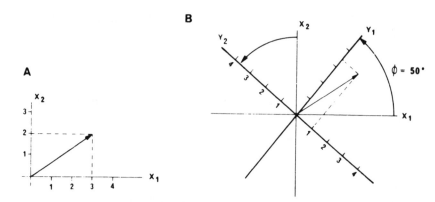

Figure 2.13. (A) The position of a row vector $x' = [3\ 2]$ given by coordinates on the orthogonal axes, x_1 and x_2. (B) The rigid rotation of axes through 50°. The coordinates of the row vector y' are given by the calculation

$$y_1 = (0.6428 \times 3) + (0.7660 \times 2) = 3.46,$$

$$y_2 = (-0.7660 \times 3) + (0.6428 \times 2) = -1.01.$$

These are the coordinates of the row vector in relation to the axes y_1 and y_2.

between y_1 and y_2 when y_3 and y_4 are eliminated. It is a measure of association between y_1 and y_2 given that y_3 and y_4 are held constant.

It should be noted that the regression coefficients b_{13} and b_{14} in [2.24] are not the same as those in [2.23]. For example, b_{13} in [2.23] represents the effect of y_3 on y_1 when the effects of y_2 and y_4 are also taken into account whereas b_{13} in [2.24] represents the effect of y_3 on y_1 when only the effect of y_4 is taken into account, that is, when y_2 is excluded. If y_2 has any effect on y_1 at all, the residual variance in [2.23] is smaller than that of [2.24].

2.9 ROTATION OF COORDINATE SYSTEMS

Systems of coordinates may be rotated by matrix operations. We shall meet several situations in which it is necessary to rotate a set of coordinate axes to new locations.

Algebra of rotation

In Fig. 2.13A, the elements of the vector x give the coordinates of a point with respect to the coordinate axes for the two variables. Suppose

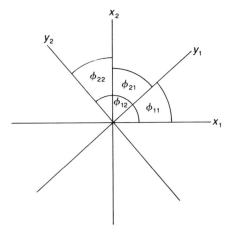

Figure 2.14. Angular relationships between a set of original coordinates, x_1 and x_2, and the rotated coordinate axes y_1 and y_2.

that for some reason we want to rotate rigidly the axes through $\phi°$ in a counterclockwise direction, as depicted in Fig. 2.13B. Our problem is then to find the coordinates of the point with reference to the new axes, Y_1 and Y_2.

From elementary trigonometry, the required coordinates are given by the equations

$$y_1 = \cos \phi x_1 + \sin \phi x_2,$$
$$y_2 = -\sin \phi x_1 + \cos \phi x_2. \qquad [2.25]$$

These two equations can be neatly written in matrix form as

$$\mathbf{y'} = \mathbf{x}\mathbf{T}, \qquad [2.26]$$

where

$$\mathbf{T} = \begin{bmatrix} \cos \phi & -\sin \phi \\ \sin \phi & \cos \phi \end{bmatrix}.$$

To extend this useful relationship to a more general case, we introduce the notation shown in Fig. 2.14. Here, the angles ϕ_{ij} are the angles between the ith old reference axis and the jth new one. For rigid rotation, and the two-dimensional scheme illustrated, we note the following angular relationships:

$$\phi_{12} = \phi_{11} + 90°,$$
$$\phi_{21} = \phi_{11} - 90°,$$
$$\phi_{22} = \phi_{11}.$$

The trigonometrical functions relating these angles are

$$\sin \phi_{11} = \sin(\phi_{21} + 90°) = \cos \phi_{21}$$
$$-\sin \phi_{11} = -\sin(\phi_{12} - 90°) = \sin(90° - \phi_{12}) = \cos \phi_{12}$$

Substituting into [2.26] yields

$$y_1 = \cos \phi_{11} x_1 + \cos \phi_{21} x_2,$$
$$y_2 = \cos \phi_{12} x_1 + \cos \phi_{22} x_2. \qquad [2.27]$$

The entire procedure can be generalized to any number of orthogonal axes and their rotated equivalents. The system of equations for p axes is

$$y_1 = \cos \phi_{11} x_1 + \cos \phi_{21} x_2 + \cdots + \cos \phi_{p1} x_p,$$
$$y_p = \cos \phi_{1p} x_1 + \cos \phi_{2p} x_2 + \cdots + \cos \phi_{pp} x_p. \qquad [2.28]$$

Putting $t_{ij} = \cos \phi_{ij}$, where i is the subscript attached to the axes of the original coordinate system and j that for the axes of the rotated system, the entire suite of equations may be summarized as

$$y' = x'T. \qquad [2.29]$$

For a collection of N row vectors in the matrix $X_{(N \times p)}$, the equation

$$Y = XT \qquad [2.30]$$

will give in Y the coordinates of all N row vectors in terms of the p rigidly rotated axes. The matrix T is termed a *transformation matrix* carrying X into Y. In order for the transformation to correspond to a rigid rotation, T must be an orthonormal matrix, that is, $T'T = I$.

Obviously, distances and angles between vectors are unchanged during rigid rotation of the reference axes.

Interpretation of rotation

Postmultiplication of a data matrix by an orthonormal matrix may be regarded as a simple, mechanical process of rigid rotation. Two features of this process are useful for providing additional insights into data manipulation by matrix operations. First, the new, rotated axes may themselves be interpreted as variables. The column vectors in Y of [2.30] are new variables that are linear combinations of the column vectors (variables) of X. The elements in the T matrix are thus seen to be the coefficients in the linear combinations. Second, the row vectors of X give the compositions of the objects in terms of the originally measured variables. The two vectors of Y give the compositions of the objects in terms of the newly derived variables.

2.10 THE STRUCTURE OF A MATRIX

The idea that variables and objects may be envisaged as vectors in N-dimensional space has been repeatedly emphasized in this chapter. The concepts of vector length, angle of separation between vectors, and distance between vectors were defined and shown to have geometrical relationships with statistical descriptors. In this section, we provide additional interpretations of the data matrix with reference to matrices derivable from it.

Matrices as products

In later sections, matrices will sometimes be treated as being the product of two or more matrices. In Section 2.6, we explained that the product of two conformable matrices $\mathbf{A}_{(s \times r)}$ and $\mathbf{B}_{(r \times t)}$ is a matrix $\mathbf{C}$ of order s by t. We now consider the reverse process of factoring a product into its components.

In scalar algebra, any number can be decomposed into factors, usually in an infinite number of ways. For example, the number 12 can be considered as the product of the factors 3 and 4, or 2 and 6, or 1 and 12, and so on. Similarly, not only can any matrix be thought of as the product of two factors but also there is an infinite number of such pairs of factors (Horst, 1963).

We shall accept the first assertion as a fundamental theorem. To demonstrate the infinity of factors, we consider a matrix $\mathbf{X}_{(N \times p)}$ and the equation

$$\mathbf{X} = \mathbf{YW},$$

where $\mathbf{Y}$ and $\mathbf{W}$ are two conformable factors of $\mathbf{X}$. We make use of the fact that there are infinitely many square orthonormal matrices of a given order (Section 2.7). Let $\mathbf{A}^*$ be any one of this infinite set that is conformable to both $\mathbf{Y}$ and $\mathbf{W}$. We may then form the two new matrices as follows:

$$\mathbf{Y}^* = \mathbf{YA}^*$$

and

$$\mathbf{W}^* = \mathbf{A}^{*\prime}\mathbf{W}.$$

Then, because $\mathbf{A}^*\mathbf{A}^{*\prime} = \mathbf{I}$, we have that

$$\mathbf{X} = \mathbf{YA}^*\mathbf{A}^{*\prime}\mathbf{W} = \mathbf{Y}^*\mathbf{W}^*.$$

Consequently, if a matrix is the product of two matrices (and it always may be so considered), it is also the product of an infinite number of pairs of matrices.

A major or minor product moment matrix is a special case of the preceding example. If $R = X'X$ and A^* is any square orthonormal matrix conformable to X, we have that

$$X^* = A^* X \quad \text{and} \quad X^{*\prime} = X'A^{*\prime},$$

then

$$R = X'A^{*\prime}A^* X = X^{*\prime}X^*.$$

This indicates that R may be considered as the minor product moment of an infinite number of matrices.

It is one of the goals of most multivariate techniques, and especially of factor analysis, to decompose a data matrix, or its major or minor product moments, into components. Since there will be an infinite number of such components, it is self-evident that some criteria will be necessary in order to find a unique pair of them that fulfill certain requirements.

Rank of a matrix

A notion that derives from the subject matter of the preceding section is of special importance in factor analysis with respect to determining the possible dimensions of the factors of a matrix and, in particular, the minimum possible dimensions of the factors.

We state by definition that "the rank of a matrix is the smallest common order among all pairs of matrices whose product is the matrix" (Horst, 1963, p. 335). The meaning of this definition is illustrated by Fig. 2.15. In Fig. 2.15A, $X_{(N \times p)}$ is seen to be the product of two matrices of common order t. Figure 2.15B shows that it is also possible in this case to derive X from matrices, the common order of which is s and $s < t$. In Fig. 2.15C, the common order of the factors is r and $r < s < t$. If we find in this case that factors cannot be found of which the common order is less than r, then the rank of X is r.

It follows from this that the rank of a matrix cannot exceed its smaller dimension; a data matrix, for example, must have a rank equal to, or less than, the number of columns or rows in it, whichever is the lesser. The rank of a product moment matrix derived from the data matrix cannot exceed the smallest dimension of the data matrix. In fact, the rank of the product moment matrix is equal to the rank of the data matrix.

One further definition relating to rank will be found useful, namely, that "a *basic matrix* is one whose rank is equal to its smallest order" (Horst, 1963, p. 337). This implies that a basic matrix cannot be expressed as the product of two matrices the common order of which is less than the smaller order of the matrix.

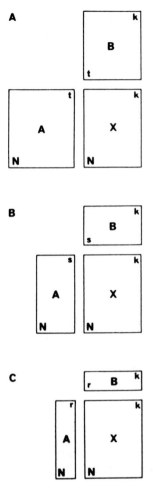

Figure 2.15. The concept of rank in terms of the factors of a matrix.

The concept of rank provides a first insight into the anatomy of factor analysis in that it suggests that a data matrix may be regarded as the product of two matrices of which the orders may be considerably less than that of the data matrix, provided that the latter is not "basic."

Linear independence and rank

The concept of rank outlined in the foregoing pages may also be approached by studying the nature of the relationships between the column (or row) vectors contained in the matrix. These relationships can be described in terms of linear dependence and linear independence.

A vector that is a scalar multiple of another vector, or a weighted or unweighted sum of a set of vectors, is said to be *linearly dependent*. For a vector y and a set of vectors $v_1, v_2, \ldots, v_m$, linear dependence is established by the relationship

$$y = a_1 v_1 + a_2 v_2 + \cdots + a_m v_m \qquad [2.31]$$

if at least one of the scalars a_i in [2.31] is nonzero. As an example, consider the vectors

$$y = \begin{bmatrix} 3 \\ 2 \\ 0 \end{bmatrix}, \quad v_1 = \begin{bmatrix} 2 \\ 1 \\ 1 \end{bmatrix}, \quad v_2 = \begin{bmatrix} 1 \\ 0 \\ 2 \end{bmatrix}.$$

Vector y is linearly dependent on v_1 and v_2 because

$$y = (2)v_1 + (-1)v_2.$$

We note that

$$(2)v_1 + (-1)v_2 - y = 0.$$

Clearly, v_1 and v_2 are not mutually dependent for there is no scalar value, other than zero, that relates one to the other. This leads us to the definition of *linear independence*.

A vector is linearly independent of a set of vectors if it is neither a scalar multiple nor a weighted or unweighted sum of any combination of the members in the set. For the relationship

$$y = a_1 v_1 + a_2 v_2 + \cdots + a_m v_{m_i}$$

all a_i $(i = 1, \ldots, m)$ must be zero in order that y be linearly independent of the set of vectors v_i. For example:

$$y = \begin{bmatrix} 1 \\ 0 \\ 0 \end{bmatrix}; \quad v_1 = \begin{bmatrix} 2 \\ 1 \\ 1 \end{bmatrix}; \quad v_2 = \begin{bmatrix} 1 \\ 0 \\ 2 \end{bmatrix}.$$

No set of scalars a_i, other than zero, satisfies the equality

$$y = a_1 v_1 + a_2 v_2.$$

For an entire set of vectors x_i $(i = 1, \ldots, p)$, such as the column vectors of a data matrix, the set is said to be linearly dependent if the relationship

$$a_1 x_1 + a_2 x_2 + \cdots + a_p x_p = 0$$

holds with at least one nonzero scalar, a_i. If all a_i $(i = 1, \ldots, p)$ are zero, the vectors are linearly independent.

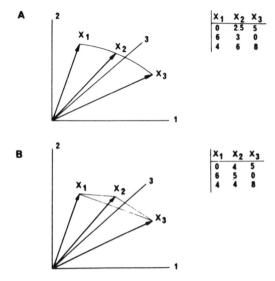

Figure 2.16. (A) A set of linearly dependent column vectors. (B) A set of linearly independent column vectors.

Geometry of linearly dependent and independent vectors

Figure 2.16A illustrates the situation for a set of linearly dependent vectors; x_2 is linearly dependent on x_1 and x_3 for $x_2 = 0.5x_1 + 0.5x_3$ (x_1 and x_3 are linearly independent). The set of vectors is coplanar. Note that any one of the vectors is linearly dependent on the other two.

Figure 2.16B illustrates a set of linearly independent vectors. These vectors are not coplanar.

Our rather simple diagrams are important for giving you an understanding of factor analysis. In both instances, the data matrix utilized was of order 3 by 3. In Fig. 2.16A, the three column vectors are embedded in the three-dimensional object space. Because there are no more than two linearly independent vectors in the set, the vectors are located in a two-dimensional subspace. The two linearly independent vectors are said to form a *basis* for this subspace.

For the general case, consider a data matrix $X_{(N \times p)}$ where only r of the p column vectors are linearly independent. Then the p column vectors will span the r-dimensional subspace of the original N-space, and the r linearly independent vectors will form a basis of this subspace. The $p - r$ linearly dependent vectors are also contained in the subspace.

The number of linearly independent vectors in a set defines the *dimensionality* of the subspace that contains the set of vectors, and each linearly independent vector corresponds to a *dimension* of the subspace. Inasmuch as there are three linearly independent vectors in Fig. 2.16B, they are contained in a three-dimensional space of which they form a basis.

The infinity of basis vectors

An important point is that, although the linearly independent vectors of a set of vectors form a basis, it is not the only possible basis. In fact, for a space of any given dimensionality, there are infinitely many linearly independent vectors that also form a basis. For example, in Fig. 2.16A, vectors x_1 and x_2 form a basis for the set of vectors shown; however, we might also consider any orthogonal vectors q_1 and q_2, lying in the plane of x_1 and x_2, as a basis for this two-dimensional space. Vectors q_1 and q_2 could be rigidly rotated to an infinity of positions within the plane of the remaining vectors and still form a basis for the space, since orthogonal vectors are linearly independent.

The foregoing definitions and concepts are used to formulate another definition of rank, to wit: A matrix of rank r contains at least one set of r linearly independent column or row vectors. Some implications of this definition are as follows:

1. Given a matrix $\mathbf{X}_{(N \times p)}$ ($p < N$) whose rank is r ($r < p$), then a set of r linearly independent vectors can be found to form the basis of the space containing the p column vectors (and also the N row vectors). Any set of r orthogonal vectors will form a suitable basis.
2. The set of column or row vectors, whichever is the lesser, of a basic matrix forms a basis for the vectors of the matrix.

The rank of product moment matrices

Finding the rank of a data matrix is one of the goals of many factor-analytical techniques for, in so doing, we establish the minimum number of linearly independent vectors that can form the basis of the space containing both the column and row vectors that make up the matrix. Many methods of factor analysis determine this rank by analyzing the product moment of the data matrix rather than the data matrix itself. It has already been pointed out that the rank of a product moment matrix is equal to the rank of its factor. That is, if $\mathbf{X}_{(N \times p)}$ is of rank r and

$$\mathbf{B}_{(p \times p)} = \mathbf{X}'_{(p \times N)} \mathbf{X}_{(N \times p)},$$

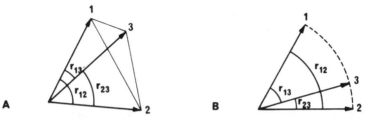

Figure 2.17. (A) Linearly independent vectors enclosed by a tetrahedron of which the volume is proportional to the determinant of the matrix. (B) A singular correlation matrix is produced by coplanar vectors. The volume of the enclosed tetrahedron is zero.

then **B** is of rank r. Similarly, if

$$\mathbf{C}_{(N \times N)} = \mathbf{X}_{(N \times p)} \mathbf{X}'_{(p \times N)},$$

C is of rank r.

The correlation matrix, in particular, is a common starting point in factor analysis. You may recall that this matrix is the minor product moment of the standardized data matrix. Moreover, the elements of this matrix contain the cosines of the angles between all possible pairs of standardized column vectors. It is possible to portray geometrically the relationships between the vectors whose cosines are represented by the correlations. In Fig. 2.17A, for example, the three vectors correspond to the correlation matrix

$$\mathbf{R}_1 = \begin{bmatrix} 1.000 & 0.500 & 0.866 \\ 0.500 & 1.000 & 0.707 \\ 0.866 & 0.707 & 1.000 \end{bmatrix}.$$

In order to maintain the consistency of the angles given by the elements of the matrix, the vectors have to be situated in 3-space. It follows that the three vectors are linearly independent and the rank of the matrix is 3. In Fig. 2.17B, the vectors of another correlation matrix, $\mathbf{R}_2$, are pictured. Here,

$$\mathbf{R}_2 = \begin{bmatrix} 1.000 & 0.500 & 0.707 \\ 0.500 & 1.000 & 0.966 \\ 0.707 & 0.966 & 1.000 \end{bmatrix}$$

and geometrical requirements force the three vectors to lie in 2-space. Thus, any one of the vectors is linearly dependent on the other two, and the rank of the matrix is 2. This example is demonstrated in MATLAB in the appendix.

We note that in Fig. 2.17A, a tetrahedron may be constructed by joining the tips of the three vectors. As recorded by Aitken (1951), the volume of this tetrahedron is proportional to the determinant of the correlation matrix. In these examples we have

$$|\mathbf{R}_1| = 0.5 - (-0.0562) + (-0.4438) = 0.1124,$$

$$|\mathbf{R}_2| = 0.0668 - (-0.0915) + (-0.1583) = 0.0.$$

Because the vectors shown in fig. 2.17B are coplanar, the volume of the tetrahedron formed by joining the vector ends is zero. (You can think of it having collapsed into a plane, as it were). The determinant of the associated correlation matrix is zero.

The relationships between linearly independent vectors, rank, determinant, and the geometrical configuration of vectors may be generalized to spaces of any dimension. The foregoing discussion of the rank of a product moment matrix suggests the following conclusions:

1. Any given matrix of which the rank r is less than its smaller dimension, p, may be conveniently thought of in terms of the two types of vectors it contains – the set of $(p - r)$ linearly dependent vectors and the set of r linearly independent vectors.
2. The rank defines the dimensionality of the space containing all the vectors, and the r linearly independent vectors form one of an infinite number of possible bases for this space.

The exact rank of a matrix may be determined by any one of several methods. We shall discuss the eigenstructure method in detail, as it offers several advantages over other methods.

2.11 EIGENVALUES AND EIGENVECTORS

You will have perceived by now that by finding the rank of a data matrix, much can be learned about the complexity of the contained data. The rank will give the dimensionality of the matrix, which determines the number of linearly independent vectors necessary to span the space containing the vectors of the matrix. The eigenvectors and eigenvalues of a matrix will not only determine the rank but will also yield a set of linearly independent basis vectors.

Eigenvalues and eigenvectors (synonyms: latent roots and vectors, characteristic roots and vectors, proper values and vectors) are used in many scientific connotations. Because of the diversity of the subjects to which they are applied, they are open to a wide variety of interpretations. We shall begin by giving you an algebraic account of them within the scope of this book.

Algebraic review

We restrict our discussion to the eigenvalues of a real symmetric matrix **R**, which might be, for example, the minor or major product moment of a data matrix, the covariance, or the correlation matrix.

An eigenvector of **R** is a vector u with elements, not all zero, such that

$$\mathbf{R}u = u\lambda, \tag{2.32}$$

where λ is an unknown scalar.

Essentially, we need to find a vector such that the vector $\mathbf{R}u$ is proportional to u. Another form of this equation is

$$\mathbf{R}u - u\lambda = 0,$$

or

$$(\mathbf{R} - \lambda\mathbf{I})u = 0, \tag{2.33}$$

where 0 is the null vector.

This implies that the unknown vector u is orthogonal to all row vectors of $(\mathbf{R} - \lambda\mathbf{I})$. Equation [2.33] represents a system of homogeneous equations. For a 2×2 matrix **R**, this is

$$(r_{11} - \lambda)u_1 + r_{12}u_2 = 0,$$
$$r_{21}u_1 + (r_{22} - \lambda)u_2 = 0. \tag{2.34}$$

If $\mathbf{R} - \lambda\mathbf{I}$ is nonsingular, one can solve [2.33] by premultiplying both sides of the equation by $(\mathbf{R} - \lambda\mathbf{I})^{-1}$; however, this would yield the trivial solution of $u = 0$. Hence, it is only possible to obtain a nontrivial solution if $(\mathbf{R} - \lambda\mathbf{I})$ is singular, that is, its determinant must be zero.

The first step in solving for u and λ requires setting the determinant of $(\mathbf{R} - \lambda\mathbf{I})$ to zero, thus

$$|\mathbf{R} - \lambda\mathbf{I}| = 0. \tag{2.35}$$

This implies that the rank of $(\mathbf{R} - \lambda\mathbf{I})$ is at least one less than its order. Again, for a 2×2 matrix, the general form of this determinantal equation, known as the *characteristic equation*, is

$$\begin{vmatrix} r_{11} - \lambda & r_{12} \\ r_{21} & r_{22} - \lambda \end{vmatrix} = 0, \tag{2.36}$$

which on expansion yields

$$(r_{11} - \lambda)(r_{22} - \lambda) - r_{12}r_{21} = 0.$$

Multiplying out and reordering terms gives

$$\lambda^2 - \lambda(r_{11} + r_{22}) + (r_{11}r_{22} - r_{12}r_{21}) = 0,$$

which is a quadratic equation, the solution of which is (remembering that $r_{12} = r_{21}$)

$$\tfrac{1}{2}\left[-(r_{11} + r_{22}) \pm \left\{(r_{11} + r_{22})^2 - 4(r_{11}r_{22} - r_{12}^2)\right\}^{1/2}\right]$$

Two values for λ, λ_1 and λ_2, will be obtained. These two roots are termed the eigenvalues of $\mathbf{R}$. It should be noted that

$$\lambda_1 + \lambda_2 = r_{11} + r_{22},$$

$$\lambda_1\lambda_2 = r_{11}r_{22} - r_{12}r_{21} = |\mathbf{R}|.$$

In the general case of $\mathbf{R}_{(p \times p)}$, the characteristic equation will be a polynomial of degree p and will yield p roots, or eigenvalues, $\lambda_1, \lambda_2, \ldots, \lambda_p$. If $\mathbf{R}$ is a real, square symmetric matrix, these eigenvalues are always real. However, these p eigenvalues may not all be different and some may be equal to zero. If two or more eigenvalues are equal, we say that this is a multiple eigenvalue; otherwise, the eigenvalue is said to be distinct.

Once the eigenvalues have been obtained, the associated eigenvectors may be found from [2.33]. A unique solution cannot be obtained for an eigenvector. That is, if u is a solution, so is cu, c being a scalar. By convention, therefore, eigenvectors are always normalized; they are constrained to be of unit length.

Associated with each distinct eigenvalue there is an eigenvector that is unique, except for its sign (i.e., it may be multiplied by -1), and eigenvectors associated with different eigenvalues are orthogonal. For multiple eigenvalues, there are many different solutions to [2.33], but one can always choose an eigenvector that is orthogonal to all other eigenvectors. Consequently, whether or not there are multiple eigenvalues present, one can always associate an eigenvector of unit length to each eigenvalue such that every eigenvector is orthogonal to all of the others.

If the eigenvalues λ_i $(i = 1, \ldots, p)$ are placed as the elements of a diagonal matrix $\mathbf{\Lambda}$ and the eigenvectors collected as columns into the matrix $\mathbf{U}$, then [2.32] becomes in matrix form

$$\mathbf{RU} = \mathbf{U\Lambda}. \qquad [2.37]$$

The matrix $\mathbf{U}$ is square orthonormal, so that

$$\mathbf{U'U} = \mathbf{UU'} = \mathbf{I},$$

that is, every eigenvector is orthonormal to all others in the set. Postmultiplication of [2.37] by $\mathbf{U'}$ gives

$$\mathbf{R} = \mathbf{U\Lambda U'}, \qquad [2.38]$$

an equation of fundamental importance in many multivariate methods. It shows that the symmetric matrix **R** may be represented in terms of its eigenvalues and eigenvectors. Another way to write this is

$$\mathbf{R} = \lambda_1 u_1 u_1' + \lambda_2 u_2 u_2' + \cdots + \lambda_p u_p u_p',$$

which illustrates that **R** is a weighted sum of matrices $u_i u_i'$ of order p by p and of rank 1. Furthermore, each term is orthogonal to all other terms since for $i \neq j$, $u_i' u_j = 0$; therefore,

$$u_i u_i' u_j' u_j = 0.$$

Premultiplication of [2.37] by **U'** gives

$$\mathbf{\Lambda} = \mathbf{U'RU}, \qquad\qquad [2.39]$$

which shows that **U** is a matrix that reduces **R** to diagonal form.

Some useful properties of eigenvalues are listed next:

1. Trace $\mathbf{\Lambda}$ = trace **R**; the sum of the eigenvalues equals the sum of the elements in the principal diagonal of the matrix. [2.40]
2. $\prod_{i=1}^{p} \lambda_i = |\mathbf{R}|$; the product of the eigenvalues equals the determinant of the matrix. Clearly, if one or more eigenvalues are zero, the determinant of **R** will be zero, thus indicating that **R** is singular. [2.41]
3. The number of nonzero eigenvalues equals the rank of **R**. [2.42]

An example of the calculations for a simple eigenvalue problem now follows.

To illustrate the calculation of eigenvalues and eigenvectors, consider the data matrix $\mathbf{X}_{(8 \times 2)}$:

$$\mathbf{X} = \begin{bmatrix} -8 & -1 \\ 6 & 10 \\ -2 & -10 \\ 8 & 1 \\ 0 & 3 \\ -6 & -6 \\ 0 & -3 \\ 2 & 6 \end{bmatrix}. \quad .$$

Step 1. Compute the minor product moment, $\mathbf{X'X} = \mathbf{C}$:

$$\mathbf{C} = \begin{bmatrix} 208 & 144 \\ 144 & 292 \end{bmatrix}.$$

Step 2. Establish the characteristic equation:

$$\begin{vmatrix} 208 - \lambda & 144 \\ 144 & 292 - \lambda \end{vmatrix} = 0,$$

$$(208 - \lambda)(292 - \lambda) - 144 \times 144 = 0,$$

$$\lambda^2 - 500\lambda + 40{,}000 = 0.$$

Step 3. Calculate the roots:

$$\lambda_1 = 400 \quad \text{and} \quad \lambda_2 = 100.$$

Step 4. Substitute these values into the next equation to obtain the eigenvectors for λ_1:

$$(\mathbf{C} - \lambda_1\mathbf{I})\mathbf{u}_1 = \mathbf{0};$$

$$\left[\begin{bmatrix} 208 & 144 \\ 144 & 292 \end{bmatrix} - \begin{bmatrix} 400 & 0 \\ 0 & 400 \end{bmatrix} \right] \begin{bmatrix} u_{11} \\ u_{21} \end{bmatrix} = \begin{bmatrix} 0 \\ 0 \end{bmatrix},$$

$$-192u_{11} + 144u_{21} = 0,$$

$$144u_{11} - 108u_{21} = 0,$$

$$-192u_{11} + 144u_{21} = 144u_{11} - 108u_{21},$$

$$-336u_{11} = -252u_{21},$$

$$u_{11} = 0.75u_{21}.$$

Setting $u_{11} = 1$, then $u_{21} = 1.33$. Normalizing to unit length.

$$u_{11} = 1/1.66 = 0.6, \qquad u_{21} = 1.33/1.66 = 0.8.$$

Thus, the first eigenvector is

$$\mathbf{u}_1 = \begin{bmatrix} 0.6 \\ 0.8 \end{bmatrix}.$$

For λ_2,

$$\left[\begin{bmatrix} 208 & 144 \\ 144 & 292 \end{bmatrix} - \begin{bmatrix} 100 & 0 \\ 0 & 100 \end{bmatrix} \right] \begin{bmatrix} u_{12} \\ u_{22} \end{bmatrix} = \begin{bmatrix} 0 \\ 0 \end{bmatrix},$$

$$108u_{12} + 144u_{22} = 0,$$

$$144u_{12} + 192u_{22} = 0,$$

$$u_{12} = -1.33u_{22};$$

Thus, the second eigenvector is

$$u_2 = \begin{bmatrix} -0.8 \\ 0.6 \end{bmatrix}.$$

Step 5. Check results,

$$Cu = \lambda u:$$

For λ_1:

$$\begin{pmatrix} 208 & 144 \\ 144 & 292 \end{pmatrix} \begin{pmatrix} 0.6 \\ 0.8 \end{pmatrix} = \begin{pmatrix} 240 \\ 320 \end{pmatrix},$$

$$400 \begin{pmatrix} 0.6 \\ 0.8 \end{pmatrix} = \begin{pmatrix} 240 \\ 320 \end{pmatrix}.$$

For λ_2:

$$\begin{pmatrix} 208 & 144 \\ 144 & 292 \end{pmatrix} \begin{pmatrix} -0.8 \\ 0.6 \end{pmatrix} = \begin{pmatrix} -80 \\ 60 \end{pmatrix},$$

$$100 \begin{pmatrix} -0.8 \\ 0.6 \end{pmatrix} = \begin{pmatrix} -80 \\ 60 \end{pmatrix}.$$

You have just learned that the rank of a square symmetric matrix is equal to the number of nonzero eigenvalues of the matrix. More precisely, if for a square matrix **R** of order p there are m zero eigenvalues, the rank of **R** is $(p - m)$. Furthermore, the set of $(p - m)$ orthogonal eigenvectors associated with the $(p - m)$ nonzero eigenvalues forms a set of orthogonal basis vectors that span the space of the $(p - m)$ linearly independent vectors of **R**.

Geometrical interpretation of eigenvalues and eigenvectors

Figure 2.18 shows a bivariate scatter diagram of observations on two variables. The data are assumed to be in standardized form and the variables normally distributed. The data points can be enclosed by equal-density contours. The inner contour surrounds 66% of the points, the outer, 95%. It is well known that these contours form ellipses for bivariate, normally distributed variables. If the variables are uncorrelated, the ellipses will describe a circle; if perfectly correlated, they will degenerate into a line. For more than two variables, the data points can be viewed as forming a p-dimensional hyperellipsoid (p variables).

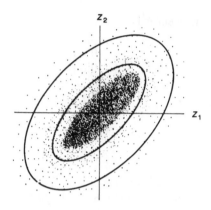

Figure 2.18. Bivariate scatter diagram with equal-density contour lines. The inner ellipse encloses 66% of the data points, the outer ellipse 95% of the data points.

Pearson (1901) and, later, Hotelling (1933) realized that the major and minor axes of the hyperellipsoid can be found from the eigenvectors of the correlation matrix. The equation of the hyperellipsoid is

$$w'R^{-1}w = c,$$

where R is the correlation of covariance matrix of the data, w is a vector of coordinates for the points on the ellipsoid, and c is a constant. The major axis of the ellipsoid is determined by the points on the ellipsoid furthest away from the centroid. To locate these points, you need only find the point on the ellipsoid whose distance to the origin is a maximum. The squared distance from the point, w, on the ellipsoid to the origin is $w'w$, so we have to maximize $w'w$ subject to $w'R^{-1}w = c$. Using simple calculus to solve this maximization problem leads us to the result:

$$Rw = \lambda w,$$

with $w'w = \lambda$. Hence, w must be an eigenvector of R, corresponding to the largest eigenvalue λ. The eigenvector w may be scaled to unit length as before; it defines the direction of the major axis of the ellipsoid. The length of the major axis, that is, the distance from the origin to the point on the ellipsoid furthest away from the origin, is $\lambda^{1/2}$.

Using a similar line of reasoning, it can be demonstrated that the second eigenvector corresponds to the largest minor axis of the ellipsoid, the third eigenvector to the third largest minor axis of the ellipsoid, and so on.

The interpretation of the eigenvalues and eigenvectors of covariance and correlation matrices leads to several useful generalizations:

1. The location of eigenvectors along the principal axes of the hyperellipsoids in effect positions the eigenvectors so as to coincide with the directions of maximum variance. The eigenvector associated with the largest eigenvalue determines the direction of maximum variance of the data points; the eigenvector associated with the second largest eigenvector locates the direction of maximum variance orthogonal to the first.

2. The location of eigenvectors may be viewed as a problem of rotation or transformation. The original axes, representing the original measured variables, are rotated to new positions according to the criteria reviewed previously.

   The rows of the matrix of eigenvectors are the coefficients that rotate the axes of the variables to positions along the major and minor axes. The elements of the vectors can be considered as the cosines of the angles of rotation from the old to a new system of coordinate axes. The columns of the matrix of eigenvectors give the direction cosines of the eigenvectors themselves.

3. Eigenvectors are linearly independent vectors that are linear combinations of the original variables. As such, they may be viewed as "new" variables with the desirable property that they are uncorrelated and account for the variance of the data in decreasing order of importance.

4. The sum of the squared projections of data points onto the eigenvectors is proportional to the variance along the eigenvectors. This variance is equal to the associated eigenvalue. The square root of the eigenvalue can therefore be used as the standard deviation of the "new" variable represented by the eigenvector.

5. It follows from the foregoing paragraph that an eigenvalue of zero indicates that the corresponding minor axis is of zero length. This suggests that the dimensionality of the space containing the data points is less than the original space – which, in turn, implies that the rank of the data matrix is less than its smaller dimension.

In summary, the eigenvectors associated with the nonzero eigenvalues of a covariance or correlation matrix represent a set of orthogonal axes to which the data points may be referred. They are a set of linearly independent basis vectors the positions of which are located so as to account for maximum variance, which variance is given by the associated eigenvalues. The dimensionality of the space is given by the number of nonzero eigenvalues.

The preceding interpretation of eigenvalues and eigenvectors may also be applied to the vectorial interpretation of correlation.

2.12 BASIC STRUCTURE OF A MATRIX AND THE ECKART–YOUNG THEOREM

We begin with a brief historical background to the Q-R-mode duality relationship. A precursor of principal component analysis was presented in a remarkable paper by the renowned English mathematician J. J. Sylvester [wrongly attributed a North American origin by Preisendorfer (1988)], who in 1889 treated the reduction of a square matrix into its singular value components.

A rigorous development of the singular value decomposition appeared later in the hands of the French mathematician Autonne (1913, 1915). The full extension to rectangular matrices came in an article by Eckart and Young (1936). The Eckart–Young partnership is interesting to contemplate. Eckart, a theoretical physicist, clearly saw the connection between the singular value decomposition of a rectangular matrix and the problem of diagonalizing quadratic forms in physics. There is a little-known sequel to the 1936 paper (Eckart and Young, 1939).

Subsequent developments seem to stem from the particular direction initiated by Hirschfeld (1935), then by Bénzécri (1973), whereby the special flavor of originally francophone work developed.

In Section 2.11, the eigenvalues and eigenvectors of square symmetric matrices were defined. Even more fundamental to the theory of factor analysis than these, as developed in this book, is the notion of the *basic structure* or *singular value decomposition* of an arbitrary rectangular matrix. You will soon perceive how we are able to unite the concepts of Q- and R-mode analyses by means of the theory of basic structure.

Let **X** be a given data matrix of order N by p, where $N > p$, and let r be the rank of **X**. In most cases, $r = p$, but the following account is quite general and holds also for $r < p$. The basic structure or singular value decomposition of **X** states that

$$\mathbf{X} = \mathbf{V}\mathbf{\Gamma}\mathbf{U}' \qquad [2.43]$$

where (1) **V** is an N-by-r matrix the columns of which are orthonormal, that is, $\mathbf{V}'\mathbf{V} = \mathbf{I}_r$; (2) **U** is a p-by-r matrix of which the columns are orthonormal, that is, $\mathbf{U}'\mathbf{U} = \mathbf{I}_r$; and (3) $\mathbf{\Gamma}$ is a diagonal matrix of order r by r with positive diagonal elements $\gamma_1, \gamma_2, \ldots, \gamma_r$, called *singular values* of **X**.

The idea of basic structure of a matrix is sometimes referred to as the *Eckart–Young theorem*. If you want to learn more about this important theorem, we refer you to the paper by Johnson (1963). The mathematical operation involved is not peculiar to statistics; it is well known in algebra, as indicated at the beginning of this section. In terms of the diagonal elements $\gamma_1, \gamma_2, \ldots, \gamma_r$ of Γ, the columns $v_1, v_2, \ldots, v_r$, of V, and the columns $u_1, u_2, \ldots, u_r$ of U, the basic structure of X, as defined in [2.43], may be written as

$$X = \gamma_1 v_1 u_1' + \gamma_2 v_2 u_2' + \cdots + \gamma_r v_r u_r', \qquad [2.44]$$

which shows that the matrix X of rank r is a linear combination of r matrices of rank 1.

The relationships between the quantities Γ, U, and V in the basic structure of X, and the eigenvalues and eigenvectors of the major and minor product moments of X, will now be established. The following fundamental relationships hold:

1. The major product moment XX', which is square symmetric and of order N by N, has r positive eigenvalues and $(N - r)$ zero eigenvalues. The positive eigenvalues of XX' are $\gamma_1^2, \gamma_2^2, \ldots, \gamma_r^2$ and the corresponding eigenvectors are $v_1, v_2, \ldots, v_r$. Thus, the singular values, $\gamma_1, \gamma_2, \ldots, \gamma_r$ are the positive square roots of the positive eigenvalues of XX', and the columns of V are the corresponding eigenvectors.

2. The minor product moment $X'X$, which is square, symmetric, and of order p by p, has r positive eigenvalues and $(p - r)$ zero eigenvalues. The positive eigenvalues are $\gamma_1^2, \gamma_2^2, \ldots, \gamma_r^2$, and the corresponding eigenvectors are $u_1, u_2, \ldots, u_r$. Thus, the singular values $\gamma_1, \gamma_2, \ldots, \gamma_r$ are the positive square roots of the positive eigenvalues of $X'X$, and the columns of U are the corresponding eigenvectors.

3. The positive eigenvalues of XX' and $X'X$ are the same, namely, $\gamma_1^2, \gamma_2^2, \ldots, \gamma_r^2$. Furthermore, if v_m is an eigenvector of XX', and u_m an eigenvector of $X'X$, corresponding to one of these eigenvalues γ_m^2, then the following relationships hold between u_m and v_m, for $m = 1, \ldots, r$:

$$v_m = (1/\gamma_m)Xu_m \quad \text{and} \quad u_m = (1/\gamma_m)X'v_m. \qquad [2.45]$$

These relationships make it possible to compute v_m from u_m, and vice versa. In compact form, the relationships are

$$V = XU\Gamma^{-1} \quad \text{and} \quad U = X'V\Gamma^{-1}. \qquad [2.46]$$

The analysis of the minor product moment $X'X$ is referred to as *R-mode analysis* and that of the major product moment XX' as *Q-mode*

analysis. The relationships discussed in (3) show how the results of a
Q-mode analysis may be obtained from those of an R-mode analysis,
and conversely. In practice, the steps in the calculations of the singular
value decomposition are as follows:

1. Compute XX' or $X'X$, whichever has the smaller order. Let us
   here assume that this is $X'X$.
2. Compute the positive eigenvalues, $\lambda_1, \lambda_2, \ldots, \lambda_r$, of $X'X$ and
   the corresponding eigenvectors, $u_1, u_2, \ldots, u_r$.
3. Compute the singular values $\gamma_1 = \sqrt{\lambda_1}$, $\gamma_2 = \sqrt{\lambda_2}, \ldots, \gamma_r$
   $= \sqrt{\lambda_r}$.
4. Compute $v_1, v_2, \ldots, v_r$ from

$$v_m = (1/\gamma_m)X u_m \qquad (m = 1, 2, \ldots, r).$$

As an illustration, consider the problem of determining the basic
structure of the data matrix X following [2.42]:

$$X = \begin{bmatrix} -8 & -1 \\ 6 & 10 \\ -2 & -10 \\ 8 & 1 \\ 0 & 3 \\ -6 & -6 \\ 0 & -3 \\ 2 & 6 \end{bmatrix}.$$

Step 1. Compute the minor product moment:

$$X'X = \begin{bmatrix} 208 & 144 \\ 144 & 292 \end{bmatrix}.$$

Step 2. Compute the eigenvalues and eigenvectors of $X'X$. These
were previously found to be as follows:

$$\lambda_1 = 400, \qquad \lambda_2 = 100;$$

$$u_1 = \begin{bmatrix} 0.6 \\ 0.8 \end{bmatrix}, \qquad u_2 = \begin{bmatrix} -0.8 \\ 0.6 \end{bmatrix}.$$

Step 3. Compute the singular values. These are the square roots of
the eigenvalues. Hence,

$$\gamma_1 = 20 \quad \text{and} \quad \gamma_2 = 10.$$

Step 4. Compute v_1 and v_2 as $v_1 = (1/\gamma_1)Xu_1$, and analogously for v_2:

$$v_1 = (1/\gamma_1)Xu_1 = 0.05 \begin{bmatrix} -8 & -1 \\ 6 & 10 \\ -2 & -10 \\ 8 & 1 \\ 0 & 3 \\ -6 & -6 \\ 0 & -3 \\ 2 & 6 \end{bmatrix} \begin{bmatrix} 0.6 \\ 0.8 \end{bmatrix}$$

$$= 0.05 \begin{bmatrix} -5.6 \\ 11.6 \\ -9.2 \\ 5.6 \\ 2.4 \\ -8.4 \\ -2.4 \\ 6.0 \end{bmatrix} = \begin{bmatrix} -0.28 \\ 0.58 \\ -0.46 \\ 0.28 \\ 0.12 \\ -0.42 \\ -0.12 \\ 0.30 \end{bmatrix},$$

$$v_2 = (1/\gamma_2)Xu_2 = 0.10 \begin{bmatrix} -8 & -1 \\ 6 & 10 \\ -2 & -10 \\ 8 & 1 \\ 0 & 3 \\ -6 & -6 \\ 0 & -3 \\ 2 & 6 \end{bmatrix} \begin{bmatrix} -0.8 \\ 0.6 \end{bmatrix}$$

$$= 0.10 \begin{bmatrix} 5.8 \\ 1.2 \\ -4.4 \\ -5.8 \\ 1.8 \\ 1.2 \\ -1.8 \\ 2.0 \end{bmatrix} = \begin{bmatrix} 0.58 \\ 0.12 \\ -0.44 \\ -0.58 \\ 0.18 \\ 0.12 \\ -0.18 \\ 0.20 \end{bmatrix}.$$

Summarizing these results, we have that

$$V = \begin{bmatrix} -0.28 & 0.58 \\ 0.58 & 0.12 \\ -0.46 & -0.44 \\ 0.28 & -0.58 \\ 0.12 & 0.18 \\ -0.42 & 0.12 \\ -0.12 & -0.18 \\ 0.30 & 0.20 \end{bmatrix}, \quad \Gamma = \begin{bmatrix} 20 & 0 \\ 0 & 10 \end{bmatrix}, \quad U = \begin{bmatrix} 0.6 & -0.8 \\ 0.8 & 0.6 \end{bmatrix}.$$

It may be verified that $\mathbf{V\Gamma U'} = \mathbf{X}$ and that the two columns of $\mathbf{V}$ are orthonormal, that is, that $\mathbf{V'V} = \mathbf{I}$.

To terminate this illustration, we shall also verify that the columns of $\mathbf{V}$ are really the eigenvectors of the major product moment $\mathbf{XX'}$, corresponding to the eigenvalues λ_1 and λ_2. The easiest way of demonstrating this is to check that $(\mathbf{XX'})\mathbf{V} = \mathbf{V\Lambda}$.

We have that

$$\mathbf{XX'} = \begin{bmatrix}
65 & -58 & 26 & -63 & -3 & 54 & 3 & -22 \\
-58 & 136 & -112 & 58 & 30 & -96 & -30 & 72 \\
26 & -112 & 104 & -26 & -30 & 72 & -30 & -64 \\
-63 & 58 & -26 & 65 & 3 & -54 & -3 & 22 \\
-3 & 30 & -30 & 3 & 9 & -18 & -9 & 18 \\
54 & -96 & 72 & -54 & -18 & 72 & 18 & -48 \\
3 & -30 & -30 & -3 & -9 & 18 & 9 & -18 \\
-22 & 72 & -64 & 22 & 18 & -48 & -18 & 40
\end{bmatrix}$$

and

$$\mathbf{X'XV} = \begin{bmatrix}
-112 & 58 \\
232 & 12 \\
-182 & -44 \\
112 & -58 \\
48 & 18 \\
168 & 12 \\
-48 & -18 \\
120 & 20
\end{bmatrix} = \mathbf{V\Lambda}.$$

2.13 LEAST-SQUARES PROPERTIES OF EIGENVALUES AND EIGENVECTORS

In Section 2.11, it was shown how every symmetric matrix $\mathbf{S}$ of order p by p can be represented in terms of its eigenvalues $\lambda_1 > \lambda_2 > \cdots > \lambda_r$ and eigenvectors $\mathbf{u}_1, \mathbf{u}_2, \ldots, \mathbf{u}_r$ as

$$\mathbf{S} = \lambda_1 \mathbf{u}_1 \mathbf{u}_1' + \lambda_2 \mathbf{u}_2 \mathbf{u}_2' + \cdots + \lambda_r \mathbf{u}_r \mathbf{u}_r', \qquad [2.47]$$

where r is the rank of $\mathbf{S}$. Similarly, it was shown in Section 2.12 that every rectangular matrix $\mathbf{X}$ of order N by p and of rank r can be represented in terms of its singular values $\gamma_1 > \gamma_2 > \cdots > \gamma_r$ and vectors $\mathbf{v}_1, \mathbf{v}_2, \ldots, \mathbf{v}_r$ and $\mathbf{u}_1, \mathbf{u}_2, \ldots, \mathbf{u}_r$ as

$$\mathbf{X} = \gamma_1 \mathbf{v}_1 \mathbf{u}_1' + \gamma_2 \mathbf{v}_2 \mathbf{u}_2' + \cdots + \gamma_r \mathbf{v}_r \mathbf{u}_r'. \qquad [2.48]$$

Still more fundamental than these two representations are the least-squares properties of the successive terms in them. This has to do with the solutions to the following two problems.

Problem 1

Given a symmetric Gramian matrix (a Gramian matrix has nonnegative eigenvalues) $S_{(p \times p)}$ of rank r, find a symmetric Gramian matrix $T_{(p \times p)}$ of a given lower rank, $k < r$, that approximates S in the least-squares sense.

Problem 2

Given a rectangular matrix $X_{(N \times p)}$ of rank r and a number $k < r$, find a matrix $W_{(N \times p)}$ of rank k that approximates X in the least-squares sense.

The solutions to these problems are

$$T = \lambda_1 u_1 u_1' + \lambda_2 u_2 u_2' + \cdots + \lambda_k u_k u_k',$$
$$W = \gamma_1 v_1 u_1' + \gamma_2 v_2 u_2' + \cdots + \gamma_k v_k u_k', \qquad [2.49]$$

respectively, which are the sums of the first k terms in each of the representations. The matrices T and W are called *lower-rank least-squares matrix approximations* of S and X, respectively. The least-squares property can be expressed by saying that T is as close to S as possible in the sense that the sum of squares of all the elements in $S - T$ is a minimum. Similarly, W is as close to X as possible in the sense that the sum of squares of all the elements of $X - W$ is a minimum. A measure of the closeness or goodness of the approximation is, in both cases,

$$\lambda_{k+1} + \lambda_{k+2} + \cdots + \lambda_r;$$

that is, this measure is the sum of the $r - k$ smallest eigenvalues of $X'X$. Sometimes, the relative measure

$$\frac{\lambda_{k+1} + \lambda_{k+2} + \cdots + \lambda_r}{\lambda_1 + \lambda_2 + \cdots + \lambda_r} \qquad [2.50]$$

or some related function is used as a measure of goodness of the approximation.

The importance of these two problems will now be discussed in the most common case of $r = p$. In Problem 1, S could be the covariance matrix of p linearly independent variables. The matrix T could also represent the covariance matrix of p variables but, since the rank of T is $k < p$, these p variables are linearly dependent on k variables. Hence, the original p variables, whose covariance matrix is S, may be generated approximately by k variables. If we consider Problem 2, these ideas will become clearer. The original matrix, X, of order N by

p, may be written as (cf. [2.43])

$$X = V\Gamma U',$$

where V is of order N by p, with orthonormal columns, Γ is a diagonal matrix of order p by p, and U is a square orthonormal matrix of order p by p. The lower-rank approximation, W, may be written as

$$W = V_k\Gamma_kU'_k,$$

where V_k consists of the first k columns of V, Γ_k consists of the first k rows of columns of Γ, and U_k is composed of the first k columns of U. Since $W \approx X$, we have that

$$X \approx V_k\Gamma_kU'_k. \qquad [2.51]$$

Postmultiplication of this matrix by $U_k\Gamma_k^{-1}$ gives (compare [2.46])

$$V_k = XU_k\Gamma_k^{-1}. \qquad [2.52]$$

The matrix $U_k\Gamma_k^{-1}$, of order p by k, represents a transformation of the rows of X from Euclidean p-space to Euclidean k-space, and [2.51] shows that there is a transformation from the N-by-p matrix X to the N-by-k matrix V_k. The matrix X represents N points in Euclidean p-space that may be approximately embedded in Euclidean k-space. The matrix V_k represents the coordinates of the N points in Euclidean k-space.

2.14 CANONICAL ANALYSIS OF ASYMMETRY

Skew-Symmetric Matrices

A quadratic matrix A is said to be *skew symmetric* (or *antimetric*) if it is equal to the negative of its transpose,

$$A = -A'.$$

Thus $a_{ik} = -a_{ki}$ and $a_{ii} = 0$. This can be seen from the following example:

$$\text{If} \quad A = \begin{bmatrix} 0 & 2 & 4 \\ -2 & 0 & -1 \\ -4 & 1 & 0 \end{bmatrix}, \quad \text{then} \quad A' = \begin{bmatrix} 0 & -2 & -4 \\ 2 & 0 & 1 \\ 4 & -1 & 0 \end{bmatrix} = -A.$$

Every quadratic matrix A can be partitioned into a symmetric and a skew-symmetric part:

$$A = M + N, \qquad [2.53]$$

where

$$\mathbf{M} = \tfrac{1}{2}(\mathbf{A} + \mathbf{A}'), \qquad [2.54]$$

$$\mathbf{N} = \tfrac{1}{2}(\mathbf{A} - \mathbf{A}'). \qquad [2.55]$$

An example demonstrates these relationships:

$$\mathbf{A} = \begin{bmatrix} 5 & 1 & -4 \\ 3 & 7 & 8 \\ -2 & 0 & 3 \end{bmatrix} = \begin{bmatrix} 5 & 2 & -3 \\ 2 & 7 & 4 \\ -3 & 4 & 3 \end{bmatrix} = \begin{bmatrix} 0 & -1 & -1 \\ 1 & 0 & 4 \\ 1 & -4 & 0 \end{bmatrix}.$$

The decomposition [2.53] gives, in statistical connections, an orthogonal break-down of sums of squares

$$\sum_{i,j} a_{ij}^2 = \sum_{i,j} m_{ij}^2 + \sum_{i,j} n_{ij}^2, \qquad [2.56]$$

which mean that $\mathbf{M}$ and $\mathbf{N}$ can be analyzed separately (Gower, 1977).

The symmetric part $\mathbf{M}$ can be analyzed by multidimensional scaling through the computation of coordinates for the p objects, which represent the symmetric values in two or three dimensions.

The canonical analysis of a skew-symmetric matrix $\mathbf{N}$ depends on its canonical decomposition into a sum of elementary skew-symmetric matrixes of rank 2. Thus,

$$\mathbf{N} = \mathbf{USV}', \qquad [2.57]$$

where $\mathbf{V} = \mathbf{UJ}$, $\mathbf{U}$ is an orthogonal matrix, $\mathbf{J}$ is the elementary block-diagonal skew-symmetric orthogonal matrix composed of 2×2 diagonal blocks

$$\begin{bmatrix} 0 & 1 \\ -1 & 0 \end{bmatrix},$$

and $\mathbf{S}$ is a diagonal matrix holding the singular values s_i. If n is odd, the final singular value s_n is zero and the final diagonal element of $\mathbf{J}$ is unity. Because $\mathbf{N}$ is skew symmetric, the singular values occur pairwise and can be arranged in descending order of magnitude. From the Eckart–Young (1936) theorem, the terms corresponding to the r largest singular values provide a least-squares fit of rank $2r$ to $\mathbf{N}$.

The pair of columns of $\mathbf{U}$ corresponding to each pair of equal singular values, when scaled by the square root of the corresponding singular values, holds the coordinates of the p specimens in a two-dimensional space (a plane) that approximates the skew symmetry. The proportion of the total skew symmetry represented by this plane is yielded by the size of the corresponding singular value. The first singular value s_1 ($= s_2$) is the largest; hence, the first two columns of $\mathbf{U}$ contain the coordinates of the p specimens in the plane accounting for the largest proportion of skew symmetry. If the first singular value is very much greater than the other singular values, the plot of the first plane can be expected to give a good approximation to the values of **N**. If this is not so, then other planes will be required in order to yield a good approximation.

Skew symmetry is nonmetric. Therefore, the value of **N** are represented in the planes by areas of triangles made with the origin. Hence, the area of triangle (P_i, P_j, P_0) defined by the points i, j, and the origin, is proportional to the i, jth value in matrix **N**. Moreover,

$$\text{Area}(P_i, P_j, P_0) = -(P_j, P_i, P_0).$$

Pairs of points lying on a line through the origin, as well as coincident pairs of points, will give zero areas. Triangles on the same base, which include a vertex at the origin and the other vertex on a line parallel to the base, are equal in area.

3 Aims, ideas, and models of factor analysis

3.1 INTRODUCTION

Factor analysis is a generic term that we use to describe a number of methods designed to analyze interrelationships within a set of variables or objects. Although the various techniques differ greatly in their objectives and in the mathematical model underlying them, they all have one feature in common, the construction of a few hypothetical variables (or objects), called factors, that are supposed to contain the essential information in a larger set of observed variables or objects. The factors are constructed in a way that reduces the overall complexity of the data by taking advantage of inherent interdependencies. As a result, a small number of factors will usually account for approximately the same amount of information as do the much larger set of original observations. Thus, factor analysis is, in this one sense, a multivariate method of data reduction.

Psychologists developed, and have made extensive use of, factor analysis in their studies of human mental ability. The method was primarily devised for analyzing the observed scores of many individuals on a large battery of psychological tests. Tests of aptitude and achievement were designed to measure various aspects of mental ability but it soon became apparent that they often displayed a great deal of correlation with each other. Factor analysis attempts to "explain" these correlations by an analysis, which, when carried out successfully, yields a small number of underlying factors, which contain all the essential information about the correlations among the tests. Interpretation of the factors has led to the theory of fundamental aspects of human ability.

In this chapter, we shall introduce the basic ideas and concepts embodied in R-mode factor analysis, that is, problems concerned with the analysis of relationships among variables. The discussion will center on two techniques, *principal component analysis* and *"true" factor analysis*. The concept of Q-mode factor analysis, introduced in Section 2.11, in which attention is directed toward studying relationships among objects, follows naturally from the ideas treated here. It will be taken up in Chapter 5.

Consistent with what was said in Chapter 2, we shall unravel the intricacies of the subject in terms of a data matrix $\mathbf{X}_{(N \times p)}$. To facilitate discussion, and in order to follow the usual statistical development of factor analysis, we often digress to the extent of considering one row of the data matrix, referred to then as a p-variate random vector of observations.

3.2 FIXED AND RANDOM CASES

Factor analysis, like many other statistical techniques, can be employed in several ways. We here call attention to the distinction between the descriptive use of the method and the inferential use.

Fixed case

This refers to the use of factor analysis when interest is solely directed toward analyzing one particular collection of data (sample) without regarding it as a sample from some statistical population of objects. *Any factors obtained in the analysis can only be interpreted with respect to the objects in the sample*, and no inference is usually made about a population larger than the actual sample.

Random case

When the sample of objects may be conceived of as a random sample from some specified population, we refer to this as the random case. This allows inferences to be made about the population from one analysis of the sample. This approach requires consideration of such questions as the nature of the multivariate distribution of the variables and the theory of sampling from the population. There is a point that must be made here. Early factor analysis was exclusively of the fixed-mode type and it is this approach that has dominated in factor studies in biology and geology, as well as in most applications in psychology up to the present day. Modern developments in the theory of multivariate analysis remained unnoticed until fairly recently, when statisticians began examining factor analysis in relation to statistical principles. Many of these results are still so new that it is still unclear just to what extent they will influence interpretations of factor analysis in the natural sciences.

3.3 DISCUSSION OF MODELS

The R-mode methods of factor analysis are based on mathematical models that present simplified, but never exact, representations of the data, based on certain assumptions. The distinctions between the various techniques of factor analysis are largely connected with differences in these assumptions. We shall present the models in two different manners: (1) at the level of variances and covariances of the observed variables; and (2) at the level of the observed variables.

By employing these two explanations, we can develop various aspects of the methods, but it is necessary to realize that the model is the same at both levels of discussion.

Model for the observed variables

Here, we develop the factor model in relation to the data matrix (fixed case). The mathematical model is

$$\mathbf{X}_{(N \times p)} = \mathbf{F}_{(N \times k)} \mathbf{A}'_{(k \times p)} + \mathbf{E}_{(N \times p)}, \qquad [3.1]$$

where $\mathbf{X}$ is the data matrix, $\mathbf{F}$ the matrix of factor scores, $\mathbf{A}$ the matrix of factor loadings, and $\mathbf{E}$ is a matrix of residuals or error terms. Here k is a scalar denoting the number of factors to be used; it may be postulated in advance of the analysis or determined by various means subsequent to the analysis. It is almost always less than p, the number of variables, and often much less.

For any given value of the data in the matrix, say that of the nth row and the ith column, the model becomes in scalar notation

$$x_{ni} = \sum_{j=1}^{k} f_{nj} a_{ij} + e_{ni}. \qquad [3.2]$$

For any particular row, x', of $\mathbf{X}$ we have

$$x' = f' \mathbf{A}' + e' \qquad [3.3]$$

where f' and e' are corresponding row vectors of $\mathbf{F}$ and $\mathbf{E}$. Finally, for convenience, we take the transpose of [3.3] to get

$$x = \mathbf{A} f + e; \qquad [3.4]$$

x is now a column vector representing one of the objects of the data matrix (it is the p-variate random vector of observations of statistics that we mentioned earlier in this section). Equation [3.4] is the fundamental model equation for all forms of R-mode factor analysis. It states

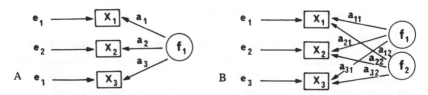

Figure 3.1. Diagrammatical representation of factor relationships.

that each observed variable (an element of x) is a weighted sum of factors plus an error term or residual. The product, $\mathbf{A}f$, produces a vector of estimates of x; the vector e represents the difference between this estimate and the observed vector. These residuals are assumed to be uncorrelated with the factors.

The column vector f contains the factors, and the elements of $\mathbf{A}$ are the weights to be attached to each factor for a particular variable.

These concepts are diagrammatically portrayed in Fig. 3.1A. We consider the simple case of three variables, denoted by squares on the diagram. We suppose that one factor will account for the variables and, if weighted by the coefficients a_1, a_2, a_3, will provide an approximation to x_1, x_2, and x_3, respectively. The e's represent the difference between the approximations and the observed values.

The model equations for this example are

$$x_1 = a_1 f_1 + e_1,$$
$$x_2 = a_2 f_1 + e_2,$$
$$x_3 = a_3 f_1 + e_3. \qquad [3.5]$$

If it be found that the approximations to the observed values are poor, a two-factor model might be postulated, as indicated schematically in Fig. 3.1B. Here, each variable is linked to both f_1 and f_2 by appropriate weighting coefficients, the a's:

$$x_1 = a_{11} f_1 + a_{12} f_2 + e_1,$$
$$x_2 = a_{21} f_1 + a_{22} f_2 + e_2,$$
$$x_3 = a_{31} f_1 + a_{32} f_2 + e_3. \qquad [3.6]$$

The significance of the terms in these model equations is not immediately obvious and they will now be discussed. Referring to [3.4] and Fig. 3.1, we remind you that any column vector of $\mathbf{X}$ is a variable on which we have N observations (i.e., the objects); a row vector of $\mathbf{X}$ represents one object on which p variables have been measured. A column vector of $\mathbf{F}$, the factor score matrix, is a hypothetical variable, or factor. Although unobservable, the N elements in this vector reflect the varying amount of the factor in each object. Hence, each factor is some unobservable attribute of the objects and $\mathbf{F}$ is the totality of such attributes for the objects. The fact that only k such attributes exist $(k < p)$ suggests that there is some redundancy in the original data matrix.

The vector f in [3.4] is a typical row vector of $\mathbf{F}$. It is a k-variate random vector describing the composition of a particular object in terms of the hypothetical variables or factors, that is, the scores of an individual object on the factors.

The factor loadings matrix, $\mathbf{A}$, contains the coefficients that must be used to combine the factors into the estimate of a particular variable. A column vector of $\mathbf{A}$ may be regarded as containing coefficients that describe the composition of the factor in terms of the original variables. This particular point of view leads to the notion that a factor is a linear combination of the observed variables, a concept that is used in principal component analysis.

For the general case of p variables and k factors, [3.6] may be generalized to the form:

$$x_i = a_{i1}f_1 + a_{i2}f_2 + \cdots + a_{ik}f_k + e_i, \qquad [3.7]$$

where x_i is the ith variable in vector x. There will be p such equations, one for each variable. The expression [3.7] represents the regression of x_i on the factors $f_1, f_2, \ldots, f_k$ with residual e_i. The difference between this and ordinary regression as outlined in Section 2.8 is that the factors are not directly observed. Otherwise, all the assumptions and results of regression hold. For example, the linear combination

$$c_i = a_{i1}f_1 + a_{i2}f_2 + \cdots + a_{ik}f_k \qquad [3.8]$$

is the best linear estimate of x_i and is called the *common part* of x_i because this part of the variable has something in common with the other variables as an outcome of their links to the factors. The residual term e_i is usually taken to be the sum of two uncorrelated parts, s_i and ϵ_i:

$$e_i = s_i + \epsilon_i, \qquad [3.9]$$

where ϵ_i is *measurement error* and s_i is referred to as the *specific part* of x_i. It contains that part of x_i unaccounted for by the factors and not due to measurement error. By measurement error, we mean the error that occurs when the same object is measured many times; it is a common observation that there will be differences in the results of these measurements. We explain the terms introduced here in more detail in the next section.

3.4 MODEL FOR THE VARIANCES AND COVARIANCES

Although it is possible to develop the solution for the factor model directly from the data matrix, it is more convenient to approach the topic from a consideration of the variances and covariances of the variables. To proceed, we combine [3.7], [3.8], and [3.9] to obtain

$$x_i = c_i + e_i = c_i + s_i + \epsilon_i. \qquad [3.10]$$

It is, of course, possible to find the total variance for variable x_i, since it is observable. Likewise, there will be a variance, unobservable for the most part, for the terms on the right-hand side of [3.10]. One of the assumptions of the factor model is that all the terms on this side are uncorrelated. Because of this assumption, and the well-known property of additivity of variances, it is possible to decompose the total variance of a variable x_i as follows:

$$\sigma_{x_i}^2 = \sigma_{c_i}^2 + \sigma_{e_i}^2 = \sigma_{c_i}^2 + \sigma_{s_i}^2 + \sigma_{\epsilon_i}^2, \qquad [3.11]$$

where:

$\sigma_{c_i}^2$ denotes the common variance, or *communality*; it represents that
　　　part of the variance of x_i that is in common with the other
　　　variables and is involved in the covariances between them;
$\sigma_{e_i}^2$ is the *residual variance*, often referred to as the *uniqueness*; it is the
　　　variance of x_i unaccounted for by the factors and is therefore
　　　not shared by other variables;
$\sigma_{s_i}^2$ is the specific variance, or *specificity*, of x_i; it denotes the variance
　　　specific to a variable;
$\sigma_{\epsilon_i}^2$ is the error variance of x_i; it is that part of the variability inherent in
　　　x_i due solely to measurement error and is possible to estimate by
　　　repeated trials.

The variance components expressed in [3.10], and the components of a variable in [3.7], may be made more understandable by means of a simple example.

We shall consider an hypothetical study on a suite of igneous rocks in which the amounts of three trace elements in a particular mineral are determined, as well as the tensile strength of the bulk rock specimen. The trace elements may be expected to be highly correlated due to their supposed mutual dependence on physical and chemical conditions at the time of mineral formation. In a factor model, they would exhibit large common parts and low specificity. The tensile strength of the rock would presumably have little in common with the trace elements and would thus contain a large specific part and low communality when all four variables are considered simultaneously. If we were to add some fifth variable closely related to tensile strength, it would be observed that this would decrease the specific part for tensile strength and increase its common part.

In a practical application, it is usually impossible to know in advance of the analysis what the different variance terms are. They must be estimated in some manner, usually, as the analysis proceeds. For example, the communality may be determined from [3.8] as follows. Since

$$c_i = a_{i1}f_1 + a_{i2}f_2 + \cdots + a_{ik}f_k,$$

we have

$$
\sigma_{c_i}^2 = \sum_{n=1}^{N} \frac{c_{ni}^2}{N} = \sum_{n=1}^{N} \frac{(a_{i1}f_{n1} + a_{i2}f_{n2} + \cdots + a_{ik}f_{nk})^2}{N}
$$

$$
= a_{i1}^2 \frac{\sum f_{n1}^2}{N} + a_{i2}^2 \frac{\sum f_{n2}^2}{N} + \cdots + a_{ik}^2 \frac{\sum f_{nk}^2}{N}
$$

$$
+ a_{i1}a_{i2}\frac{\sum f_{n1}f_{n2}}{N} + a_{i1}a_{i3}\frac{\sum f_{n1}f_{n3}}{N}
$$

$$
+ \cdots + a_{ik-1}a_{ik}\frac{\sum f_{nk-1}f_{nk}}{N}. \tag{3.12}
$$

The terms $\sum f_{nj}^2/N$ are the variances of the factors, which we can assume are unity, that is, the factors will be in standardized form. Furthermore, the terms $\sum f_{nj}f_{ne}/N$ are the correlations between the factors. Thus, [3.12] becomes

$$
\sigma_{c_i}^2 = a_{i1}^2 + a_{i2}^2 + \cdots + a_{ik}^2 + a_{i1}a_{i2}\phi_{12}
$$

$$
+ \cdots + a_{ik-1}a_{ik}\phi_{k-1k}, \tag{3.13}
$$

where the ϕs are correlations between the factors. If the factors are uncorrelated, we have that

$$
\sigma_{c_i}^2 = a_{i1}^2 + a_{i2}^2 + \cdots + a_{ik}^2. \tag{3.14}
$$

In this case, the communality of variable x_i is the sum of squared elements in the ith row of the factor loadings matrix **A**.

Another way of approaching the factor model is from the consideration of the covariance matrix. This will allow us to develop the relationship between the data matrix and the population. To simplify the development, we shall assume that the data matrix is in deviate form (Section 2.8).

The factor model ([3.1]) is

$$
\mathbf{Y} = \mathbf{FA'} + \mathbf{E} \tag{3.15}
$$

and

$$\mathbf{Y}' = \mathbf{A}\mathbf{F}' + \mathbf{E}'. \tag{3.16}$$

In terms of variances and covariances:

$$\frac{1}{N}\mathbf{Y}'\mathbf{Y} = \mathbf{A}\left[\frac{1}{N}\mathbf{F}'\mathbf{F}\right]\mathbf{A}' + \mathbf{A}\left[\frac{1}{N}\mathbf{F}'\mathbf{E}\right] + \left[\frac{1}{N}\mathbf{E}'\mathbf{F}\right]\mathbf{A}' + \frac{1}{N}\mathbf{E}'\mathbf{E}. \tag{3.17}$$

The matrix $\mathbf{Y}$ can be regarded as a sample (consisting of N row vectors) from a p-variate population. The sample size could be augmented by including more row vectors (objects) and this would, in general, lead to variances and covariances more likely to be closer to the population values than those of the smaller sample. Thus, if we let the sample size N increase indefinitely, each term in [3.17] will converge toward its population value. Hence,

$$\frac{1}{N}\mathbf{Y}'\mathbf{Y} \to \boldsymbol{\Sigma}, \qquad \frac{1}{N}\mathbf{F}'\mathbf{F} \to \boldsymbol{\Phi}, \qquad \frac{1}{N}\mathbf{F}'\mathbf{E} \to \mathbf{0}, \qquad \frac{1}{N}\mathbf{E}'\mathbf{E} \to \boldsymbol{\Psi},$$

and we have

$$\boldsymbol{\Sigma} = \mathbf{A}\boldsymbol{\Phi}\mathbf{A}' + \boldsymbol{\Psi}. \tag{3.18}$$

In this formula, $\boldsymbol{\Sigma}$ is the p-by-p population covariance matrix of the observed variables, $\mathbf{A}$ is the p-by-k matrix of factor loadings, $\boldsymbol{\Phi}$ is the k-by-k covariance matrix for the factors (if the factors are in standardized form, this is a correlation matrix with 1s in the diagonal), and $\boldsymbol{\Psi}$ is the p-by-p residual covariance matrix.

You will have noticed that here and elsewhere in this book, we try to use Greek letters for population quantities and Roman letters for the sample counterparts, although in a few cases, this has not been observed for reasons of expediency. The Greek–Roman convention is well established in the statistical literature.

Equation [3.18] presents, in matrix form, the complete factor model for the variances and covariances of the observed variables. The model holds for the two major R-mode factor methods, and we now specify and contrast the subtle but important differences between the two.

3.5 PRINCIPAL COMPONENT VERSUS TRUE FACTOR ANALYSIS

The fundamental differences between principal component and factor analysis depend upon ways in which factors are defined and upon assumptions concerning the nature of the residuals.

In principal component analysis, factors are determined so as to account for *maximum variance* of all the observed variables. In true factor analysis, the factors are defined to account maximally for the *intercorrelations* of the variables. Thus, principal component analysis can be said to be variance-oriented, whereas true factor analysis is correlation-oriented.

The residual terms e_i are assumed to be small in principal component analysis, whereas this is not so in true factor analysis. Basically, this implies that component analysis accepts that a large part of the total variance of a variable is important and in common with other observed variables. On the other hand, factor analysis allows for a considerable "amount of uniqueness" to be present in the data and utilizes only that part of a variable that takes part in correlation with other variables.

In both methods, the residuals are assumed to be uncorrelated with the factors. However, in principal component analysis, there is no assumption about correlations among the residuals, whereas in true factor analysis, it is assumed that they are uncorrelated among themselves.

The constraints imposed on the covariance matrix Σ of [3.18] differ for the two methods. In principal component analysis, because the residual covariance matrix Ψ is assumed to be small, then $\Sigma \approx A\Phi A'$. This implies that the p-by-p matrix Σ is approximately of rank k. In true factor analysis, Ψ is a diagonal matrix because interresidual correlations are zero. Hence, Σ in [3.18] is the sum of $A\Phi A'$, a matrix of rank k exactly, and a diagonal matrix Ψ. The off-diagonal elements of Σ are the same as the off-diagonal elements of $A\Phi A'$. The diagonal of $A\Phi A'$ contains the communalities of the variables. Thus, in components analysis, the aim is to reproduce, or fit, both the diagonal and the off-diagonal elements of Σ, whereas in true factor analysis, the goal is to do this for the off-diagonal elements only, that is, they account for covariances but not variances.

3.6 TRANSFORMATIONAL INDETERMINACY OF FACTORS

The factor models specified by [3.1], [3.4], and [3.18] contain an inherent indeterminancy if $k > 1$; without additional constraints, a unique solution to the equations is not possible. This is due to the fact that any set of factors may be transformed linearly to another set of factors having the same properties. This fact was mentioned in Section 2.10.

We shall now demonstrate what we mean when we state that factor-analytical models are indeterminate. Let us suppose that a vector f and a matrix A have been found that satisfy [3.4]. Now consider a nonsingular, linear transformation of the factors f to another set f^* by means of the transformation matrix **T**:

$$f^* = \mathbf{T}f. \tag{3.19}$$

Moreover, if **A*** is defined as

$$\mathbf{A}^* = \mathbf{A}\mathbf{T}^{-1} \tag{3.20}$$

and we use the factor equation $x = \mathbf{A}f + e$, then simple substitution leads to

$$x = \mathbf{A}\mathbf{T}^{-1}\mathbf{T}f + e = \mathbf{A}^*f^* + e. \tag{3.21}$$

For the data matrix representation with $\mathbf{F}^* = \mathbf{F}\mathbf{T}'$ and $\mathbf{A}^* = \mathbf{A}\mathbf{T}^{-1}$, we have that

$$\mathbf{X} = \mathbf{F}\mathbf{A}' + \mathbf{E} = \mathbf{F}\mathbf{T}'\mathbf{T}'^{-1}\mathbf{A}' + \mathbf{E} = \mathbf{F}^*\mathbf{A}^{*\prime} + \mathbf{E}. \tag{3.22}$$

For the covariance matrix representation of [3.9], we let $\mathbf{\Phi}^* = \mathbf{T}\mathbf{\Phi}\mathbf{T}'$ and let **A*** be as before, which leads to

$$\mathbf{\Sigma} = \mathbf{A}\mathbf{\Phi}\mathbf{A}' + \mathbf{\Psi} = \mathbf{A}\mathbf{T}^{-1}\mathbf{T}\mathbf{\Phi}\mathbf{T}'\mathbf{T}'^{-1}\mathbf{A}' + \mathbf{\Psi} = \mathbf{A}^*\mathbf{\Phi}^*\mathbf{A}^{*\prime} + \mathbf{\Psi}. \tag{3.23}$$

From the foregoing equations it is clear that f and f^* have the same properties and without further restrictions are indistinguishable. In principal component analysis, f and f^* account for the same total amount of variance. In true factor analysis, they account for the intercorrelations equally well. It is therefore necessary to impose constraints on the solution in order to obtain a particular set of factors.

The transformation matrix **T**, used to transform from f to f^*, is square and of order k by k. This means that the factors are only determined up to an arbitrary nonsingular linear transformation. Hence, it is only the Euclidean space in which the factors are embedded which is determined. To determine a unique set of factors, one must define a set of basis vectors in this space so that the factors can be referred to these vectors. The bias vectors may be chosen to be orthogonal to each other, in which case we say that the factors are orthogonal, or uncorrelated; alternatively, some of them may be chosen so as to be oblique, in which case we say that factors are oblique, or correlated. Oblique solutions are not very common in the natural sciences, although they are widely employed in psychology.

In both the orthogonal and oblique situations, the basis vectors are usually chosen so that the factors obtained have certain variance and correlation properties. Once a set of factors has been determined, they may be transformed to another set of factors in order to facilitate interpretation, should this be deemed advisable. The interpretation of the factors is usually made by inspection of the factor loading matrix **A**.

3.7 FACTOR PATTERN AND FACTOR STRUCTURE

The matrix **A** has been referred to as the factor loading matrix. More properly, it should be called the *factor pattern*. This holds for both the orthogonal and oblique solutions.

In the orthogonal case, in which the factors are uncorrelated, the factor pattern contains elements a_{ij} that are the covariances between the variables and the factors; a_{ij} is the covariance between variable x_i and factor f_j. If both the variables and the factors are standardized, then this covariance is a correlation.

In the oblique case, the factors are themselves intercorrelated and this must be taken into account when computing the covariance between factors and variables. The matrix giving the covariances between variables and factors is termed the *factor structure* and is given by **AΦ**. In the oblique case, both the factor pattern and the factor structure are necessary for a full description of the solution. In this case, the elements of **A** are not covariances; they can only be interpreted as regression coefficients. In the orthogonal case, when **Φ** = **I**, the factor pattern and the factor structure are both given by **A**. The preceding statements can be most easily demonstrated by using the deviate data matrix representation [3.15]. Then, the covariances between variables and factors are the elements of

$$\frac{1}{N}\mathbf{Y'F} = \frac{1}{N}(\mathbf{AF'} + \mathbf{E'})\mathbf{F}$$

$$= \mathbf{A}\frac{1}{N}\mathbf{F'F} + \frac{1}{N}\mathbf{E'F}.$$

When N increases, the second term tends to zero and the first term to **AΦ**.

3.8 EXAMPLE OF TRUE FACTOR ANALYSIS

Before proceeding to the actual mechanics of solving the factor model equations, it may be helpful to take you through the results of a simple numerical example.

Suppose that six variables, $x_1, x_2, \ldots, x_6$, have been measured on a large population of organisms and their environment. We shall assume that variables x_1, x_2, and x_6 are various measures of morphological variability of some organism and that x_3, x_4, and x_5 are measures of some sort of environmental conditions that are suspected to have influenced the morphology of these organisms. For simplicity, the variables have been transformed to standardized form. The correlations between variables are as follows:

	x_1	x_2	x_3	x_4	x_5	x_6
x_1	1.000	0.720	0.378	0.324	0.270	0.270
x_2	0.720	1.000	0.336	0.288	0.240	0.240
$\Sigma = x_3$	0.378	0.336	1.000	0.420	0.350	0.126
x_4	0.324	0.288	0.420	1.000	0.300	0.108
x_5	0.270	0.240	0.350	0.300	1.000	0.090
x_6	0.270	0.240	0.126	0.108	0.090	1.000

Inasmuch as we are dealing with a very large sample, we denote an element of this matrix as ρ_{ij} and take it to represent a population value.

N.B. This matrix is a matrix of correlations, which by rights should be designated **P** (capital rho). However, for the purposes of illustration, we use the designation Σ, normally reserved for the covariance matrix.

True factor analysis

We determine first the factors of the foregoing correlation matrix according to [3.18], and we use the assumptions for the true factor model. Factor analysis sets about explaining these correlations by introducing factors $f_1, f_2, \ldots, f_k$ that account for them. One way of viewing the operation is by posing the following question. Is there a factor, f_1, such that if it be partialed out, no intercorrelation remains between the variables? If so, the partial correlations (cf. Section 2.9) between any pair of variables, x_i and x_j, must vanish after f_1 has been eliminated; that is, according to Section 2.9, we would have

$$x_i = a_i f_1 + e_i \quad \text{and} \quad x_j = a_j f_1 + e_j, \qquad [3.24]$$

with e_i and e_j uncorrelated for all $i \neq j$. The correlation ρ_{ij} would then be

$$\rho_{ij} = a_i a_j, \qquad i \neq j,$$

so that the correlations in rows i and j would be proportional. An inspection of Σ reveals that this is not so and therefore it must be concluded that one factor is insufficient to account for all the intercorrelations in Σ and that two or more factors are needed to do so.

In the present problem, we assume that the factors are uncorrelated, that is, $\Phi = I$, and we can write [3.18] as

$$\Sigma = AA' + \Psi. \qquad [3.25]$$

Thus, the major product of **A** should reproduce exactly the off-diagonal elements of Σ, the correlation matrix. In expanded form,

$$\rho_{ij} = a_{i1}a_{j1} + a_{i2}a_{j2} + \cdots + a_{ik}a_{jk}. \qquad [3.26]$$

When such a matrix A has been found, the communality for variable x_i is given by

$$\sigma_{c_i}^2 = a_{i1}^2 + a_{i2}^2 + \cdots + a_{ik}^2. \qquad [3.27]$$

Utilizing methods to be described later, it was found that the following A matrix satisfied [3.26] exactly:

$$A = \begin{matrix} & \text{Factor 1} & \text{Factor 2} \\ x_1 \\ x_2 \\ x_3 \\ x_4 \\ x_5 \\ x_6 \end{matrix} \begin{bmatrix} 0.889 & -0.138 \\ 0.791 & -0.122 \\ 0.501 & 0.489 \\ 0.429 & 0.419 \\ 0.358 & 0.349 \\ 0.296 & -0.046 \end{bmatrix}.$$

For example,

$$\rho_{12} = a_{11}a_{21} + a_{12}a_{22} = 0.889 \times 0.791 + (-0.138) \times (-0.122)$$
$$= 0.7032 + 0.0168$$
$$= 0.7200.$$

Also,

$$\sigma_{c1}^2 = 0.889^2 + (-0.138)^2 = 0.81,$$

the communality of the first variable.

We have thus established the fact that, for these data, two uncorrelated factors, f_1 and f_2, exist such that the representation

$$x_1 = 0.889f_1 - 0.138f_2 + e_1,$$
$$x_2 = 0.791f_1 - 0.122f_2 + e_2,$$
$$x_3 = 0.501f_1 + 0.489f_2 + e_3,$$
$$x_4 = 0.429f_1 + 0.419f_2 + e_4,$$
$$x_5 = 0.358f_1 + 0.349f_2 + e_5,$$
$$x_6 = 0.296f_1 - 0.046f_2 + e_6 \qquad [3.28]$$

holds with $\rho(e_i, e_j) = 0$ for all i and j.

The resemblance between equations [3.28] and multiple regression (Section 2.8) should be noted. The fs are unobservable variables used to predict the values of the observed variables, the x's.

The factors f_1 and f_2 are, however, not the only two factors that satisfy [3.25]. For example, the following two factors also do this:

$$f_1^* = 0.988f_1 - 0.153f_2 \quad \text{and} \quad f_2^* = -0.153f_1 + 0.988f_2, \quad [3.29]$$

which, in terms of f_1 and f_2, becomes

$$f_1 = 0.988f_1^* + 0.153f_2^* \quad \text{and} \quad f_2 = -0.153f_1^* + 0.988f_2^* \quad [3.30]$$

In doing the operation of [3.30], we are, in fact, using an orthonormal transformation matrix:

$$\mathbf{T} = \begin{bmatrix} 0.988 & -0.153 \\ 0.153 & 0.988 \end{bmatrix}.$$

Applying now [3.20], we obtain the new factor loadings, $\mathbf{A}^*$, that is, the factor pattern is now

$$
\mathbf{A}^* = \begin{array}{c} \\ x_1 \\ x_2 \\ x_3 \\ x_4 \\ x_5 \\ x_6 \end{array}
\begin{array}{c} \text{Factor 1} \quad \text{Factor 2} \\ \begin{bmatrix} 0.90 & 0 \\ 0.80 & 0 \\ 0.42 & 0.56 \\ 0.36 & 0.48 \\ 0.30 & 0.40 \\ 0.30 & 0 \end{bmatrix} \end{array}.
$$

By substituting [3.30] into [3.28], we obtain the following equations in terms of the original variables:

$$
\begin{aligned}
x_1 &= 0.90f_1^* & & + e_1, \\
x_2 &= 0.80f_1^* & & + e_2, \\
x_3 &= 0.42f_1^* + 0.56f_2^* & & + e_3, \\
x_4 &= 0.36f_1^* + 0.48f_2^* & & + e_4, \\
x_5 &= 0.30f_1^* + 0.40f_2^* & & + e_5, \\
x_6 &= 0.30f_1^* & & + e_6. \quad [3.31]
\end{aligned}
$$

It is apparent that the same relationship holds for $\mathbf{A}^*$ as for $\mathbf{A}$, that is,

$$\rho_{12} = 0.90 \times 0.80 + 0 \times 0 = 0.72$$

and

$$\sigma_{c_i}^2 = 0.90^2 + 0^2 = 0.81.$$

Although $\mathbf{A}$ and $\mathbf{A}^*$ both satisfy the factor model equally well, $\mathbf{A}^*$ exhibits a kind of simplicity in that some of its elements are zero. The factor pattern of $\mathbf{A}^*$ approaches what factor analysts in psychology call *simple structure*; it contains many zero elements. The variables will be seen to have been grouped into two sets.

In general, greater simplicity can be achieved by allowing the factors to become correlated, which is done by using oblique rather than orthogonal factors. For example, the two correlated factors

$$f_1^{**} = f_1^* \quad \text{and} \quad f_2^{**} = 0.6f_1^* + 0.8f_2^*$$

with $\rho(f_1^{**}, f_2^{**}) = 0.6$, that is,

$$\Phi = \begin{bmatrix} 1.0 & 0.6 \\ 0.6 & 1.0 \end{bmatrix},$$

also reproduce the intercorrelations among variables. The transformation from f^* to f^{**} is given by

$$T = \begin{bmatrix} 1 & 0 \\ 0.6 & 0.8 \end{bmatrix}.$$

The inverse of T is

$$T^{-1} = \begin{bmatrix} 1 & 0 \\ -0.75 & 1.25 \end{bmatrix}.$$

Using this T^{-1}, we find the factor pattern A^{**} corresponding to f^{**} from $A^{**} = A^* T^{-1}$ to be

$$
A^{**} = \begin{array}{c} \\ x_1 \\ x_2 \\ x_3 \\ x_4 \\ x_5 \\ x_6 \end{array}
\begin{array}{cc} \text{Factor 1} & \text{Factor 2} \\ \begin{bmatrix} 0.9 & 0 \\ 0.8 & 0 \\ 0 & 0.7 \\ 0 & 0.6 \\ 0 & 0.5 \\ 0.3 & 0 \end{bmatrix} \end{array}.
$$

The corresponding factor structure matrix, $A\Phi$, is

$$
\begin{array}{c} \\ x_1 \\ x_2 \\ x_3 \\ x_4 \\ x_5 \\ x_6 \end{array}
\begin{array}{cc} \text{Factor 1} & \text{Factor 2} \\ \begin{bmatrix} 0.90 & 0.54 \\ 0.80 & 0.48 \\ 0.42 & 0.70 \\ 0.36 & 0.60 \\ 0.30 & 0.50 \\ 0.30 & 0.18 \end{bmatrix} \end{array}.
$$

The factor pattern matrix A^{**} now illustrates the breakdown of the variables into two groups. Factor 1 contains a factor that combines the morphological measures only, whereas Factor 2 combines environmental controls only.

3.9 REPRESENTATION OF RESULTS

Using the geometrical notions of vector and matrix operations, to which you were introduced in Chapter 2, we can illustrate the results of the factor analysis of our example.

For the R-mode model, we adopt the vector representation of correlation (Section 2.8). Here, the variables as vectors are viewed as being located in N-dimensional object space with their angular separations determined by the correlation coefficients. The factors provide a basis for these vectors. In the present example, the results indicate a dimensionality of 2.

In the orthogonal cases given by $\mathbf{A}$ and $\mathbf{A}^*$, the factor axes may be represented as two orthogonal vectors of unit length. An element of the matrix, a_{ij}, gives the correlation, and thus the angle between the vector of variables, i, and the jth factor axis. A graph of the variables in relation to the factor axes is therefore a convenient way of illustrating the relationships between the variables. Fig. 3.2A shows such a plot for the $\mathbf{A}$ matrix; Fig. 3.2B displays the plot for matrix $\mathbf{A}^*$. In Fig. 3.2A, the reference axes represent the system (f_1, f_2). The variables have moderately large projections on both factors. Figure 3.2B demonstrates the effects of a rigid rotation of (f_1, f_2) in a clockwise direction to the system (f_1^*, f_2^*); that is, an angle of $90°$ is maintained between the axes. Here, variables 1, 2, and 6 lie along the first axes and have, therefore, zero loadings on factor 2. Variables 3, 4, and 5 have nonzero loadings on both factors.

The transformation from (f_1^*, f_2^*) to (f_1^{**}, f_2^{**}) corresponds to a rotation of axis 2 so that it passes through variables 3, 4, and 5, but keeping the first axis in the same position as in Fig. 3.2B. The angle between the axes corresponds to a cosine of 0.6, the correlation between the oblique factors (Fig. 3.2C).

3.10 TRUE FACTOR ANALYSIS SUMMARIZED

The basic principles of true factor analysis have been demonstrated in the foregoing example. We shall now summarize these.

The principle of conditional linear independence.   This principle expresses the idea that factors shall account for all linear relationships among the variables. Once the factors have been partialed out, no correlation shall remain between the variables. In this sense, factor analysis is a method for classification of linear independence.

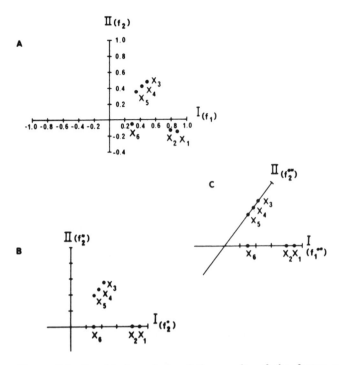

Figure 3.2. Graphical analysis of the results of the factor analysis. (A) Plot of the unrotated results. (B) Effects of a rigid clockwise rotation of the reference axes. (C) The effect of rotating the second reference axis.

The principle of simple structure. Once a set of k factors has been found that account for the intercorrelations of the variables, these may be transformed to any other set of k factors that account equally well for the correlations. Because there are infinitely many sets of factors, this represents a great indeterminacy in the model. However, Thurstone (1947) put forward the idea that only those factors for which the variables have a very simple representation are meaningful, which is to say that the matrix of factor loadings should have as many zero elements as possible. If $a_{ij} = 0.0$, the jth factor does not enter into the ith variable. According to Thurstone again, a variable should not depend on all common factors but only on a small part of them. Moreover, the same factor should only be connected with a small portion of the variables. Such a matrix is considered to yield simple structure.

In our example, we have put forth three possible solutions. It is evident that the matrix **A** does not conform to the principle of simple

Table 3.I. Communalities and uniquenesses for factor example

Variable	Communality	Uniqueness
1	0.81	0.19
2	0.64	0.36
3	0.49	0.51
4	0.36	0.64
5	0.25	0.75
6	0.09	0.91
Total	2.64	3.36

structure. A^* gives a simple structure with two orthogonal factors. This matrix might be interpreted as showing one general factor, f_1^*, that is involved in all six variables, and an additional factor, f_2^*, uncorrelated with f_1^*, but connected only to variables x_3, x_4, and x_5.

A^{**} is an oblique, simple structure solution. Here, f_1^{**} is interpretable as being solely morphological, and f_2^{**} as solely environmental. These factors correlate at 0.6 in the population studied. In many aspects, the interpretation based on A^{**} seems to be conceptually more reasonable and scientifically appealing.

Our example also demonstrates that communalities and unique variances are unaffected by linear transformations of factors.

The general formula for finding communalities [3.8] is given by the following summation expression (for two factors):

$$\sigma_{c_i}^2 = a_{i1}^2 + a_{i2}^2 + a_{i1}a_{i2}\rho(f_1, f_2). \qquad [3.32]$$

If the factors are uncorrelated, the correlation term in [3.32] disappears, leaving the sum of squares in the ith row of A.

Using this equation (with zero correlation between factors) on matrix A^*, we obtained the results listed in Table 3.I. The same result was also obtained when we applied the formula to matrix A^{**}, thus showing that the communalities and the uniqueness are unaffected by linear transformations of factors. This property is quite general and the unique factors $e_1, e_2, \ldots, e_p$ are unaffected by transformations of the common factors.

4 *R*-mode methods

4.1 INTRODUCTION

In this chapter we describe various methods suitable for studying the interrelationships among variables – so-called *R*-mode methods. In contrast, we treat in Chapter 5 *Q*-mode methods, which are designed for the study of interrelationships among objects.

In Chapter 3, we discussed the purposes, concepts, models, and terminology of factor analysis, a discussion that was put forward in terms of the population properties of the models. In practice, the data we have at our disposal consist of a sample drawn from the population. In the data matrix X, of order $N \times p$, the rows constitute a random sample of observations from the population. From X one can compute the sample covariance matrix S, as we showed in Chapter 2. The statistical problem then is how does one go about estimating the population model from the data in X or S? In terms of the concepts of Chapter 3, we have the random case. The approach taken is that of fitting the population model to the data of the sample. However, even in the fixed case we need to fit a model to the data, since X or S will be approximated by matrices of lower rank. Some of the results are the same for the fixed and random cases. In the fixed case we are usually interested in estimating both the factor loadings matrix A and the factor scores matrix F, whereas in the random case, we are usually only interested in A and perhaps the covariance matrix Φ of the factors. We now consider principal component analysis and factor analysis separately. In the ensuing text, a "hat" (^) on a population symbol denotes that this is a derived sample estimate. Special considerations are reviewed in later sections, including the graphical procedure of *ordination*.

4.2 PRINCIPAL COMPONENT ANALYSIS

Fixed case

We start by considering again (cf. Section 3.1) the model for the data matrix Y in deviate form:

$$Y = FA' + E. \tag{4.1}$$

For convenience we assume that all the elements of **Y** have been divided by $\sqrt{N}$ so that the covariance matrix **S** is simply given by

$$S = Y'Y,$$

the minor product moment of **Y**. Here, as in [3.1], **F** is a matrix of factor scores, **A** is the matrix of factor loadings, and **E** is a matrix of residuals.

In the fixed case, we may fit this model by applying the *method of least squares* to the data matrix **Y**. This means that we have to determine the matrices $F_{(N \times k)}$ and $A_{(p \times k)}$, for a given $k < p$, such that the sum of squares of all the elements of the matrix

$$E = Y - FA' \tag{4.2}$$

is as small as possible. As stated in Section 2.13, the solution to this problem is obtained as

$$\hat{F}\hat{A}' = \gamma_1 v_1 u_1' + \gamma_2 v_2 u_2' + \cdots + \gamma_k v_k u_k',$$

the k terms of the singular value decomposition of **Y** corresponding to the k largest singular values $\gamma_1, \gamma_2, \ldots, \gamma_k$. Let

$$V_k = [v_1, v_2, \ldots, v_k], \qquad U_k = [u_1, u_2, \ldots, u_k] \quad \text{and}$$
$$\gamma_k = \text{diag}(\gamma_1, \gamma_2, \ldots, \gamma_k).$$

Then the solution is

$$\hat{F}\hat{A}' = V_k \Gamma_k U_k'.$$

Although this tells us what the product of **FA'** is, it does not give a unique solution for **F** and **A** themselves. Clearly there are infinitely many solutions for **F** and **A**, since we may replace **F** by $F^* = FT$ and **A** by $A^* = AT^{-1}$, where **T** is any nonsingular matrix. This corresponds to the transformation of factors discussed in Chapter 3.

We shall now consider two different choices of solutions for **A** and **F** and discuss the advantages and disadvantages of each of these. It should be emphasized that these two solutions are essentially the same; they differ only in the way the factors are scaled.

Solution 1

$$\hat{F} = V_k, \qquad \hat{A} = U_k \Gamma_k. \tag{4.3}$$

It is easily shown that as a consequence of the fact that **Y** has column sums equal to zero, those of V_k will also be zero. Therefore, **F** will be in deviate form and the covariance matrix of the factors (in the sample) will be

$$\hat{F}'\hat{F} = V_k'V_k = I,$$

that is, the factors are uncorrelated and standardized. For the factor loadings matrix we have

$$\hat{A}'\hat{A} = \Gamma_k U_k' U_k \Gamma_k = \Gamma_k^2 = \Lambda_k,$$

where Λ_k is a diagonal matrix of order $k \times k$ whose diagonal elements are $\lambda_1 = \gamma_1^2, \lambda_2 = \gamma_2^2, \ldots, \lambda_k = \gamma_k^2$, the eigenvalues of S, the covariance matrix. Furthermore, if E in [4.1] is small so that we have approximately

$$Y \approx \hat{F}\hat{A}',$$

the covariance matrix S is approximately

$$S = Y'Y \approx \hat{A}\hat{F}'\hat{F}\hat{A}' = \hat{A}\hat{A}',$$

that is, $\hat{A}\hat{A}'$ approximates S.

To compute this solution, one proceeds as follows, assuming that $N \ge p$:

1. Compute the covariance matrix S.
2. Compute the k largest eigenvalues Λ_k and the corresponding eigenvectors U_k of S.
3. Compute $\hat{A} = U_k \Lambda_k^{1/2}$. This amounts to scaling each eigenvector so that its squared length equals the eigenvalue.
4. Compute $\hat{F} = Y\hat{A}\Lambda_k^{-1}$. This means first postmultiplying Y by $\hat{A}$ and then scaling the columns by the reciprocals of the eigenvalues.

Solution 2

$$\hat{A} = U_k, \qquad \hat{F} = V_k \Gamma_k. \tag{4.4}$$

With this choice, the covariance matrix of the factors is

$$\hat{F}'\hat{F} = \Gamma_k V_k' V_k \Gamma_k = \Gamma_k^2 = \Lambda_k,$$

which is diagonal, but with diagonal elements equal to the eigenvalues of S in descending order of magnitude. Hence, the factors are still uncorrelated, but they have different variances. For the factor loadings matrix A we have that

$$\hat{A}'\hat{A} = U_k' U_k = I,$$

that is, the columns of $\hat{A}$ are orthonormal. Furthermore, in this case we have

$$S = Y'Y \approx \hat{A}\hat{F}'\hat{F}\hat{A}' = \hat{A}\Lambda_k \hat{A}'.$$

The computation of this solution is as follows, assuming as before that $N \geq p$:

1. Compute the covariance matrix $\mathbf{S}$.
2. Compute the k largest eigenvalues Λ_k and the corresponding eigenvectors $\mathbf{U}_k$ of $\mathbf{S}$.
3. Then $\hat{\mathbf{A}} = \mathbf{U}_k$.
4. Compute $\hat{\mathbf{F}} = \mathbf{Y}\hat{\mathbf{A}}$.

Note that in this case no scaling of columns is needed and therefore this choice is computationally somewhat simpler than the other.

We shall now discuss the interpretation of the results in the two solutions. As noted in Chapter 3, the matrix $\mathbf{A}$, in general, represents the regression of the variables on the factors. In this sense, the factor loading a_{im} represents the effect of factor m on variable i. It is usually the factor loading matrix $\mathbf{A}$ that the investigator inspects when interpreting the results of a factor analysis, but certain facts must then be remembered. In Solution 1, all the factors are uncorrelated and have the same variance, unity, so that the columns of $\hat{\mathbf{A}}$ are directly comparable. In this solution, the matrix $\hat{\mathbf{A}}$ also represents the covariances between the variables and the factors, that is, the factor structure. In particular, to get the correlations between variables and factors, one only needs to divide each row of $\hat{\mathbf{A}}$ by the standard deviation of the variables. Hence, the matrix of correlations between variables and factors is

$$\hat{\mathbf{C}} = \mathbf{D}_s^{-1}\hat{\mathbf{A}} = \mathbf{D}_s^{-1}\mathbf{U}_k\Lambda_k^{1/2},$$

where $\mathbf{D}_s$ is a diagonal matrix with diagonal elements equal to the standard deviations of the variables.

In Solution 2, on the other hand, the factors are also uncorrelated, but they have different variances, so the columns of $\hat{\mathbf{A}}$ are not directly comparable. Because the covariance matrix of the factors is Λ_k, the covariances between variables and factors are given by $\hat{\mathbf{A}}\Lambda_k$ and the corresponding correlations are given by

$$\hat{\mathbf{C}} = \mathbf{D}_s^{-1}\hat{\mathbf{A}}\Lambda_k\Lambda_k^{-1/2} = \mathbf{D}_s^{-1}\mathbf{U}_k\Lambda_k^{1/2},$$

which is the same as for Solution 1. Hence, in Solution 2, both rows and columns of $\hat{\mathbf{A}}$ must be scaled to get the correlations between variables and factors.

To illustrate principal component analysis in the fixed case, we make use of a set of artificial data, consisting of 10 hypothetical objects on which 5 characters have been measured. The data matrix, of order 10×5, was constructed, so as to have approximately rank 2. This matrix is listed in Table 4.I, which also supplies the means and standard deviations of each variable. The covariance matrix **S** for these five variables is given in Table 4.II. Note that since N is small here, we have computed **S** as

$$(1/(N-1))\mathbf{Y'Y}.$$

The eigenvalues of **S** are $\lambda_1 = 114.83459$, $\lambda_2 = 17.72064$, $\lambda_3 = 0.00588$, $\lambda_4 = \lambda_5 = 0.00000$. Taking $k = 2$, the eigenvectors corresponding to the two largest eigenvalues are

$$\mathbf{U}_2 = \begin{bmatrix} 0.17 & 0.32 \\ 0.75 & 0.19 \\ 0.21 & 0.10 \\ -0.01 & 0.86 \\ -0.61 & 0.35 \end{bmatrix}.$$

The least-squares solutions for **A** and **F** are as follows.

Solution 1

$$\hat{\mathbf{A}} = \begin{bmatrix} 1.86 & 1.34 \\ 8.01 & 0.81 \\ 2.25 & 0.40 \\ -0.13 & 3.60 \\ -6.50 & 1.46 \end{bmatrix}, \quad \hat{\mathbf{F}} = \begin{bmatrix} 0.71 & -1.73 \\ 1.00 & 1.61 \\ -1.88 & 0.40 \\ 0.88 & -0.04 \\ 0.19 & -1.31 \\ -0.85 & -0.45 \\ 0.65 & 0.82 \\ 0.28 & -0.30 \\ 0.36 & 0.70 \\ -1.34 & 0.31 \end{bmatrix}.$$

Here, $\hat{\mathbf{F}}'\hat{\mathbf{F}} = (N-1)\mathbf{I} = 9\mathbf{I}$.

Solution 2

$$\hat{\mathbf{A}} = \begin{bmatrix} 0.17 & 0.32 \\ 0.75 & 0.19 \\ 0.21 & 0.10 \\ -0.01 & 1.86 \\ -0.61 & 0.35 \end{bmatrix}, \quad \hat{\mathbf{F}} = \begin{bmatrix} 7.61 & -7.29 \\ 10.67 & 6.77 \\ -20.20 & 1.69 \\ 9.44 & -0.18 \\ 2.05 & -5.49 \\ -9.08 & -1.90 \\ 6.97 & 3.45 \\ 2.97 & -1.28 \\ 3.89 & 2.94 \\ -14.33 & 1.30 \end{bmatrix}.$$

Here, $\hat{\mathbf{F}}'\hat{\mathbf{F}} = 9\mathbf{\Lambda}_k$.

Table 4.I. *The artificial data*

(A) Data matrix **X**

Object	Variable				
	x_1	x_2	x_3	x_4	x_5
1	5.0	25.0	15.0	5.0	5.0
2	10.0	30.0	17.0	17.0	8.0
3	3.0	6.0	10.0	13.0	25.0
4	7.5	27.2	16.0	11.0	6.5
5	4.6	21.9	14.0	6.6	9.0
6	3.8	13.6	12.0	9.8	17.0
7	8.3	26.6	15.9	14.2	9.1
8	6.1	22.7	14.6	10.2	9.9
9	7.6	24.2	15.2	13.8	10.8
10	3.9	10.3	11.2	12.6	21.3
Mean	5.98	20.75	14.09	11.32	12.16
Standard deviation	2.29	8.04	2.29	3.61	6.66

(B) Deviation score matrix **Y**

Object	Variable				
	y_1	y_2	y_3	y_4	y_5
1	-0.98	4.25	0.91	6.32	-7.16
2	4.02	9.25	2.91	5.68	-4.16
3	-2.98	-14.75	-4.09	1.68	12.84
4	1.52	7.15	1.91	-0.32	-5.66
5	-1.38	0.45	-0.09	-4.72	-3.16
6	-2.18	-7.15	-2.09	-1.52	4.84
7	2.32	5.85	1.81	2.88	-3.06
8	0.12	1.95	0.51	-1.12	-2.26
9	1.62	3.45	1.11	2.48	-1.36
10	-2.08	-10.45	-2.89	1.28	9.14

(C) Covariance matrix **S**

	x_1	x_2	x_3	x_4	x_5
x_1	5.2573				
x_2	15.9933	64.7517			
x_3	4.7387	18.3650	5.2410		
x_4	4.5849	1.9111	1.1769	13.0062	
x_5	-10.1498	-50.8122	-14.0438	6.0676	44.3049

Table 4.II. *The artificial data*

(A) Standardized data matrix **Z**

Object	Variable				
	z_1	z_2	z_3	z_4	z_5
1	−0.43	0.53	0.40	−1.75	−1.08
2	1.75	1.15	1.27	1.57	−0.62
3	−1.30	−1.83	−1.79	0.47	1.93
4	0.66	0.89	0.83	−0.09	−0.85
5	−0.60	0.06	−0.04	−1.31	−0.47
6	−0.95	−0.89	−0.91	−0.42	0.73
7	1.01	0.73	0.79	0.80	−0.46
8	0.05	0.24	0.22	−0.31	−0.34
9	0.71	0.43	0.48	0.69	−0.20
10	−0.91	−1.30	−1.26	0.35	1.37

(B) Correlation matrix $\mathbf{R} = (1/9)\mathbf{Z}'\mathbf{Z}$

	x_1	x_2	x_3	x_4	x_5
x_1	1.0000				
x_2	0.8668	1.0000			
x_3	0.9028	0.9969	1.0000		
x_4	0.5545	0.0659	0.1426	1.0000	
x_5	−0.6650	−0.9487	−0.9216	0.2528	1.0000

For both solutions, the correlations between variables and factors are

$$\hat{\mathbf{C}} = \begin{bmatrix} 0.81 & 0.58 \\ 0.99 & 0.10 \\ 0.98 & 0.18 \\ -0.04 & 1.00 \\ -0.98 & 0.22 \end{bmatrix}$$

From this matrix it is seen that factor 1 is highly associated with variables 1, 2, and 3 and negatively with variable 5, whereas factor 2 is highly associated with variable 4 and little with the other variables.

The technique just described is called *principal component analysis* and in the fixed case this is applied to the data matrix directly. Our illustration was based on the deviation score matrix **Y**, but the technique can also be applied to the standardized data matrix **Z** and, as will now be demonstrated, a different result will be obtained. The standardized data matrix **Z** is given in Table 4.IIA and the correlation matrix $\mathbf{R} = (1/N)\mathbf{Z}'\mathbf{Z}$ in Table 4.IIB. The eigenvalues of **R** are $\lambda_1 = 3.68237$,

$\lambda_2 = 1.31747$, $\lambda_3 = 0.00016$, $\lambda_4 = \lambda_5 = 0.00000$. The eigenvectors corresponding to the two largest eigenvalues are

$$
\mathbf{U}_2 = \begin{bmatrix}
0.48 & 0.34 \\
0.52 & -0.10 \\
0.52 & -0.04 \\
0.09 & 0.86 \\
-0.47 & 0.37
\end{bmatrix}.
$$

The least-squares Solution 1 is given by

$$
\hat{\mathbf{A}} = \begin{bmatrix}
0.92 & 0.39 \\
0.99 & -0.12 \\
1.00 & -0.04 \\
0.18 & 0.98 \\
-0.90 & 0.42
\end{bmatrix}, \qquad
\hat{\mathbf{F}} = \begin{bmatrix}
0.32 & -1.84 \\
1.32 & 1.36 \\
-1.75 & 0.80 \\
0.84 & -0.25 \\
-0.09 & -1.31 \\
-0.92 & -0.26 \\
0.82 & 0.66 \\
0.21 & -0.35 \\
0.51 & 0.61 \\
-1.24 & 0.59
\end{bmatrix}.
$$

Here, $\hat{\mathbf{F}}'\hat{\mathbf{F}} = 9\mathbf{I}$.

In this case, when both variables and factors are standardized, the matrix $\hat{\mathbf{C}}$ of correlations between variables and factors is identical to $\hat{\mathbf{A}}$. Even though the eigenvalues and eigenvectors of $\mathbf{R}$ are quite different from those of $\mathbf{S}$, it will be seen that the matrix $\hat{\mathbf{C}}$ of correlations between variables and factors is roughly the same in the two cases. The fact that they are not identical means that the result of a principal component analysis is scale-dependent, a point that we shall discuss later in this chapter.

Random case

In the random case, the model is the same as in [4.1]. With uncorrelated factors, this model in the population implies

$$
\mathbf{\Sigma} = \mathbf{A}\mathbf{A}' + \mathbf{\Psi}. \qquad [4.5]
$$

In principal component analysis, $\mathbf{E}$ and $\mathbf{\Psi}$ are assumed to be small. Therefore, a reasonable criterion for fitting the model to the data is to make $\mathbf{S} - \mathbf{A}\mathbf{A}'$ as small as possible by some suitable method. In component analysis, this is usually done by choosing $\mathbf{A}$ so that the sum of squares of all the elements of $\mathbf{S} - \mathbf{A}\mathbf{A}'$ is minimized. This is called the method of *unweighted least squares* (hereinafter abbreviated as ULS).

As demonstrated in Chapter 2, a solution to this problem is

$$\hat{\mathbf{A}}\hat{\mathbf{A}}' = \lambda_1 u_1 u_1' + \lambda_2 u_2 u_2' + \cdots + \lambda_k u_k u_k', \qquad [4.6]$$

or

$$\hat{\mathbf{A}}\hat{\mathbf{A}}' = \mathbf{U}_k \mathbf{\Lambda}_k \mathbf{U}_k',$$

where, as before, $\mathbf{\Lambda}_k$ is a diagonal matrix whose diagonal elements are the k largest eigenvalues of $\mathbf{S}$, and the columns of $\mathbf{U}_k$ are the corresponding eigenvectors.

The solution for $\mathbf{A}$ may be chosen as

$$\mathbf{A} = \mathbf{U}_k \mathbf{\Lambda}_k^{1/2}, \qquad [4.7]$$

that is, the columns of $\hat{\mathbf{A}}$ are the eigenvectors of $\mathbf{S}$ corresponding to the k largest eigenvalues and scaled so that the sum of squares equals the corresponding eigenvalue. This solution for $\mathbf{A}$ is the same as Solution 1 in the fixed case. We note that

$$\hat{\mathbf{A}}'\hat{\mathbf{A}} = \mathbf{\Lambda}_k. \qquad [4.8]$$

The contribution of the ith component to the total variance of all variables is the sum of squares of all the elements in the ith column of $\hat{\mathbf{A}}$, which is the same as λ_i, the ith largest eigenvalue. The components defined by the solution [4.7] are called *principal components*. They have the following properties.

> The first principal component is that linear combination of the variables which accounts for maximum variance.
>
> The second principal component is uncorrelated with the first and accounts for maximum variance among all components that are uncorrelated with the first principal component.
>
> The third principal component is uncorrelated with the first two principal components and accounts for the maximum variance among all components that are uncorrelated with the first two principal components, and so on.

We hasten to point out that the principal component result [4.7] is not the only solution to the minimization problem presented earlier. The general solution of this is given by

$$\hat{\mathbf{A}} = \mathbf{U}_k \mathbf{\Lambda}_k^{1/2} \mathbf{T}, \qquad [4.9]$$

where $\mathbf{T}$ is an arbitrary orthogonal matrix of order $k \times k$. The solution in [4.9] corresponds to an orthogonal transformation of the principal components as described in Chapter 3. Although the transformed components are still uncorrelated, they may not have the property of accounting successively for maximum variance. As also was pointed out in Chapter 3, the principal components may be transformed to a set of correlated components. Whether the k new components are uncorrelated or not, they *together* account for as much of the variance as do the k principal components together, but their internal variance and correlation properties are changed by the transformation.

The total variance accounted for by all the principal components is given by $\operatorname{tr} \Lambda_k$, that is, the sum of the variances of all the linear combinations:

$$\operatorname{tr} \Lambda_k = \lambda_1 + \lambda_2 + \cdots + \lambda_k.$$

This will be the same as the *total variance* of all the variables if $k = p$, because

$$\mathbf{S} = \mathbf{U}\Lambda\mathbf{U}' \quad \text{and} \quad \operatorname{tr} \mathbf{S} = \operatorname{tr}(\mathbf{U}\Lambda\mathbf{U}') = \operatorname{tr} \Lambda.$$

The variance left unexplained by the components, that is, the total *residual variance*, is therefore

$$\lambda_{k+1} + \lambda_{k+2} + \cdots + \lambda_p.$$

Consequently, the total variance of all the variables is the sum of the variance explained by the k components and the residual variance.

How many components?

Up to now, we have assumed that k, the number of components extracted, is known. The normal situation, however, is that k is unknown, and one would like to determine it in such a manner as to account for a "satisfactory" amount of the total variation. There is no exact solution to this problem, but the following guidelines are often useful.

A useful way to start is to compute the cumulative percentage variance contribution obtained for successive values of $k = 1, 2, \ldots$ and to stop when this is sufficiently large, for example, larger than 75%, 90%, or 95%. How far one can get usually depends to a large extent on the nature of the data, that is, the degree of collinearity and redundancy in it. The cumulative percentage variance contribution for a given value of k is computed as

$$100 = \frac{\sum_{i=1}^{k} \lambda_i}{\operatorname{tr} \mathbf{S}}.$$

The eigenvalues, $\lambda_1, \lambda_2, \ldots, \lambda_k$, should be well separated and significantly positive, which means that the falloff in size from one of them to the next should be appreciable. Approximative confidence intervals can sometimes be computed to see if the λ's are sufficiently different from zero (see the section on confidence intervals on p. 100).

The residual eigenvalues $\lambda_{k+1}, \ldots, \lambda_p$, should be small and approximately equal. In order to discover the existence of such a situation, it is often informative to plot the values of λ_i against their order and to look for the flattening out of the tail. A variant of this is to plot the percentage of the trace. The so-called Bartlett's test for equality of last roots is often invoked, but it is seldom really useful, because it is invariably overoptimistic in that too many small roots are selected. Krzanowski (1987a) has given this question close attention using a cross-validatory choice, prediction sum of squares (PRESS is the acronym). This method is discussed in Section 4.5, which deals with atypical observations. Further references are Jolliffe (1986, p. 99) and Jackson (1991).

Scores for principal components

Principal-components scores are calculated as follows. Consider an individual observation vector, x. Then, if the k principal components fit reasonably well, we should have, approximately,

$$x \approx \hat{\mathbf{A}} f, \qquad [4.10]$$

where f is the vector of principal-component scores corresponding to the observational vector x. Premultiplying this equation by $\hat{\mathbf{A}}'$ yields

$$\hat{\mathbf{A}}'x \approx \hat{\mathbf{A}}'\hat{\mathbf{A}} f,$$

or, from [4.8], $\hat{\mathbf{A}}'x \approx \Lambda_k f$. Thus $f \approx \Lambda_k^{-1}\hat{\mathbf{A}}'x$, or from [4.7],

$$f \approx \Lambda_k^{-1/2} \mathbf{U}_k' x. \qquad [4.11]$$

Taking the transpose of this expression, and writing it for all observational vectors of the sample, gives

$$\hat{\mathbf{F}} \approx \mathbf{X}\mathbf{U}_k \Lambda_k^{-1/2}. \qquad [4.12]$$

Thus, the scores are the product of the first k eigenvectors with the data matrix, scaled by the inverse square root of the eigenvalues. This is again the same solution as for Solution 1 in the fixed case. Note that [4.11] may be used, not only to compute the factor scores for the objects in the sample but also for a new object.

Scale dependence

Changing the unit of measurement in a variable corresponds to multiplying all the observed values for that variable by a scale factor. If each variable is multiplied by a scale factor d_{ii}, this results in the matrix **DSD** instead of the matrix **S**, where **D** is a diagonal matrix of the scale factors. The difficulty attaching to this is that the principal components based on **DSD** cannot be directly related to those of **S**. This is due to the fact that the unweighted least-squares criterion used to fit the model requires that all residuals be equally weighted and this equal weighting may be appropriate in one metric but not in another.

Principal component analysis should therefore be used only when it can be expected that the residual variances are the same for all the variables. This will usually be the case if all variables are measured in the same unit of measurement and the measurements are of the same order of magnitude. If this does not hold, it is difficult to justify the use of principal components.

A common way around the problem posed by scaling differences is to use **R**, the correlation matrix, instead of the covariance matrix **S**. Statistically, this option is not ideal, because the sampling theory available for the covariance matrix can only be applied to **R** in a few cases. This has the disadvantage that many developments, such as tests of significance, confidence intervals of eigenvalues, and the like cannot be applied to the correlation matrix.

Confidence intervals for eigenvalues of the covariance matrix

In practice, it transpires that tests of significance in factor analysis are not really very useful. Statisticians tend to be interested in developing such tests for reasons of completeness of the theoretical framework. However, it is possible to compute approximate confidence intervals for eigenvalues and eigenvectors of the covariance matrix, and these may be useful in some studies.

If the sample size is large and the data are multivariate normally distributed, it is possible to construct asymptotic confidence intervals for the eigenvalues of **S**, the covariance matrix. If $t_{1/2}$ denotes the pertinent upper percentage point of the standard normal distribution, the confidence interval for the ith eigenvalue is

$$\frac{\hat{\lambda}_i}{1 + t_{1/2}\sqrt{2/n}} \leq \lambda_i \leq \frac{\hat{\lambda}_i}{1 - t_{1/2}\sqrt{2/n}}, \qquad [4.13]$$

where λ_i is the population counterpart of the estimated sample eigenvalue $\hat{\lambda}_i$, and $n = N - 1$. Usually, one computes the 95% confidence interval, for which $t_{1/2} = 1.96$ (Anderson, 1963). Empirical confidence intervals can be obtained as part of an analysis involving cross-validation. An example is given in the section dealing with influential observations (Section 4.5). This example also shows the practical usefulness of the "bootstrapping" technique (Efron, 1979).

The role of the smallest principal components

Principal components with small nontrivial variances can be considered to express near-constant relationships that may be of considerable interest. Reyment (1978) could relate "small components" to specific situations pertaining in biological and chemical data.

The smallest principal components have also direct statistical applications. One of these concerns their use in regression for selecting a subset of informative variables. Another use concerns the identification of atypical observations in a sample, a question that is discussed in detail by Jolliffe (1986). The subject of atypical observations is taken up in detail in Section 4.5, which deals with the use of cross-validation and other techniques for uncovering the presence of atypical and influential values in a sample.

Equality of eigenvalues of a covariance matrix

The tests applied in this connection are not very useful and, for many situations, they may even give misleading results. We mention them only for completeness.

One test sometimes mentioned in the literature is one for testing the hypothesis of the equality of a certain number of the smallest eigenvalues of the covariance matrix. For example, in the three-dimensional case, if all the eigenvalues are equal, we are dealing with a spherical distribution. If the largest is different, and the two smallest equal, the cross section in the plane of the second and third axes is a circle. In the general p-dimensional case, if the $p - k$ smallest eigenvalues are equal, there is no use taking out more than k principal components, since the remaining may not be distinguishable. The test for the equality of the $p - k$ smallest eigenvalues is due to Bartlett (1954). For further results and discussion see Lawley (1956) and Lawley and Maxwell (1971).

Test that an eigenvector equals a specified vector

The test that some eigenvector of a covariance matrix is equal to a specified vector is a result of Anderson (1963). This test was designed to ascertain whether the eigenvector u_i associated with the distinct eigenvalue λ_i of the population covariance matrix Σ is equal to some specified vector b; it applies only to large samples and has the following construction, where S is a large-sample covariance matrix:

$$\chi^2 = n\left(\lambda_i b' S^{-1} b + \frac{1}{\lambda_i} b' S b - 2\right). \qquad [4.14]$$

This criterion is distributed asymptotically as chi-squared with $p - 1$ degrees of freedom when the null hypothesis of equality of vectors is true. As before, $n = N - 1$. This is a useful test in work involving growth studies. The same problem occurs in meteorology (Preisendorfer, 1988).

4.3 TRUE FACTOR ANALYSIS

In "true" factor analysis, the model equation is the same as [4.5]. The difference now is that Ψ is a diagonal matrix and its diagonal elements are not necessarily small. The diagonal elements of Ψ, which are the unique variances, must be estimated from the data, together with the factor loadings matrix $\mathbf{A}$. In accordance with the approach of the foregoing section, a reasonable criterion for fitting the model to the data would be to minimize the sum of squares of all the elements of $\mathbf{S} - \mathbf{A}\mathbf{A}' - \Psi$, that is, minimizing

$$\text{tr}(\mathbf{S} - \mathbf{A}\mathbf{A}' - \Psi)^2. \qquad [4.15]$$

If Ψ were known, one could fit $\mathbf{A}\mathbf{A}'$ to $\mathbf{S} - \Psi$ by the same method as in Section 4.2, thereby choosing the columns of $\mathbf{A}$ as eigenvectors of $\mathbf{S} - \Psi$ corresponding to the k largest eigenvalues and scaled so that the sum of squares in each column equals the corresponding eigenvalue. In practice, Ψ is not known but has to be estimated from the data. The act of selecting Ψ is equivalent to choosing the communalities, as the diagonal elements of $\mathbf{S} - \Psi$ are estimates of the communalities. Several methods of guessing communalities have been tried in the past. Traditionally, the most widely used method has been that of choosing the communality of each variable in relation to the squared multiple correlation coefficient (SMC) of that variable with all other variables (see, for example, Harman, 1967). Guttman (1956) showed that the SMC provides the "best possible" lower bound for the communality. As noted in Section 2.14, this can be shown to amount to choosing

$$\hat{\Psi} = \left(\text{diag}\,\mathbf{S}^{-1}\right)^{-1}, \qquad [4.16]$$

that is, the ψ's are set equal to the reciprocals of the diagonal elements of the inverse of $\mathbf{S}$. With this choice of Ψ, one can then proceed to estimate $\mathbf{A}$ as outlined previously. This is called the *principal-factor method* and gives estimates that are systematically biased. It has therefore been suggested that this method can be improved upon by iteration in the following way. Once the estimate $\hat{\mathbf{A}}$ of $\mathbf{A}$ has been determined, we can select a new, and hopefully better, estimate of $\boldsymbol{\Psi}$ as

$$\hat{\boldsymbol{\Psi}} = \text{diag}(\mathbf{S} - \hat{\mathbf{A}}\hat{\mathbf{A}}').$$

This can then be used to furnish the basis for a new estimate of $\hat{\mathbf{A}}$, and so on. This process is iterated until the estimate of $\hat{\mathbf{A}}$ converges to a stable value. This is known in the literature as the *iterated principal-factor method*.

These methods have no more than historical interest and they are no longer used by statisticians. More efficient methods are now available for minimizing the function [4.15]. One of these is the MINRES method of Harman (see Harman, 1967), which minimizes only the sum of squares of all the off-diagonal elements of $\mathbf{S} - \mathbf{A}\mathbf{A}'$ and therefore avoids the communalities. Another method is Jöreskog's (1976) ULS method, which, by elimination of $\mathbf{A}$, reduces the function to a function of $\boldsymbol{\Psi}$ only and then minimizes this by the Newton–Raphson procedure.

Scale-free estimation of A and $\boldsymbol{\Psi}$

One of the difficulties of the ULS approach is that it does not yield a scale-free solution. A scale-free method is one such that if $\mathbf{S}$ is analyzed to give $\mathbf{A}$ and $\boldsymbol{\Psi}$, then an analysis of $\mathbf{DSD}$, where $\mathbf{D}$ is a diagonal matrix of scale factors, should give estimates $\mathbf{DA}$ and $\mathbf{D}^2\hat{\boldsymbol{\Psi}}$. This is a desirable property in factor analysis, inasmuch as the units of measurement in the variables are sometimes arbitrary or irrelevant. The ULS method is not scale-free and, in practice, one usually employs the correlation matrix $\mathbf{R}$ instead of $\mathbf{S}$.

A scale-free method of estimating $\mathbf{A}$ and $\boldsymbol{\Psi}$ may be developed as follows. Let us assume that $\boldsymbol{\Psi}$ is known and that the model fits well. In this case, we have approximately that

$$\mathbf{S} - \boldsymbol{\Psi} \approx \mathbf{A}\mathbf{A}'.$$

Pre- and postmultiplication with $\boldsymbol{\Psi}^{-1/2}$ yields

$$\boldsymbol{\Psi}^{-1/2}\mathbf{S}\boldsymbol{\Psi}^{-1/2} - \mathbf{I} \approx \boldsymbol{\Psi}^{-1/2}\mathbf{A}\mathbf{A}'\boldsymbol{\Psi}^{-1/2}.$$

Therefore, with $\boldsymbol{\Psi}$ known, we can compute $\boldsymbol{\Psi}^{-1/2}\mathbf{S}\boldsymbol{\Psi}^{-1/2} - \mathbf{I}$ and fit $\mathbf{A}^*\mathbf{A}^{*\prime}$ to this, where $\mathbf{A}^* = \boldsymbol{\Psi}^{-1/2}\mathbf{A}$. Having obtained the estimate $\hat{\mathbf{A}}^*$ from the eigenvalues and eigenvectors of $\boldsymbol{\Psi}^{-1/2}\mathbf{S}\boldsymbol{\Psi}^{-1/2} - \mathbf{I}$, we then compute $\hat{\mathbf{A}}$ as $\boldsymbol{\Psi}^{1/2}\hat{\mathbf{A}}^*$. With the choice of $\hat{\boldsymbol{\Psi}}$ as in [4.16], this amounts to computing the eigenvalues and eigenvectors of

$$\left(\text{diag}\, \mathbf{S}^{-1}\right)^{1/2}\mathbf{S}\left(\text{diag}\, \mathbf{S}^{-1}\right)^{1/2} - \mathbf{I}, \qquad [4.17]$$

Table 4.IIIA. *Correlation matrix S for eight variables*

	1	2	3	4	5	6	7	8
1	1.000							
2	0.466	1.000						
3	0.456	0.311	1.000					
4	0.441	0.296	0.185	1.000				
5	0.375	0.521	0.184	0.176	1.000			
6	0.312	0.286	0.300	0.244	0.389	1.000		
7	0.247	0.483	0.378	0.121	0.211	0.210	1.000	
8	0.207	0.314	0.378	0.341	0.153	0.289	0.504	1.000

a matrix that can be proved to be invariant under transformations of scale in the variables. Harris (1962) related this approach to Guttman's (1953) *image analysis* and to the *canonical factor method* of Rao (1955).

Since the estimate Ψ in [4.16] is systematically too large, Jöreskog (1963) proposed that it be replaced by

$$\hat{\Psi} = \theta \left(\text{diag } S^{-1} \right)^{-1},$$

where θ is an unknown scalar, less than 1, to be estimated from the data. Let $\lambda_1, \lambda_2, \ldots, \lambda_p$ be the eigenvalues of

$$S^* = \left(\text{diag } S^{-1} \right)^{1/2} S \left(\text{diag } S^{-1} \right)^{1/2}. \qquad [4.18]$$

The least-squares estimate of θ is then

$$\hat{\theta} = \frac{1}{p-k} \sum_{m=k+1}^{p} \lambda_m, \qquad [4.19]$$

that is, the average of the $p - k$ smallest eigenvalues of S^*. Furthermore, let U_k be the matrix of order p by k, the columns of which are the orthonormal eigenvectors of S^*, corresponding to the k largest eigenvalues, and let Λ_k be the diagonal matrix of these eigenvalues. Then the least-squares estimate of A is given by

$$\hat{A} = \left(\text{diag } S^{-1} \right)^{-1/2} U_k \left(\Lambda_k - \hat{\theta} I \right)^{1/2}. \qquad [4.20]$$

This amounts to first scaling the columns of U_k so that the sum of squares of the jth column equals $\lambda_j - \hat{\theta}$ and then scaling the ith row by $1/\sqrt{s^{ii}}$, where s^{ii} is the ith diagonal element of S^{-1}.

The computations involved in this method will now be illustrated using the correlation matrix of Table 4.IIIA. Here we take this to be the covariance matrix S. The diagonal elements of S^{-1} and various related quantities are shown in Table 4.IIIB. The elements of S^* are computed

Table 4.IIIB. *Elements s^{ii} and related quantities*

	s^{ii}	$\sqrt{s^{ii}}$	$\dfrac{1}{\sqrt{s^{ii}}}$	$\dfrac{1}{s^{ii}}$	$1 - \dfrac{1}{s^{ii}}$	$\sqrt{1 - \dfrac{1}{s^{ii}}}$
1	1.785	1.336	0.748	0.560	0.440	0.663
2	1.911	1.382	0.723	0.523	0.477	0.690
3	1.506	1.227	0.815	0.664	0.336	0.580
4	1.422	1.193	0.838	0.703	0.297	0.545
5	1.553	1.246	0.802	0.644	0.356	0.597
6	1.323	1.150	0.869	0.756	0.244	0.494
7	1.687	1.299	0.670	0.593	0.407	0.638
8	1.623	1.274	0.785	0.616	0.384	0.620

Table 4.IIIC. *The matrix* S^*

	1	2	3	4	5	6	7	8
1	1.785							
2	0.861	1.911						
3	0.748	0.528	1.506					
4	0.703	0.488	0.271	1.422				
5	0.624	0.898	0.281	0.262	1.553			
6	0.479	0.455	0.423	0.335	0.558	1.323		
7	0.429	0.867	0.603	0.187	0.342	0.314	1.687	
8	0.352	0.553	0.591	0.518	0.243	0.423	0.834	1.623

from the elements of S and the elements $\sqrt{s^{ii}}$ in column 3 of Table 4.IIIB according to the formula

$$s_{ij}^* = \sqrt{s^{ii}} \, s_{ij} \sqrt{s^{jj}} .$$

The matrix S^* is given in Table 4.IIIC. This matrix would have been the same if we had started with the covariance matrix instead of the correlation matrix. In fact, it is completely invariant under changes of scales in the variables.

The eigenvalues λ_i of S^* are 5.281, 1.809, 1.507, 1.199, 1.152, 0.703, 0.625, 0.534 with sum equal to 12.810 = tr S^*. To determine the number of factors k, we use a rough rule as follows. We regard only those eigenvalues as significant that are as much above unity as the smallest eigenvalue is below unity. The smallest eigenvalue in this case is 0.534, which is 0.466 below unity. The number of eigenvalues greater than 1.466 is 3. Hence we take $k = 3$. To compute the factor loadings matrix we must first compute the estimate $\hat{\theta}$ of θ. This is equal to the arithmetic mean of the five smallest eigenvalues, which is

$$\hat{\theta} = \tfrac{1}{5} \sum_{i=4}^{8} \hat{\lambda}_i = 0.843.$$

Subtracting $\hat{\theta}$ from the three largest eigenvalues, we get the numbers

$$\lambda_m - \hat{\theta} = 4.438, 0.966, 0.664.$$

The square roots of these numbers are

$$\sqrt{\lambda_m - \hat{\theta}} = 2.107, 0.983, 0.815.$$

The eigenvectors of S^* corresponding to the three largest eigenvalues are

$$\mathbf{U}_3 = \begin{bmatrix} 0.41 & -0.37 & 0.36 \\ 0.47 & -0.15 & -0.42 \\ 0.33 & 0.20 & 0.26 \\ 0.27 & -0.15 & 0.56 \\ 0.32 & -0.44 & -0.42 \\ 0.27 & -0.13 & 0.05 \\ 0.37 & 0.52 & -0.33 \\ 0.34 & 0.54 & 0.17 \end{bmatrix}.$$

Finally, the factor loadings matrix $\mathbf{A}$ is computed from [4.20] as

$$\hat{a}_{im} = \frac{u_{im}\sqrt{\lambda_m - \hat{\theta}}}{\sqrt{s^{ii}}},$$

where u_{im} is taken from $\mathbf{U}_3$, and $1/\sqrt{s^{ii}}$ is taken from Table 4.IIIB, column 4. The factor loading matrix $\mathbf{A}$ is given in Table 4.IIID.

More efficient methods of estimation

Still more efficient methods, which are also scale-free, are available today for computing the estimates of $\mathbf{A}$ and $\mathbf{\Psi}$. These are the methods of *generalized least squares* (GLS), which minimizes

$$\text{tr}(\mathbf{S}^{-1}\mathbf{\Sigma} - \mathbf{I})^2, \qquad [4.21]$$

and the *maximum-likelihood method* (ML), which minimizes

$$\log|\mathbf{\Sigma}| + \text{tr}(\mathbf{S}\mathbf{\Sigma}^{-1}) - \log|\mathbf{S}| - p. \qquad [4.22]$$

In [4.21] and [4.22], we have written $\mathbf{\Sigma}$ for $\mathbf{AA}' + \mathbf{\Psi}$ for reasons of clarity. Both functions have to be minimized with respect to $\mathbf{A}$ and $\mathbf{\Psi}$.

Table 4.IIID. *The factor loadings matrix A*

i	a_{i1}	a_{i2}	a_{i3}
1	0.65	0.27	0.22
2	0.71	−0.11	−0.25
3	0.57	0.16	0.17
4	0.47	−0.12	0.38
5	0.55	−0.35	−0.28
6	0.50	−0.11	0.03
7	0.59	0.40	−0.20
8	0.56	0.42	0.11

In these methods, one must assume close adherence to the multivariate normal distribution. Even although they are based on the assumption of multivariate normality, they are fairly robust to deviations therefrom. Both functions are minimized by first eliminating **A** and then minimizing the function with respect to **Ψ** by a numerical iterative procedure. The computational procedure is too complicated to be described here but a computer program, EFAP (Jöreskog and Sörbom, 1980), is available for anyone to use. For further information about these methods, we refer you to Jöreskog (1967, 1976), Lawley and Maxwell (1971), and Jöreskog and Goldberger (1972).

Assessing the fit of the model

One advantage of the ML and GLS methods as compared to ULS and other methods is that there is a statistical measure available for assessing the goodness of fit of the model. This measure is $(N - 1)$ times the minimum value of [4.21] or [4.22]. If you are fitting k factors to your model and k is the correct number of factors in the population, then this measure is approximately distributed in large samples as χ^2 (chi-squared) with degrees of freedom equal to $\frac{1}{2}[(p - k)^2 - (p + k)]$.

A small value of χ^2, relative to the degrees of freedom, means that the model "fits too well" and that it should be simplified and k decreased. On the other hand, a large value of χ^2 indicates that too few factors have been selected and k should therefore be increased. Overfitting is generally regarded as more serious than underfitting. If your model overfits, it means that you are "capitalizing on chance" and some of your factor loadings may not have a real meaning. If you have underfitting, you are not utilizing the information in your sample fully.

A practical procedure for determining the number of common factors goes as follows. For each successive value of k from a lower bound k_L (for example, 0, 1, or some other number that may be regarded as a safe underestimate of the number of factors) and up, one computes the value of χ^2, say χ_k^2, and the corresponding degrees of freedom, say d_k. One also computes the difference $\chi_k^2 - \chi_{k+1}^2$ and regards this as an approximate χ^2-variate with $d_k - d_{k+1}$ degrees of freedom. If this χ^2-variate is significant continue to increase k; otherwise the current value of k is an appropriate estimate of the number of factors since no significant improvement of the fit can be obtained by increasing k. An illustration of this procedure follows next.

To illustrate the maximum-likelihood method and the use of χ^2, we use some real data from Sampson (1968). These data consist of 122 analyses of subsurface brines from the Arbuckle Group (Cambrian–Ordovician) in Kansas. Eleven variables were measured on each sample:

1. calcium, in parts per million (ppm)
2. magnesium, in ppm
3. sodium, in ppm
4. bicarbonate, in ppm
5. sulphate, in ppm
6. chloride, in ppm
7. total dissolved salts, in ppm
8. specific gravity
9. temperature, degrees celsius
10. electrical resistivity, in ohms per square meter per meter
11. pH

The correlation matrix for the 11 variables is given in Table 4.IVA. This correlation matrix was analyzed with $k = 1, 2, 3, 4,$ and 5 factors. Table 4.IVB gives the various χ^2-values obtained and illustrates the decision procedure. It is seen that for $k = 1, 2,$ and 3 the drop in χ^2 from k to $k + 1$ is considerable but from $k = 4$ to $k = 5$ the drop in χ^2 is only 7.63 with seven degrees of freedom. Hence we should stop with four factors. The maximum-likelihood solution for $k = 4$ is shown in Table 4.IVC. To achieve simple structure, these factor loadings were rotated by the varimax method (see Section 7.8), which means that an orthogonal rotation of the four coordinate axes is made such that the variance of the squared loadings in each column is maximized. This is expected to give more loadings close to 1 (or minus 1) and zero and fewer in-between loadings. The varimax solution is shown in Table 4.IVD.

Inspection of the columns of Table 4.IVD leads to the following interpretation of the factors. Factor 1 represents the relationship existing between dissolved salts and the resistivity. The second factor treats covariation in Ca and Mg and could be a factor for waters emanating

Table 4.IVA. *Correlation matrix for 11 geochemical variables; N = 122 (from Sampson, 1968)*

1	1.000					
2	0.7881	1.000				
3	0.1346	0.0768	1.000			
4	−0.2140	−0.0747	0.0216	1.000		
5	0.0192	0.0312	0.1579	0.0122	1.000	
6	0.7861	0.5916	0.2044	−0.1051	0.1365	1.000
7	0.4866	0.4828	0.2549	−0.1287	0.2969	0.6524
8	0.7055	0.5312	0.1781	−0.0856	0.1533	0.8716
9	0.0249	−0.0282	0.0262	0.0243	−0.1856	−0.0421
10	−0.5286	−0.3680	−0.2647	0.2732	−0.2845	−0.5465
11	−0.3542	−0.3384	−0.0137	−0.0312	−0.1537	−0.3318

1					
2					
3					
4					
5					
6					
7	1.000				
8	0.5897	1.000			
9	−0.0866	−0.0621	1.000		
10	−0.6898	−0.4893	−0.0818	1.000	
11	−0.2380	−0.2296	−0.0544	0.0422	1.000

Table 4.IVB. χ^2-*Values for* ML *solutions for Sampson's data*

k	χ_k^2	d_k	$\chi_k^2 - \chi_{k+1}^2$	$d_k - d_{k+1}$
1	185.08	44	73.55	10
2	112.53	34	64.67	9
3	47.86	25	26.66	8
4	21.20	17	7.63	7
5	13.57	10		

from limey country rock; it is significant that this is the only factor in which pH enters (logically, in a negative association), albeit with a rather low loading. Factor 3 is not easy to interpret because all loadings are low, but it does seem as though it represents a general reflection of resistivity variation. The fourth factor expresses a connection between chloride concentration and specific gravity. You will see that some of the variables determined, such as temperature, are not important for the analysis and it would be advisable in a further study to think about dropping these.

Table 4.IVC. *Unrotated factor loadings*

Variable	Factor			
	1	2	3	4
1 Ca^{++}	−0.529	0.844	−0.056	0.001
2 Mg^{++}	−0.368	0.700	−0.054	−0.309
3 Na^+	−0.265	0.001	0.121	−0.044
4 HCO_3^-	0.273	−0.069	0.174	−0.090
5 SO_4^{--}	−0.284	−0.144	0.171	−0.205
6 Cl^-	−0.547	0.623	0.519	0.056
7 Total salts	−0.690	0.170	0.370	−0.405
8 Specific gravity	−0.490	0.561	0.484	0.060
9 Temperature	−0.081	−0.031	−0.156	0.174
10 Resistivity	1.000	0.001	0.000	−0.000
11 pH	0.042	−0.401	−0.119	0.229

Table 4.IVD. *Varimax-rotated factor loadings*

	1	2	3	4
1 Ca^{++}	0.100	0.807	0.361	0.452
2 Mg^{++}	0.168	0.807	0.100	0.186
3 Na^+	0.259	0.042	0.053	0.123
4 HCO_3^-	−0.092	−0.098	−0.315	−0.025
5 SO_4^{--}	0.412	−0.009	−0.056	0.013
6 Cl^-	0.338	0.479	0.032	0.784
7 Total salts	0.766	0.376	−0.046	0.274
8 Specific gravity	0.303	0.422	0.022	0.723
9 Temperature	−0.053	−0.063	0.235	−0.007
10 Resistivity	−0.765	−0.171	−0.573	−0.239
11 pH	−0.036	−0.427	0.160	−0.141

Extensions of factor analysis

The type of factor analysis described in this chapter is called *exploratory factor analysis*. It is used to discover or detect the most important sources of variation and covariation in the observed data. As more knowledge is gained about the nature of the important factors and how to measure them, exploratory factor analysis may not be so useful anymore. Other types of models to account for variation and covariation have also been developed. We mention briefly that exploratory factor analysis has been extended to *confirmatory factor analysis* (Jöreskog, 1969), to *simultaneous factor analysis in several populations* (Jöreskog, 1971), to more general *covariance structures* (Jöreskog, 1973, 1981), and to *mean structures* (Sörbom, 1974). All these models may be estimated using the GLS or ML fit functions in [4.21] and [4.22], using the same computer program (Jöreskog and Sörbom, 1989).

4.4 ROBUST PRINCIPAL COMPONENT ANALYSIS

Robust methods in multivariate analysis have begun to attract attention in recent years. There are now robust methods for several of the better known methods, such as canonical variates, discriminant functions, and also principal component analysis (Campbell, 1979). A useful reference for robust methods of estimation is the book by Hampel et al. (1986, p. 271); others are Gnanadesikan (1977) and Seber (1984).

A principal component analysis of the covariance matrix S (or associated correlation matrix R) chooses a linear combination, which for present purposes we write as

$$y_m = u'x_m,$$ [4.23]

of the original variables x_m such that the usual sample variance of the y_m is a maximum. As we have already shown in Section 2.11, the solution is given in terms of the usual eigenproblem,

$$S = UAU';$$

the eigenvectors u_i of U define the linear combinations, and the corresponding diagonal elements λ_i of the diagonal matrix Λ are the variances of the principal components.

The directions u_i should not be decided by one or two atypical values in the sample. The obvious procedure for robustifying the analysis is to replace S by the robust estimator S*. In terms of this approach, an observation is weighted in accordance with its total distance from the robust estimate of location. However, this distance can be decomposed into constituents along each eigenvector. Thus an observation may have a large component along one direction and small components along the remaining directions, with the result that it will not be downweighted.

The method of *M*-estimators (Hampel et al., 1986; Seber, 1984, p. 171) can be applied in an *ad hoc* manner to the mean and variance of each principal component. The direction cosines can then be chosen to maximize the robust variance of the resulting linear combinations.

Hence the aim is to identify those observations that have undue influence on the eigenvectors (i.e., the directions) and to determine directions that are little influenced by atypical observations. This can be specified more succinctly as follows (discussion excerpted from

Campbell and Reyment, 1980):

1. Take as the initial estimate of u_1 the first eigenvector of matrix **S**.
2. Form the principal component scores in the usual manner.
3. With the median and $\{0.74 \text{ (interquartile range)}\}^2$ of the y_m as initial robust estimates, find the M-estimator of the mean and variance. [N.B. The value 0.74 is $(2 \times 0.675)^{-1}$, and 0.675 is the 75% quantile for the $N(0, 1)$ distribution.] Then use these weights to determine

$$\bar{x}^w = \frac{\sum_{m=1}^n w_m x_m}{\sum_{m=1}^n w_m},$$ [4.24]

and hence

$$\mathbf{S}^w = \frac{\sum_{m=1}^n w_m^2 (x_m - \bar{x}^w)(x_m - \bar{x}^w)'}{\left(\sum_{m=1}^n w_m^2 - 1\right)}.$$ [4.25]

4. After the first iteration it is desirable to take the weights w_m as the minimum of the weights for the current and previous iterations in order to prevent the solution from oscillating.
5. Now determine the first eigenvalue and eigenvector u^* of $\mathbf{S}^w$.
6. Repeat until successive estimates of the eigenvalue are sufficiently close.

To ascertain successive directions u_i^*, $2 < i < p$, project the data onto the space orthogonal to that spanned by the previous eigenvectors $u_1^*, \ldots, u_{i-1}^*$ and repeat the iterative steps, taking as the initial estimate the second eigenvector from the last iteration for the previous eigenvector. The sequence of events is as follows:

7. $x_{im} = (\mathbf{I} - \mathbf{U}_{i-1}^* \mathbf{U}_{i-1}^{*\prime}) x_m,$ [4.26]
   where $\mathbf{U}_{i-1}^* = (u_1^*, \ldots, u_{i-1}^*)$.
8. Repeat the iterations 2–5 with x_{im} replacing x_m, and determine $\tilde{u}_i$. The covariance matrix based on the x_{im} will be singular, with rank $p - i + 1$. This poses no special difficulty, since only the first eigenvalue and eigenvector are needed.
9. The *principal component scores* are yielded by

$$\tilde{u}' x_{im} = \tilde{u}_i (\mathbf{I} - \mathbf{U}_{i-1} \mathbf{U}_{i-1}') x_m.$$ [4.27]

Steps 7, 8, and 9 are repeated until all p eigenvalues and eigenvectors, with associated weights, have been obtained, or some specified proportion of variation has been extracted. When the principal component analysis is based on the correlation matrix, the data must be standardized to unit variance for successive eigenextractions. Two pos-

Table 4.V. *Comparison of robust and standard principal component analysis for measures on the ostracod carapace, using the covariance matrix (number of dimensions = 5; number of observations = 102)*

(A) Covariance matrix (upper triangle) and correlation matrix (lower triangle)

	1	2	3	4	5
1	0.0110	0.0054	0.0080	0.0065	0.0012
2	0.9686	0.0029	0.0040	0.0032	0.0006
3	0.9863	0.9629	0.0060	0.0048	0.0009
4	0.9714	0.9447	0.9666	0.0041	0.0007
5	0.6990	0.7326	0.7194	0.6877	0.0002

(B) Eigenvalues and eigenvectors pairwise for robust (R) and *usual* (U) computations

		Eigenvectors				
		Variables				
	Eigenvalues	1	2	3	4	5
R(1)	0.023582	0.6802	0.3402	0.5002	0.4074	0.0735
U(1)	0.023582	0.6802	0.3402	0.5002	0.4074	0.0735
R(2)	0.000802	−0.1094	0.7480	−0.0741	−0.4377	0.4811
U(2)	0.000202	−0.0433	−0.5598	−0.1208	0.7480	−0.3326
R(3)	0.000105	−0.6574	0.1016	0.2700	0.6263	0.3039
U(3)	0.000153	0.4776	−0.4773	0.1719	−0.5209	−0.4931
R(4)	0.000741	0.1978	0.1983	−0.8185	0.4976	0.0634
U(4)	0.000153	−0.2920	−0.4698	0.6892	−0.0502	0.4653
R(5)	0.000067	0.2326	−0.5245	−0.0383	−0.0504	0.8166
U(5)	0.000125	0.4713	−0.3498	−0.4802	−0.0225	0.6514

sibilities exist: (1) the data can be standardized by the robust estimates of standard deviation found for S_j^* or (2) the robust estimates of standard deviation after determining u_1^* (i.e., from S^w) can be employed.

A robust estimate of the correlation or covariance matrix can be found from $U^* \Lambda^* U^*$ to provide an alternative robust estimate. This method of estimation gives a positive definite matrix as a result, whereas robust estimation of each entry on its own does not always achieve this condition.

Example. The data analyzed here are means of species of the ostracod genus *Echinocythereis*. The material derives from the Lutetian (Eocene) of Aragon, Spain. There are 102 specimens on which 5 measures were taken on the shells, namely, length, anterior height,

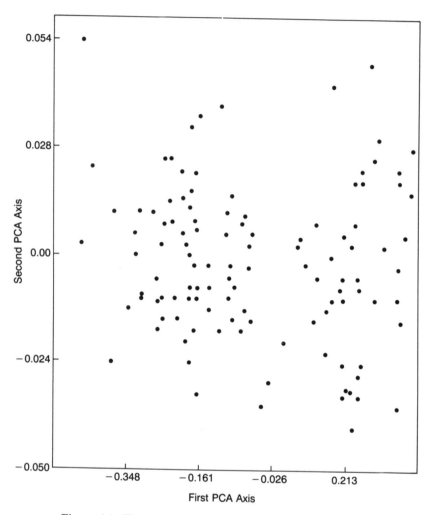

Figure 4.1. The *Echinocythereis* data: Heterogeneity in the plot of the scores of the first two principal components, the result of a speciation event.

posterior height, length of posterior margin, and distance to the adductor muscle scar from the anterior margin. The data were thought to harbor several species with somewhat different properties. An outcome of this could be expected to be instability in the principal components.

The results of the robust principal component analysis are listed in Table 4.V. It will be seen that the effects of "robustifying" the covariance matrix by this procedure do not affect the first component at all, but there are strong differences in all other vectors. These are of an order such as to invalidate an analysis relying on the usual method. More than 97% of the trace of the covariance matrix is covered by the first eigenvalue, which saps most of the variability in the material. This situation is not unusual in crustaceans, but in the present case, it is in part, at least, connected with the quite pronounced heterogeneity in the data (Fig. 4.1), illustrated in the plot of the raw scores for the first two principal components. This heterogeneity could be proved to be due to the presence of two species, arising from a speciation event.

4.5 PRINCIPAL COMPONENT ANALYSIS AND CROSS-VALIDATION

Krzanowski (1988, and references therein) has made a synthesis of techniques for improving the stringency of a principal component analysis. The method of cross-validation forms a major element in the methodology. This approach has not been widely embraced by statisticians in the past and has achieved most success in chemometrics (Wold, 1976). In our opinion, cross-validatory techniques have proven their worth in applied multivariate statistics, and for this reason we now include the subject in the present text. The questions addressed by his group of methods are as follows:

1. How many principal components can be usefully retained?
2. Can any variables be excluded on the grounds of redundancy?
3. Which observations deviate in some multivariate manner from the main body of the data?

Krzanowski distinguished between two types of divergent individuals (Krzanowski, 1987a, 1987b). An *atypical observation* is one that differs markedly from the rest of the sample with respect to the measures on the set of traits. An *influential observation* is one that causes a marked change in the analysis when it is excluded, but the measures on which show no obvious divergencies.

The simplest way to approach cross-validatory procedures is perhaps by going through the steps required to perform an analysis, and then to look at a worked example. Cross-validation is basically an exploratory method that looks for interesting patterns and scans a data matrix for redundancy. The following summary is extracted from Reyment (1990).

Step 1. Compute the principal components of the covariance matrix **S** or the correlation matrix **R** of data matrix **X**:

$$\mathbf{S} = \mathbf{V\Lambda U'}$$

and

$$\mathbf{U'U} = \mathbf{I}.$$

Alternatively, the singular value decomposition of **X** can be used with computational advantage. Thus [2.43],

$$\mathbf{X} = \mathbf{V}\mathbf{\Gamma}\mathbf{U}',$$

where

$$\gamma_i^2 = (N - 1)\lambda_i.$$

Step 2. Compute the scores of the principal components:

$$\mathbf{Z} = \mathbf{X}\mathbf{U}.$$

Step 3. Determine the criterion W_m: This criterion is computed from the *average squared discrepancy* between the actual and predicted values of the data matrix. This is done by the method of cross-validation as outlined next:

    a. Divide **X** into several groups.
    b. Delete each group in turn from the data matrix and compute the values of the predictor from the remaining data, and then predict the deleted values. In practice, the deleted group can be conveniently made to be just one row of **X** (i.e., an individual). The manner in which the method is usually applied involves deletion of variables as well as individuals (i.e., columns as well as rows of the data matrix).

Step 4. Determine the *influence* of each object: Proceed by computing *critical angles* between subspaces of a common data space. The critical angle is a measure of influence of each individual in the sample, with $t = \arccos \tau$ (where τ denotes the smallest element of the diagonal matrix $\mathbf{\Gamma}$ of step 1). Large values of the critical angle denote highly *influential* observations in the sample. There is no test available as yet, and experience must be the guide for any decision.

Step 5. Informative variables: Variables are deleted, one by one, and the residual sums of squares scrutinized. Small residual sums of squares have slight effects on a principal component analysis, and such a variable could well be a candidate for removal from the analysis.

Step 6. Selection of the "best" number of principal components: This problem has been considered elsewhere in this book (Section 4.2). It is usually considered an important item in factor analysis. The method proposed by Krzanowski (1987b) is now summarized. If the jth elements of the matrices **X** and **U** are denoted by x_{ij} and u_{ij}, respectively, the singular value decomposition is in terms of its elements

$$x_{ij} = \sum_{t=1}^{p} v_{it} \gamma_t u_{tj}. \qquad [4.28]$$

The data structure for m dimensions ($p - m$ deleted) is modeled by

$$x_{ij} = \sum_{t=1}^{m} v_{it} \gamma_t u_{tj} + e_{ij}, \qquad [4.29]$$

where e_{ij} is an error term.

Associated with a given number m of components we have the predictor

$$\hat{x}_{ij}^{(m)} = \sum_{t=1}^{m} v_{it} \gamma_t u_{tj} \qquad [4.30]$$

for reconstructing the elements of the data matrix. The average squared discrepancy between actual and predicted values is given by the relationship

$$\text{PRESS}(m) = \frac{1}{np} \sum_{i=1}^{n} \sum_{j=1}^{p} \left[\hat{x}_{ij}^{(m)} - x_{ij} \right]^2. \qquad [4.31]$$

Consider now fitting components sequentially in Equation [4.29]. This is done by defining the relationship

$$W_m = \frac{\text{PRESS}(m-1) - \text{PRESS}(m)}{D_m} + \frac{\text{PRESS}(m)}{D_r}. \qquad [4.32]$$

where D_m is the number of degrees of freedom required to fit the mth component and D_r is the number of degrees of freedom remaining after fitting the mth component: $D_m n + p - 2m$. (PRESS is an acronym for prediction sum of squares.) Note that D_r can be obtained by successive subtraction, given $(n-1)p$ degrees of freedom in the mean-centered matrix $\mathbf{X}$. W_m represents the *increase* in predictive information in each of the remaining components. Hence "important" components should yield values of W_m greater than unity. In many applications, we have found, however, that this cutoff criterion tends to be too severe.

For many purposes, it is expedient to divide $\mathbf{X}$ into several groups, as noted earlier, and to delete variables as well as individuals according to several steps. Denote by $\mathbf{X}_{(-j)}$ the result of deleting the jth column of $\mathbf{X}$. Then by the singular value decomposition we have that

$$\mathbf{X}_{(-j)} = \tilde{\mathbf{V}} \tilde{\mathbf{\Gamma}} \tilde{\mathbf{U}}',$$

where the symbols have the following meaning:

$$\tilde{\mathbf{V}} = (\tilde{v}_{st}), \qquad \tilde{\mathbf{U}} = (\tilde{u}_{st}), \qquad \tilde{\mathbf{\Gamma}} = \mathrm{diag}(\tilde{\gamma}_1, \ldots, \tilde{\gamma}_{p-1}).$$

Consider the following predictor, in which the right-hand element is obtained from the singular value decomposition of **X**, after omitting *either* the *i*th row or the *j*th column:

$$\hat{x}_{ij}^{(m)} = \sum_{t=1}^{m} \left(\tilde{v}_{it}\sqrt{\tilde{\gamma}_t} \right)\left(\tilde{u}_{tj}\sqrt{\tilde{\gamma}_t} \right). \tag{4.33}$$

Variances of Principal Components

Krzanowski (1987b, 1987c) applies the jackknife method of estimation for finding empirical standard errors for principal components. The *i*th variance can be expressed as

$$\frac{\left[(n-1)\sum_{j=1}^{n}\left(\overline{\lambda}_{ij} - \hat{\lambda}_i \right)^2 \right]^{1/2}}{n}, \tag{4.34}$$

where $\overline{\lambda}_{ij}$ is the *i*th largest eigenvalue when the *j*th row of **X** is deleted and

$$\hat{\lambda}_i = \frac{1}{n}\sum_{j=1}^{n}\overline{\lambda}_{ij} \tag{4.35}$$

Influence of each of the sample individuals. Let $\overline{\mathbf{U}}$ contain the principal component coefficients when the *i*th row of **X** is deleted. A comparison of this matrix with **U** will indicate the influence that the *i*th individual, say, has on the outcome of the analysis. The comparison is made by computing the *critical angles* between principal component subspaces of a common data space,

$$\theta = \cos^{-1}(d),$$

where *d* is the smallest element of **D** in the singular value decomposition of the appropriate matrix $\mathbf{V}'_{(m)}\mathbf{V}_{(m)}$.

Example. An example drawn from chemical oceanology. Hubberton et al. (1991) analyzed the means for 26 deep-sea sample sites cored as a part of the activities associated with Legs 113 and 114 of the Ocean Drilling Project (ODP). The determinations were for SiO_2, TiO_2, Al_2O_3, FeO, MgO, MnO, CaO, K_2O, and Na_2O, nine traits in all. The published work reported on these analyses and related them to the stratigraphical levels Quaternary, Late Pliocene, and Early Pliocene.

Table 4.VI. *First five principal components for the covariance matrix of the chemical oceanological data of Hubberton et al. (1991)*

	Eigenvalues				
	1	2	3	4	5
	6.6688	1.3601	0.4101	0.3024	0.1657
	Eigenvectors				
	1	2	3	4	5
SiO_2	0.3767	−0.1707	0.1214	−0.0309	0.1408
TiO_2	−0.3146	0.4168	0.1807	−0.0518	0.6951
Al_2O_3	−0.2068	0.7088	−0.0330	0.0349	−0.2092
FeO	−0.3767	−0.1538	−0.0757	0.0791	0.1308
MnO	−0.3230	−0.2014	−0.1195	−0.8917	0.0062
MgO	−0.3728	0.0077	−0.1171	0.0686	−0.5721
CaO	−0.3828	−0.0427	0.0635	0.2070	−0.1074
K_2O	0.2981	0.3092	−0.8078	−0.1177	0.0519
Na_2O	0.3107	0.3649	0.5110	−0.3630	−0.3076

The question posed was whether the data contained any essential information concerning oceanographical conditions during the Neogene. We summarize, briefly, the salient features of the cross-validational analysis of the data.

Table 4.VI contains the results of the first five principal components obtained for the covariance matrix (for reasons that become apparent further on). Most of tr S is concentrated in the first principal component, more than two-thirds of the variability. The first component represents covariation in silica, sodium, and potassium poised against the other elements. The roles of each of the variables were examined by the method of successive deletion, as outlined in the foregoing part of this section. It was found that variables 3, 8, and 9 are the most important in the analysis (aluminum, potassium, and sodium).

The effect of deleting individuals from the data matrix indicates that four or, possibly, five samples are atypical in some manner. The table of maximum angles indicates that all of these individuals affect the variance and not the correlational structure of the data and hence are probably atypical (though not incorrect) observations. The results for these specimens are listed in Tables 4.VIIA and 4.VIIB. Table 4.VIIA (for the critical angles) is interpreted as follows: The columns headed "1", "2", and "3" denote the largest critical angle for each specimen, deleted in turn, for one, two, and three principal components, respectively, and in decreasing order of magnitude of the associated eigenvalue. The figures printed in bold type derive from those units that

Table 4.VIIA. *Maximum angles for principal component planes: the Ocean Drilling Project (ODP) data*

Unit deleted	Variance sensitive			Correlation sensitive		
	1	2	3	− 3	− 2	− 1
Q 114-701A-1	2.29	2.41	2.44	1.28	1.22	0.07
Q 114-701A-2	3.69	3.33	2.83	3.91	0.34	0.34
Q 114-701A-3	2.20	1.47	1.45	2.06	4.00	0.22
EP 114-701B-4A	1.97	3.98	**12.45**	5.01	4.99	1.19
EP 114-701B-4B	0.42	2.05	2.93	1.26	1.15	0.26
EP 114-701B-5	4.11	8.78	**27.25**	3.07	2.49	0.58
LM 114-701C-6	1.61	1.94	2.17	2.37	2.52	0.09
LM 114-701C-7A	0.50	0.66	8.62	3.62	5.60	0.40
EM 114-701C-7B	1.28	1.32	**20.15**	4.38	5.04	0.91
EM 114-701C-8	1.26	3.74	**33.20**	2.68	2.27	0.45
EP 113-695A-9	0.70	0.70	4.79	5.13	4.95	0.25
EP 113-695A-10	1.83	3.89	7.69	2.52	3.34	0.30
EP 113-695A-11A	0.04	1.42	1.48	0.13	0.13	0.01
EP 113-695A-11B	1.48	1.65	1.69	2.51	0.43	0.13
EP 113-695A-12A	4.00	4.19	**19.27**	2.68	2.03	0.02
EP 113-695A-12B	1.17	1.86	2.30	0.80	0.20	0.02
EP 113-695A-13	1.35	2.36	2.73	8.17	6.47	0.96
EP 113-695A-14A	0.90	1.55	3.25	0.46	0.10	0.07
EP 113-695A-14B	0.34	0.34	0.44	0.37	0.15	0.02
EP 113-696B-15	3.33	2.92	5.87	0.44	0.28	0.08
LP 113-696A-16	0.16	0.64	1.50	1.27	1.97	0.08
LP 113-696A-17	1.62	2.77	3.99	4.30	0.78	0.08
LP 113-696A-18	0.07	1.28	2.39	0.51	0.05	0.05
EP 113-697B-19	0.21	2.26	8.92	2.32	1.43	0.21
EP 113-697B-20	0.13	0.63	2.84	2.28	2.28	0.67
EP 113-697B-21	0.12	0.99	2.91	0.58	0.54	0.13

Key to abbreviations: Q = Quaternary; LP = Late Pliocene; EP = Early Pliocene; LM = Late Miocene; EM = Early Miocene.

when deleted from the analysis caused the greatest disturbances. These units are outliers of location or dispersion. The three columns bearing negative signs are for the three smallest principal components. The smallest principal components often signify outliers produced by correlations; in this present example, there are no such divergencies.

It now remains to see how the test for significant principal components fared using the PRESS method (predictor sum of squares). The succession of values is 10.90, 3.55, 0.24, 0.82, 0.66, ... In our experience, we should be inclined to accept four significant principal components. The falloff after the fifth test value is rapid.

Table 4.VIIB. *Size of the component spaces being compared*

Variable deleted	Residual sum of squares		
	PC1	PC1 + PC2	PC1 + PC2 + PC3
SiO_2	1.25	1.14	1.13
TiO_2	1.63	2.38	2.48
Al_2O_3	1.04	5.10	**5.04**
FeO	1.22	1.13	1.14
MnO	1.43	1.69	1.82
MgO	1.18	1.18	1.23
CaO	1.08	1.07	1.09
K_2O	1.45	2.39	**16.69**
Na_2O	1.56	2.56	**6.70**

This example is covered by Procedures `validate.m`, `var_info.m`, and `spe_infl.m` on the accompanying diskette by L. F. Marcus.

4.6 PRINCIPAL COMPONENT ANALYSIS OF COMPOSITIONAL DATA

Frequencies, proportions, and percentages occur commonly in the natural sciences as well as in many other areas of research and application in which it is required to express a data array in relation to some constant sum. It has seemed natural, and uncomplicated, to employ the usual methods of statistics for such data and, in fact, most people, including professional statisticians, hardly expend a single thought on eventual computational complications or interpretational malchance. The analysis of compositional data is not given adequate consideration in even the most modern of texts. For example, Jackson (1991) gives the subject short shrift in his otherwise monographic treatment of applications of principal component analysis.

We begin this section with a review of the special quandary represented by compositional data. The notation we use is that of Aitchison (1986) in order to maintain the special distinction between the statistics of usual space and those of constrained (simplex) space.

Any vector x with nonnegative elements $x_1, \ldots, x_D$ representing proportions of some whole is subject to the closure constraint

$$x_1 + \cdots + x_D = 1,$$

where all elements are greater than zero.

Failure to incorporate the unit-sum constraint into a statistical model leads to a fallacious analysis with results that are seldom unchallengeable. The following are some examples of naturally occurring compositional data in the natural sciences:

a. geochemical compositions of rocks;
b. analyses of blood-group frequencies (as well as all other polymorphic determinations);
c. frequencies of species in an ecological connection;
d. morphometric measures expressed in relation to some whole, such as maximum length of the organism;
e. pollen frequency data;
f. sedimentary compositions;
g. chemical analyses expressed as parts per liter, percentages of total weight, etc.

In more general terms, each row of the data matrix corresponds to a single specimen (i.e., a *replicate*, a single experimental or observational unit).

Let there be N replicates of D-part composition so that the unit-sum property of rows (replicates) imposes the constraint

$$\sum x_{ri} = 1 \qquad (r = 1, \ldots, N; i = 1, \ldots, p). \qquad [4.36a]$$

Denoting the complete data array as the $N \times D$ matrix $\mathbf{X}$, the constraints [4.36a] can be summarized as

$$\mathbf{X} j_D = j_D, \qquad [4.36b]$$

where j_D is a $D \times 1$ vector with unit entries.

Compositional data space

A restricted part of real space, the *simplex* plays a fundamental role in the analysis of compositional data. In referring to compositional data, the "variables" are called the *parts*. Thus in a rock analysis, the major oxides are the parts. The components are the percentages of each oxide. The parts are the *labels* identifying the *constituents* of the whole, and the components are the *numerical proportions* in which the individual parts occur.

Definitions. It is appropriate at this point to introduce some definitions of some of the salient points of compositional data analysis.

1. The d-dimensional *simplex* is the set defined by

$$\mathbb{S}^d = \{(x_1, \ldots, x_d); x_1 > 0, \ldots, x_d > 0; x_1 + \cdots + x_d < 1\}.$$

This definition is not useful in practical situations because x_D is as important as any other component. A more useful version for applied work is as follows:

2. The d-dimensional simplex embedded in D-dimensional real space is the set defined by

$$\mathbb{S}^d = \{(x_1,\ldots,x_D): x_1 > 0,\ldots, x_D > 0; x_1 + \cdots + x_D = 1\}.$$

Both forms lead to the same set of all possible D-part compositions. The difference between the two representations lies with their geometrical properties.

The two definitions are complementary. The first definition above expresses the correct dimensionality by portraying the simplex $\mathbb{S}^1$ as a subset of the one-dimensional real space $\mathbb{R}^1$, the real line, and $\mathbb{S}^2$ as a subset of two-dimensional real space $\mathbb{R}^2$, the real plane.

The second definition requires the device of embedding the d-dimensional sample space $\mathbb{S}^d$ in a real space of higher dimension, to wit, $\mathbb{R}^D$. This has the consequences that (1) $\mathbb{S}^1$ is a line segment, hence a one-dimensional subset within $\mathbb{R}^2$, and (2) $\mathbb{S}^2$ is an equilateral triangle, a two-dimensional subset, within $\mathbb{R}^3$. This latter property is useful. It is clearly identified as the well-known *ternary diagram* of many branches of the natural sciences, including sedimentology, serology, analytical chemistry, and soil science.

Covariances and correlations

The possibility of producing nonsense correlations has been known since the earliest days of biometry. Pearson (1897) succinctly defined the concept of *spurious correlation* with particular reference to the constraint introduced by the use of indices and ratios. The problem becomes acute where there are elements common to numerator and denominator, such as

$$w_1/(w_1 + w_2 + w_3).$$

Readers will doubtless recognize the architecture of this expression from their own spheres of research, for example, chemistry, biological systematics, and ecology.

The crude covariance. The almost universal manner for assessing interdependence of the components of a D-part composition x is by means of the product moment covariances or correlations of the crude components $x_1,\ldots, x_D$:

$$\kappa_{ij} = \mathrm{cov}(x_i, x_j) \qquad (i,j = 1,\ldots, D),$$

with $D \times D$ crude covariance matrix

$$\mathbf{K} = \{\kappa_{ij} : i, j = 1, \ldots, D\}$$

and crude correlations defined by

$$\rho_{ij} = \kappa_{ij} / (\kappa_{ii} \kappa_{jj})^{1/2}$$

There are several difficulties connected with interpreting compositions in terms of their crude covariance structure. Some of these are the following:

1. The negative bias difficulty – Correlations are not free to range over the usual interval $(-1, 1)$.
2. Subcomposition difficulty – There is no simple relationship between the crude covariance of a subcomposition and that of the full composition.

A particular source of wonderment is the erratic way in which the *crude covariance* associated with two specific parts can fluctuate in sign as one passes from full compositional status to subcompositions of lower and lower dimension.

3. Null correlation difficulty – Attempts at solving this problem have involved artificial constructions, the lack of a satisfactory treatment of the non-null case, and the introduction of the imaginary basis for which compositional data do not exist.

Logratio variance

In this section we consider a useful solution due to Aitchison (1986). For any two parts i and j of a D-part composition x, the logratio variance is

$$\tau_{ij} = \mathrm{var}\{\log(x_i/x_j)\}.$$

The D^2 logratio variances τ_{ij} of a D-part composition x satisfy the conditions

$$\tau_{ii} = 0 \quad (i = 1, \ldots, D),$$

$$\tau_{ij} = \tau_{ji} \quad (i = 1, \ldots, d; j = i + 1, \ldots, D)$$

and so are determined by the $\frac{1}{2}dD$ values

$$\tau_{ij} = (i = 1, \ldots, d; \, j = i + 1, \ldots, D).$$

Covariance structure of a D-part composition x is defined as the set of all

$$\sigma_{ij,\, kl} = \text{cov}\{\log(x_i/x_k), \log(x_j/x_l)\}$$

as i, j, k, l run through the values $1, \ldots, D$. The covariance structure of a d-dimensional composition $(x_1, \ldots, x_D)$ is similar to the covariance structure of a d-dimensional vector in $\mathbb{R}^d$ in its requirement of the specification of $\frac{1}{2}dD$ quantities.

Matrix specifications of covariance structures. The method of analysis developed by Aitchison (1986) relies on defining the covariances between compositional variables. The logratio covariance matrix is the fundamental concept. To paraphrase Rabbi Hillel, all else is commentary.

Logratio covariance matrix. For a D-part composition x the $d \times d$ matrix

$$\Sigma = [\sigma_{ij}] = \left[\text{cov}(\log(x_i/x_D), \log(x_j/x_D)): i, j = 1, \ldots, d\right] \quad [4.37]$$

is termed the logratio covariance matrix. It determines the covariance structure by the relationships

$$\sigma_{ij,\, kl} = \sigma_{ij} + \sigma_{kl} - \sigma_{il} - \sigma_{jk}$$

The matrix Σ is a covariance matrix of a d-dimensional random vector y defined by

$$y_i = \log(x_i/x_D) \qquad (i = 1, \ldots, d),$$

or in more compact notation,

$$y = \log(x_{-D}/x_D). \qquad [4.38]$$

Vector y is valid in $\mathbb{R}^D$ because the logratio transformation is one-to-one

$$x \in \mathbb{S}^d \to y = \log(x_{-D}/x_D) \in \mathbb{R}^d.$$

With this specification, the inadvertency caused by negative bias does not arise. The logratio transformation transports the constraints of $\mathbb{S}^d$ into the "freedom" of $\mathbb{R}^d$.

Centered logratio covariance matrix. The centered logratio co-variance matrix, which we shall denote as Γ, possesses the statistically desirable property of *symmetry* of all *D*-parts. There is a drawback, however – the matrix is singular. The matrix is computed by replacing the single divisor x_D by the geometric mean

$$g(x) = (x_1, \ldots, x_D)^{1/D}$$

of all *D* components. Hence,

$$\Gamma = [\gamma_{ij}] = \text{cov}\left[\log x_i/g(x), \log x_j/g(x)\right] \qquad (i, j = 1, \ldots, D).$$
$$[4.39]$$

This matrix determines the *covariance structure* by the relationships

$$\sigma_{ij,kl} = \gamma_{ij} + \gamma_{kl} - \gamma_{il} - \gamma_{jk}$$

The singularity of Γ means that there are *D* restrictions on its elements; these can be most satisfactorily identified as the requirement that all row sums be zero.

Logcontrast principal component analysis

The centered logratio covariance matrix Γ is the most suitable point of departure for constrained principal component analysis. The usual method of extraction of eigenvalues and eigenvectors is used. It is sometimes argued, albeit without algebraic support, that a "crude" principal component analysis of compositional data will give approximately the same result as the more involved procedure advocated in the preceding pages, at least as far as graphical purposes are concerned. There is no doubt that this can be true for some sets of data, but it can also be a complete disaster for other material. There is just no way of being sure what you are going to get, apart from cases where curvilinearity is a main cause of the deviation.

A suitable method of principal component extraction has then to be nonlinear and sufficiently flexible to accommodate approximate nonlinearity within its range.

Let the *d* positive eigenvalues of the centered logratio covariance matrix Γ be $\lambda_1 > \cdots > \lambda_d$; the standardized eigenvectors $a_1, \ldots, a_d$ satisfy

$$(\Gamma - \lambda_i I)a_i = 0 \qquad (i = 1, \ldots, d).$$

The logcontrast $a_i' \log x$ is then the *i*th *logcontrast principal component*. The main properties of logcontrast principal components are those of usual principal components.

Table 4.VIII. *Correlations for the blood polymorphisms (Iberian Peninsula)*

	1	2	3	4	5
1	1.0000	0.8010	0.5370	0.8964	−0.9468
2	−0.3624	1.0000	0.4501	0.7935	−0.8830
3	−0.2285	0.1282	1.0000	0.4180	−0.6984
4	−0.0427	−0.3485	−0.5181	1.0000	−0.3645
5	−0.7816	0.2209	0.0988	−0.3645	1.0000

The upper triangle contains the *simplex* values, the lower triangle the "raw" correlations in bold type.

The proportion of the total variability that is contained in the first c logcontrast principal components is given by the quotient

$$\sum_{i=1}^{c} \lambda_i / \mathrm{tr}(\boldsymbol{\Gamma}).$$

Example Principal component analysis of Rhesus alleles.

An important human serological category is that of the Rhesus alleles. The general notation for designating the permutations of alleles is $c/C : d/D : e/E$. The data are obtained in the form of vectors of compositions, such as is generally the case in polymorphic material. The present example consists of average values sampled at 23 sites for *CDe*, *cDE*, *cDe*, *cde* and the sum of all remaining combinations. The data are listed in Reyment (1991, Table 1).

The first step in the calculations was to compute the simplex correlation matrix. This is listed in Table 4.VIII, together with the "raw correlations", that is, the correlations obtained by the usual method. The marked differences in values need hardly be commented upon. It is quite clear that the logratio correlations produce a completely different picture from that obtained by ignoring the constraint. The inappropriate selection of statistical procedures can hardly do other than lead to erroneous conclusions.

The next step is to perform a principal component analysis. For the purposes of this example, we have also included the "usual" analysis of the raw correlations. The results are listed in Table 4.IX. Again you will see that there is very little in common between the two. A fuller treatment of the Rhesus data is given in Reyment (1991). Suffice it here to indicate that the important role played by *cDe* in the second eigenvector. It is an important marker for past African importations into Iberian populations, the demographic effects of which are still much in evidence.

Table 4.IX. *Principal component analyses for the correlation matrix of the*
Rhesus serological data (Iberian Peninsula)

(A) Simplex space

	Eigenvalues				
	1	2	3	4	5
(trace = 5)	3.9957	0.6832	0.2381	0.0830	0
Percentage of trace	79.914	13.664	4.763	1.660	

	Eigenvectors			
	1	2	3	4
CDe	0.4766	− 0.1457	0.3719	− 0.7358
cDE	0.4469	− 0.2379	− 0.8283	0.0054
cDe	0.3334	0.9019	− 0.0063	0.1247
cde	0.4607	− 0.3280	0.4189	0.6656
Remainder	− 0.4998	− 0.0530	− 0.0041	− 0.0001

(B) Usual space

	Eigenvalues				
	1	2	3	4	5
	2.2324	1.3247	0.8275	0.6254	0

	Eigenvectors			
	1	2	3	4
CDe	− 0.5124	0.4979	− 0.1574	0.3289
cDE	0.3981	0.0539	− 0.8518	− 0.2591
cDe	0.3491	0.5269	0.4417	− 0.5676
cde	− 0.4023	− 0.5926	0.0493	− 0.5292
Remainder	0.5433	− 0.3471	0.2283	0.4722

4.7 PATH ANALYSIS AND WRIGHTIAN FACTOR ANALYSIS

The method of *path analysis* was introduced by Sewall Wright (1932, 1968) for analyzing covariance structures. It is well to remember that path analysis was not developed in competition with standard multivariate statistics; it was a product of the times and many of today's most popular methods of multivariate analysis were not available, or were so vaguely formulated as to be unusable in applied work. Wright's idea was to produce a technique for assessing the direct causal contribution of one variable to another in a nonexperimental connection. The problem reduces to one of estimating the coefficients of a set of linear structural equations that represent the cause-and-effect relationships hypothesized by the investigator. (The LISREL package (Jöreskog and Sörbom, 1989) provides a straightforward way of estimating a path analysis model by unweighted least squares for directly observed variables.

It is strange that path analysis has not caught on to the same extent as other multivariate methods, despite attempts to make it known in genetics (Li, 1955; Kempthorne, 1957) and econometrics (Fornell, 1982). The thirties were an innovative period in the history of multivariate analysis. This is when such methods as principal components, discriminant functions, generalized distances, canonical variates, canonical correlation, the generalization of the *t*-test, and so on were developed in answer to a growing demand from workers in various fields of data analysis. Path analysis has a long history in the social sciences. The variant described next is known to social scientists as "direct, indirect, and total effects" – see Jöreskog and Sörbom (1989, pp. 119–23 and p. 145). In preparing the following account, we have drawn on the biologically oriented presentation by Bookstein et al. (1985).

Juxtaposition of path analysis and regression

In order to place *path analysis* in a useful perspective, we note that ordinary multiple regression is concerned with finding a linear combination of the following kind:

$$a_1 X_1 + a_2 X_2 + \cdots + a_n X_n$$

of predictors $X_1, X_2, \ldots, X_n$ that best correlate with a dependent variable Y. The quantities a_i are the coefficients for which the variance of the linear combination

$$Y - a_1 X_1 - \cdots - a_n X_n$$

is a minimum. The step to path analysis can be made as follows. Let there be, say, three predictors X_1, X_2, X_3, each of unit variance and a dependent variable Y, also of unit variance. The regression of Y on the X_i can be illustrated in the form of a *path diagram*.

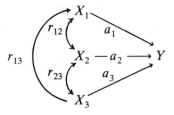

Here a single arrow means "contributes to." A double-headed arrow means "is correlated with."

The covariances between all variables are assumed to be known. The task is to compute the unknowns a_i from some criterion implicit in the path diagram.

The effect of a *unit* increase in, say, X_1 can be distributed over the *paths* as follows:

1. a *direct* effect a_i upon Y;
2. an *indirect* effect of $r_{21}a_2$ via X_2; that is, an expected increase of one unit in X_i brings about an expected increase of r_{21} units in X_2, which has, in its turn, a direct effect a_2 upon Y for each unit increase in X_2;
3. an indirect effect, by the same token, of $r_{31}a_3$ via X_3.

The sum of these three effects, one direct and two indirect, must result in the same expected change in Y per unit increment in X_1.

Summing the information in 1, 2, and 3 gives the expression

$$r_{Y1} = a_1 + a_2 r_{21} + a_3 r_{31}.$$

This gives one equation for the three unknowns a_1, a_2, a_3. The same process applied for unit changes in X_2 and X_3 results in two similar equations, to wit:

$$r_{Y2} = a_2 + a_1 r_{12} + a_3 r_{32},$$
$$r_{Y3} = a_3 + a_1 r_{13} + a_2 r_{23}.$$

These three equations are the normal ones generated for minimizing

$$\text{var}(Y - a_1 X_1 - a_2 X_2 - a_3 X_3)$$

by equating its derivatives to zero.

Each equation is a consistency criterion. Each asserts the consistency between a simple regression and the net outcome of the multiple regression with respect to one of the Xs.

Factors and correlation

The algebra of factor analysis can be deduced from a path diagram of the explanation for which it stands. Suppose two variables X_1 and X_2 (with unit variance) have a correlation coefficient r, which is attributable to a third variable Y, also with unit variance. Thus, in Wright's notation,

$$X_1 = a_1 Y + e_1$$

and

$$X_2 = a_2 Y + e_2,$$

where e_1 and e_2 denote random variation, uncorrelated with each other and uncorrelated with Y. One path linking the variables can be given as follows:

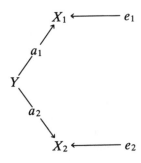

under the assumption that variances are all 1,

$$\text{var}(e_i) = 1 - a_i^2,$$

and the correlation between X_1 and X_2 is a_1a_2. Now, if it be found that the observed correlation is something other than this product of the individual regression coefficients on a third variable, then the e_i must be correlated. If Y is the common cause of these variables, and we want to remove its contribution to their correlation, then the residual covariance $r_{12} - a_1a_2$ measures the functional constraint.

If there are three variables with observed correlations, the number of parameters to be estimated equals the number of correlations and the following model has a unique solution:

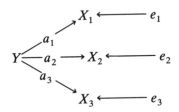

that is, $X_i = a_iY + e_i$. This solution is

$$a_1 = (r_{12}r_{13}/r_{23})^{1/2},$$

$$a_2 = (r_{12}r_{23}/r_{13})^{1/2},$$

$$a_3 = (r_{13}r_{23}/r_{12})^{1/2}.$$

This model fails if any $a_i > 1$ (i.e., the error terms e_i are correlated) or if just one of the observed r_{ij} is negative.

The *covariance* matrix of the *explained* part of each variable can be exemplified for three variables as follows:

$$
\begin{pmatrix}
a_1^2 & r_{12} & r_{13} \\
r_{21} & a_2^2 & r_{23} \\
r_{31} & r_{32} & a_3^2
\end{pmatrix}
\tag{4.40}
$$

Each diagonal element is *reduced* by the error variance of the corresponding X_i. These are the communalities (or "self-correlations") we have met earlier on in factor analysis.

If we are interested in defining a measure of "general size" in a growth study, this can be done as follows (case of three variables again):

$$
Y = \frac{a_1 \hat{X}_1 + a_2 \hat{X}_2 + a_3 \hat{X}_3}{a_1^2 + a_2^2 + a_3^2}.
\tag{4.41}
$$

This (unmeasurable) Y has unit variance and covariances a_i ($i = 1, 2, 3$) with $\hat{X}_i$.

As an alternative to the usual approach to the estimation of the a_i for any number of variables (e.g., maximum likelihood estimation), it can be instructive to use Wright's criterion of *consistency*. Hence, we require the vector *a* to be an eigenvector of the correlation matrix with the unit elements of the diagonal replaced by the communalities, as was done for three variables in [4.40].

Wright's way of finding secondary factors can be put as follows. Instead of determining the eigenvectors, one can seek a vector $(a_1, \ldots, a_p)$ that is an eigenvector of the matrix [4.42]

$$
\begin{pmatrix}
a_1^2 & a_1 a_2 & r_{13} & r_{14} & \cdots & r_{1p} \\
a_1 a_2 & a_2^2 & r_{23} & r_{24} & \cdots & r_{2p} \\
r_{13} & r_{23} & a_3^2 & a_3 a_4 & \cdots & r_{3p} \\
\vdots & \vdots & \vdots & \vdots & \vdots & a_{p-1} a_p \\
r_{1p} & r_{2p} & r_{3p} & r_{4p} & \cdots & a_p^2
\end{pmatrix}
\tag{4.42}
$$

In this matrix, not only the diagonals but also the postulated residual factor spaces have been filled with terms exactly fitting the model. In other words, referring to the first example, the 2×2 blocks

$$
\begin{pmatrix}
a_1^2 & r_{12} \\
r_{21} & a_2^2
\end{pmatrix},
\begin{pmatrix}
a_3^2 & r_{34} \\
r_{43} & a_4^2
\end{pmatrix},
\ldots,
\begin{pmatrix}
a_5^2 & r_{56} \\
r_{65} & a_6^2
\end{pmatrix}
$$

Table 4.X. *The correlations for the six characters measured on Wright's leghorns [after Wright (1968)]*

	1	2	3	4	5	6
1	1.000	0.584	0.615	0.601	0.570	0.600
2		1.000	0.576	0.530	0.526	0.555
3			1.000	0.940	0.875	0.878
4				1.000	0.877	0.886
5					1.000	0.924
6						1.000

Table 4.XI. *Single-factor solution for the leghorns*

Single-factor solution
0.665   0.615   0.953   0.942   0.923   0.941

Residuals

$$\begin{bmatrix}
0.557 & 0.175 & -0.019 & -0.026 & -0.044 & -0.026 \\
0.175 & 0.622 & -0.010 & -0.049 & -0.042 & -0.024 \\
-0.019 & -0.010 & 0.092 & 0.043 & -0.004 & -0.019 \\
-0.026 & -0.049 & 0.043 & 0.113 & 0.008 & -0.001 \\
-0.044 & -0.042 & -0.004 & 0.008 & 0.148 & 0.055 \\
-0.026 & -0.024 & -0.019 & -0.001 & 0.055 & 0.114
\end{bmatrix}$$

for, say, six variables have been extracted from the reduced correlation matrix.

Example. The classical example of path analysis is that of the six skeletal characters measured on leghorn hens. This was used by Wright (1968) to exemplify his procedure. The characters measured on 276 animals are (1) skull length, (2) skull breadth, (3) humerus, (4) ulna, (5) femur, and (6) tibia. The biologists will see that these characters fall into three disjunct groups, each of two traits. The question that arises is whether meaningful correlations can be found betwen them. [Data taken from Wright (1968).] The program of Bookstein et al. (1985) was used for the present calculations.

The correlation matrix is listed in Table 4.X: All of these correlations are high to fairly high, and a normal step would be to examine the partial correlation coefficients. The solution obtained by extracting a single factor is given in Table 4.XI, together with the resulting residuals. Note that a few of the entries in the off-diagonal positions are large, but most are small and some close to zero; also note that the factor is

Table 4.XII. *Four-factor solution for the leghorn data*

Primary path coefficients

0.636   0.583   0.957   0.946   0.914   0.933

Residual correlations

$$\begin{bmatrix} 0.595 & 0.213 & 0.006 & -0.001 & -0.012 & 0.007 \\ & 0.660 & 0.018 & -0.022 & -0.007 & 0.001 \\ & & 0.084 & 0.034 & 0.000 & -0.015 \\ & & & 0.104 & 0.012 & 0.003 \\ & & & & 0.165 & 0.072 \\ & & & & & 0.130 \end{bmatrix}$$

Table 4.XIII. *Path coefficients for the leghorn data using four factors*

	Trait			
Variable	Primary	Head	Wing	Leg
Skull length	0.636	0.461		
	0.034	*(0.074)*		
	18.59	**6.26**		
Skull width	0.583	0.461		
	0.034			
	17.31			
Humerus	0.957		0.182	
	0.047		*0.264*	
	20.22		**0.690**	
Ulna	0.947		0.182	
	0.048			
	19.65			
Femur	0.914			0.269
	0.046			*0.169*
	19.98			**1.59**
Tibia	0.932			0.269
	0.046			
	20.45			

In each group of three figures, the top value is the unweighted least-squares estimate, that beneath in italics is the corresponding standard deviation, and the third, in bold type, is the value of t (obtained by LISREL).

really the first principal component of the matrix. The four-factor solution is listed in Table 4.XII.

The largest residual occurs for res_{12}, the correlation between length and breadth of the skull. The final phase of the analysis is done by computing the path coefficients. Most of these are near to zero and so cannot be given useful interpretations. There are, however, two clearly significant values. The results are summarized in Table 4.XIII. (Note that the system LISREL is a convenient way of making a path analysis.) The conclusion indicated is that the observed correlations have been "explained" by a primary size-factor and three secondary factors. The mesh of relationships can be neatly illustrated by a path diagram:

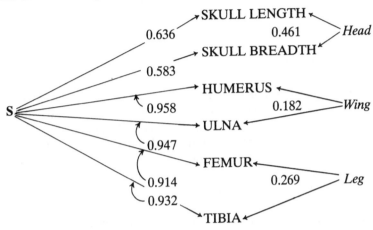

The same data were run in LISREL to yield the same results, as to be expected, but also the standard deviations of the estimates and a corresponding value of Student's *t*. This information is given in Table 4.XIII. We remark that the unweighted least-squares estimate for the head is statistically significant, but not the other two, owing to their large standard deviations.

5 Q-mode methods

5.1 INTRODUCTION

An objective of many investigations is to classify a sample of objects (rock specimens, fossils, water samples, environments, etc.) on the basis of several properties. Q-mode analysis, as first applied in geology by Imbrie (1963), is a valuable aid in doing this in cases where there are a large number of objects and, especially, where there is little a-priori knowledge of the significance of the constituents. The concept of Q-mode analysis was first developed by psychologists and then by biologists.

Q-mode analyses are designed to portray interrelationships between objects, just as in R-mode analyses interrelationships between variables are analyzed. To a certain extent, factor scores derived from R-mode analysis provide a means of describing interobject relationships; however, these associations are not usually based on a suitable measure of interobject similarity. That is, the covariance or correlation may not be the best criterion by which to judge the degree of similarity between two objects.

The mainstay of Q-mode factor analysis lies with the definition of interobject similarity. Once a suitable mathematical definition of this similarity coefficient has been established, it is possible to assemble an $N \times N$ similarity matrix containing the degree of similarity between all possible pairs of N objects. When N is large, this matrix will contain many elements and finding its rank by eigenanalysis may provide a means of adequately describing the objects in terms of fewer basic dimensions than original variables.

We describe here two methods of Q-mode analysis. The Imbrie Q-mode method defines similarity with respect to the proportions of constituents. Gower's method of principal coordinates utilizes the Euclidean distance between objects as a similarity index whereas Bénzécri's method of correspondence analysis uses a contingency table (Fisher, 1940) for assessing similarity between objects (Chapter 6).

Once the "similarity" or association matrix has been constructed, the procedures follow similar steps to principal component analysis, with or without rotation of the results. The resulting matrices are interpreted quite differently from those of R-mode analyses. The main interest in

136

doing a Q-mode analysis is graphical. Useful references are Everitt (1978), Jolliffe (1986), and Jackson (1991). Aitchison (1986) indicated a Q-mode version of principal coordinates (Section 5.2).

5.2 PRINCIPAL COORDINATES

One of the earlier special applications of factor analysis was developed in quantitative biology in the hands of numerical taxonomists. Faced with the need of uncovering possible heterogeneity in a set of data, the idea of "factor-analyzing" a matrix of similarity measures was born. That is, it was desired to ascertain whether or not a multivariate sample could reasonably be regarded as coming from a single statistical population, or whether it was composed of material from more than one population, that is, that the sample is mixed or heterogeneous. Should the material turn out to be heterogeneous, it is usual to sort the specimens into the groups disclosed by the analysis and to study their statistical properties under the assumption of homogeneity of each of the new subsamples. There is little doubt that there is a very real need for this kind of "shotgun" first step in studying data about which we know little to start off with. The method of principal coordinates considered in this section is a reasonable alternative to Q-mode factor analysis regarded as an ordinating procedure. You should be quite clear, however, that the aims of these two methods are only partly coincident. Q-mode factor analysis is concerned with obtaining answers to several questions, of which the graphical appraisal of the data is only one, whereas the main objective of principal coordinates is the graphical treatment of the data points.

The term "principal coordinate analysis" was introduced by Gower (1966a). However, the same type of technique has been used by psychologists and psychometricians for many years (see, for example, Torgerson, 1958, Chapter 11).

Description of the method

You can think of principal coordinates as a Q-mode principal component analysis of a similarity matrix of some kind. The basis of the procedure consists of extracting the first k eigenvalues and eigenvectors of an N-by-N matrix of associations (usually some sort of distance measure) derived from our usual raw-data matrix, $\mathbf{X}$. The variables need not all be quantitative, and some may be suitably scored qualitative variables and some may be dichotomous, that is to say, variables of the present-or-absent, plus-or-minus, $0, -1$, kind.

The association matrix, $\mathbf{H}_{(N \times N)}$, whose mnth element is a coefficient of association between the mth and the nth individuals of the sample, is formed using any of the manifold measures of association or similarity available. An exhaustive discussion of these kinds of coefficients is to be found in Sneath and Sokal (1973).

Distances between objects. In order to introduce you to the special features of principal coordinate analysis, we shall once again begin our presentation with a consideration of the data matrix.

As we have defined the data matrix before, the N points P_i ($i = 1, 2, \ldots, N$) are demarcated in p-space as

$$
\begin{array}{c}
P_1 \\
P_2 \\
\vdots \\
P_N
\end{array}
\begin{bmatrix}
x_{11} & x_{12} & \cdots & x_{1p} \\
x_{21} & x_{22} & \cdots & x_{2p} \\
\vdots & \vdots & \cdots & \vdots \\
x_{N1} & x_{N2} & \cdots & x_{Np}
\end{bmatrix}.
$$

The coordinates of point P_i are, therefore,

$$
(x_{i1}, x_{i2}, \ldots, x_{ip}).
$$

In principal component analysis, certain studies are directed toward using the plots of the transformed variates, defined as $u_1 z_1 + u_2 z_2 + \cdots + u_p z_p$ for the first eigenvector of standardized variables, for example, for purposes of distance-oriented analysis. It is important to understand that the plots of the sample points are separated by statistical distances. Thus, the distance between the projected points Q_m and Q_n, as we shall denote the projected counterparts of, say, points P_m and P_n, is an approximation to the distance between the original sample points.

The Euclidean distance between points P_m and P_n in p-space is defined by the equation

$$
d_{mn}^2 = \sum_{r=1}^{p} (x_{mr} - x_{nr})^2. \tag{5.1}
$$

This distance measure has the defects that it does not take into account the correlations between variables, nor is it invariant to the scales of measurement of the different variables. It is similar to measures of distance commonly employed in numerical taxonomy.

Using Fig. 5.1, we shall now give a sample geometrical interpretation of the relationship between any point P_i and its projection Q_i onto a new set of orthogonal axes. Let Q_i in Fig. 5.1 be the projection of the original data point P_i on the first eigenvector of $\mathbf{X'X}$, where $\mathbf{X}$ is as

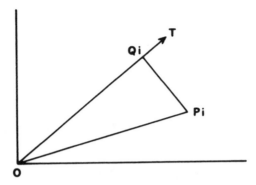

Figure 5.1. The geometrical relationship between a data point P_i and the projection of it, Q_i.

usual the data matrix. Considering now all of the i points we have, by Pythagoras' theorem, that

$$\Sigma(OP_i)^2 = \Sigma(P_iQ_i)^2 + \Sigma(OQ_i)^2.$$

By the criterion of best fit of principal component analysis, $\Sigma(P_iQ_i)^2$ is a minimum and, since $(OP_i)^2$ is a constant, $\Sigma(OQ_i)^2$ must be a maximum.

By the definition of principal components in Chapter 3, if the points are centered at the origin, $\Sigma(OQ_i)^2$ is the variance of the linear combination with the elements of the eigenvector corresponding to the greatest eigenvalue of $\mathbf{X'X}$ as coefficients. The same reasoning applies to the second eigenvector, and so on.

It follows from the foregoing geometrical illustration that any dimension with a small eigenvalue necessarily contributes little to the distances.

Q-mode development. Up to now, we have been thinking of distances between points as expressed in R-mode principal components. We shall now consider Gower's expansion of this distance concept to the Q-mode.

Suppose a measure h_{mn} of association or similarity between objects has been defined for every pair of objects in a sample (data matrix). Gower (1966a) shows that coordinates of the points $P_1, P_2, \ldots, P_N$ can be found relative to principal axes such that the squared interpoint distances are given by

$$\{\Delta(P_m, P_n)\}^2 = h_{mm} + h_{nn} - 2h_{mn}. \qquad [5.2]$$

If h_{mn} is a measure of similarity, then the matrix $\mathbf{H}$ will have diagonal elements $h_{mm} = 1$ for $m = 1, \ldots, N$, as the similarity between an object and itself is 1. This is true for any of the similarity measures cited by Sneath and Sokal (1973) and for the coefficient proposed by Gower (1966a, 1967):

$$h_{mn} = \left(1 - \sum_{i=1}^{p} |x_{mi} - x_{ni}|/R_i\right), \qquad [5.3]$$

where R_i is the range of variable i. In this situation, the squared interpoint distance produced is

$$\{\Delta(P_m, P_n)\}^2 = 2(1 - h_{mn}).$$

If $h_{mn} = -\frac{1}{2}d_{mn}^2$, where d_{mn} is a real Euclidean interpoint distance between objects m and n (e.g., [5.1]), then $h_{mm} = 0$ for $m = 1, \ldots, N$, so that

$$\{\Delta(P_m, P_n)\}^2 = d_{mn}^2.$$

Gower (1966a) shows that if d_{mn} is the Pythagorean distance, then principal coordinate analysis is the dual of principal component analysis and the principal coordinates are the transformed observational vectors of principal components.

As N points can be fitted into $N - 1$ dimensions, principal coordinate analysis must yield at least one zero eigenvalue and the other eigenvalues must be positive to give a configuration of the points in a real space. The transformation to matrix $\mathbf{H}^*$ such that

$$h_{mn}^* = h_{mn} - h_{m\cdot} - h_{\cdot n} + h_{\cdot\cdot} \qquad [5.4]$$

ensures this for Euclidean distances and many similarity measures; here, $h_{\cdot\cdot}$ denotes the grand mean of $\mathbf{H}$, and $h_{m\cdot}$ and $h_{\cdot n}$ the mean of row m and column n, respectively. $\mathbf{H}^*$ is positive semidefinite, and a diagonal element, h_{mm}^*, is the squared distance of the point P_m from the centroid O; an off-diagonal element, h_{mn}^*, is the cosine of the angle between the vectors from the centroid O to P_m and P_n times the product of distances OP_m and OP_n. If we denote the rank of $\mathbf{H}^*$ as r, the coordinates of the points $P_1, P_2, \ldots, P_N$ can be represented by the matrix $\mathbf{A}_{(N \times r)}$, the columns of which are the eigenvectors of $\mathbf{H}^*$, scaled so that their squared lengths equal the eigenvalues, in decreasing order of magnitude, that is,

$$\mathbf{A} = \mathbf{U}\mathbf{\Lambda}^{1/2} \qquad [5.5]$$

and $\mathbf{H}^* = \mathbf{A}\mathbf{A}'$.

Usually, one is interested in representing the points in a space of low dimensionality, say k. This is done by using only the first k columns of

A. That is, we approximate $\mathbf{H}^*$ by

$$\hat{\mathbf{H}}^* = \hat{\mathbf{A}}_k \hat{\mathbf{A}}'_k.$$

The rows of $\hat{\mathbf{A}}_k$ represent the coordinates of the N points, $Q_1, Q_2, \ldots, Q_N$, the projections of $P_1, P_2, \ldots, P_N$ onto the best-fitting subspace of dimensionality k. If the approximation to rank k is reasonably good, then the interpoint distances between points Q_m are approximately the same as those of P_m, $m = 1, 2, \ldots, N$. Another property is that the points Q_m are centered at the origin, that is, the column sums of $\hat{\mathbf{A}}_k$ are zero. This is a consequence of the fact that the row and column sums of $\mathbf{H}^*$ are all zero.

We note that any transformation of $\mathbf{H}$ will result in a transformation in the distance between Q_i and Q_j and the configuration of objects produced by principal coordinate analysis will be correspondingly distorted. Parabola effects etc. may result, and these are especially common when $\mathbf{H}$ has been formed from presence–absence data (the "horseshoe effect").

When $\mathbf{H}$ consists of measures that are not real Euclidean distances, a complication arises. In such situations, the matrix $\mathbf{H}^*$ needs no longer be positive semidefinite. This means that $\mathbf{H}^*$ can have negative eigenvalues. However, if these negative eigenvalues are small, we can still extract the eigenvectors corresponding to the k largest eigenvalues and approximate $\mathbf{H}^*$ as before, provided that these k eigenvalues are positive. The points Q_m are represented by $\hat{\mathbf{A}}_k$ as before.

If λ_i is small, then the contribution $\lambda_i(u_{mi} - u_{ni})^2$ to the distance between the points Q_m and Q_n will be small (the sum of squares along axis i is, by definition, λ_i). Even if λ_i is large, but the elements of the eigenvector corresponding to it are not very different, then the distance will be small. Consequently, the only coordinate vectors that contribute much to the distances are those for large eigenvalues that have a wide range of variation in the loadings of their corresponding eigenvectors.

The sum of squares of the residuals (perpendiculars onto the reduced k-dimensional representation) is given by the formula

$$\operatorname{tr} \mathbf{H}^* - \sum_{i=1}^{k} \lambda_i.$$

Finally, we wish to draw your attention to the fact that data matrices can easily exceed the storage capacities of many computers when used in the calculations of principal coordinates. Inasmuch as the Q–R duality only applies exactly for squared interpoint distances in some Euclidean space, Gower (1968) has shown how to add extra points to a plot; see also Wilkinson (1970). A useful further reference is Gordon (1981).

Principal coordinates versus Q-mode factor analysis

As noted in the introduction to this section, principal coordinate analysis has some points in common with Q-mode factor analysis. Both produce plots of transformed data points; however, the latter is concerned with examining a wide range of hypotheses in which the rotation of axes plays an important role. One might say that the graphical representation of the transformed data points is incidental to the analysis. In principal coordinates, the main reason for doing the analysis is to produce a plot of projected points; the distances between points can be related directly to other multivariate statistical procedures under certain assumptions; for example, the canonical variates application of principal coordinates and generalized statistical distances employed as similarity coefficients (Gower, 1966b).

The original development of principal coordinates by Gower leaves no place for the rotation of axes as the loadings of the eigenvectors are not amenable to reification. It is, however, worth noting that three-dimensional models of the plots for the first three axes often clarify relationships between groups.

Example We use here some artificial data of Imbrie (1963) of rank 3 to illustrate the main steps in computing principal coordinates. The data matrix is, for 10 objects (rows) and 10 variables,

x_1	x_2	x_3	x_4	x_5	x_6	x_7	x_8	x_9	x_{10}
5.0	25.0	15.0	5.0	5.0	20.0	10.0	5.0	5.0	5.0
10.0	30.0	17.0	17.0	8.0	8.0	5.0	4.0	1.0	0.0
3.0	6.0	10.0	13.0	25.0	15.0	13.0	8.0	5.0	2.0
7.5	27.9	16.0	11.0	6.5	14.0	7.5	4.5	3.0	2.5
4.6	21.2	14.0	6.6	9.0	19.0	10.6	5.6	5.0	4.4
3.8	13.6	12.0	9.8	17.0	17.0	11.8	6.8	5.0	3.2
8.3	26.6	15.9	14.2	9.1	11.1	6.8	4.6	2.2	1.2
6.1	22.7	14.6	10.2	9.9	15.4	9.1	5.3	3.8	2.9
7.6	24.2	15.2	13.8	10.8	11.8	7.6	5.0	2.6	1.4
3.9	10.3	11.2	12.6	21.3	14.8	11.9	7.3	4.6	2.1

The objects 4 to 10 were constructed by mixing objects 1, 2, and 3 in various proportions, to give the result shown in Table 5.I. Note that the rows sum to a constant.

The ranges for the variables of this matrix are as follows (these values are needed for computing [5.3]):

$$x_1 = 7 \qquad x_2 = 24 \qquad x_3 = 7 \qquad x_4 = 12 \qquad x_5 = 20$$

$$x_6 = 12 \qquad x_7 = 8 \qquad x_8 = 4 \qquad x_9 = 4 \qquad x_{10} = 5.$$

Table 5.I. *Percentage of end-members in each of the objects of the artificial data*

Object	End-member		
	1	2	3
1	100	0	0
2	0	100	0
3	0	0	100
4	50	50	0
5	80	0	20
6	40	0	60
7	20	70	10
8	50	30	20
9	20	60	20
10	10	10	80

The association matrix H computed from [5.3] is given next, with only the lower triangle of this symmetric matrix listed:

	x_1	x_2	x_3	x_4	x_5	x_6	x_7	x_8	x_9	x_{10}
x_1	1.000									
x_2	0.377	1.000								
x_3	0.440	0.183	1.000							
x_4	0.687	0.690	0.425	1.000						
x_5	0.888	0.362	0.552	0.657	1.000					
x_6	0.664	0.273	0.776	0.568	0.776	1.000				
x_7	0.566	0.800	0.384	0.861	0.561	0.473	1.000			
x_8	0.772	0.540	0.554	0.835	0.813	0.733	0.740	1.000		
x_9	0.624	0.718	0.465	0.863	0.626	0.555	0.918	0.813	1.000	
x_{10}	0.531	0.287	0.885	0.540	0.643	0.864	0.488	0.665	0.569	1.000

Using [5.4], the matrix H^* was computed next. The first two eigenvalues of this matrix were extracted; these are $\lambda_1 = 1.599$ and $\lambda_2 = 0.792$.

Owing to the transformation [5.3], there are seven nonzero remaining eigenvalues (0.314, 0.189, 0.146, 0.129, 0.089, 0.080, and 0.061) and one zero eigenvalue with eigenvector 1 (this is because the rows of H^* sum to zero). These account for 70% of the trace of H^*. (N.B. The transformation [5.3] has changed the rank from that of the data matrix; for [5.1], there are only two significant roots.)

The coordinates corresponding to roots 1 and 2 are computed as $U\Lambda^{1/2}$, that is, the product of the matrix of eigenvectors by the square root of the eigenvalue matrix.

	Coordinates	
Objects (projected data points)	Axis 1 a_{i1}	Axis 2 a_{i2}
Q_1	-0.074	0.514
Q_2	0.618	-0.258
Q_3	-0.555	-0.378
Q_4	0.319	0.048
Q_5	-0.199	0.417
Q_6	-0.463	-0.008
Q_7	0.458	-0.137
Q_8	0.034	0.178
Q_9	0.335	-0.087
Q_{10}	-0.473	-0.290

The plot of these points is shown in Fig. 5.2. We use these data again (p. 178) in illustrating correspondence analysis. Comparative discussion is deferred until then. Note that the sum of the elements of each factor is zero, as it should be (the centroid property of p. 140).

5.3 LOGCONTRAST PRINCIPAL COORDINATE ANALYSIS

A dual of logcontrast principal components, principal coordinate analysis, can be easily obtained (Aitchison, 1986). This aspect of the duality relationship derives from the fact that the covariance between components can be paralleled with the "angle" between their respective vectors in the replicate space, and the dual concept of similarity of compositions can be identified with negative "distance" between compositional representations in the sample space. The notation of this section is that appropriate to simplex space (cf. Section 4.6).

Using the logratio covariance structure [4.39] for compositions, the definition of *distance* between two compositions x_1 and x_2 is

$$\sqrt{\sum_{i=1}^{D} \left[\log \frac{x_{1i}}{g(x_1)} - \log \frac{x_{2i}}{g(x_2)}\right]^2}$$

For practical applications, the duality can be expressed in terms of the data matrix $\mathbf{U}$ in which the logarithms of the x_{ri} in $\mathbf{X}$, the raw-data matrix, are centered so that all row sums and column sums are zero:

$$\mathbf{U} = \mathbf{G}_N(\log \mathbf{X})\mathbf{G}_D. \qquad [5.6]$$

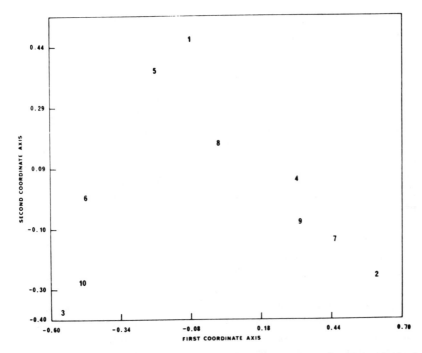

Figure 5.2. Plot of the first two coordinate axes for the principal coordinate analysis of the artificial data, $p = 10$, $N = 10$. Note that this plot "distorts the data," a result of using the coefficient [5.3]. An "exact" representation is yielded if the Pythagorean distance [5.1] is used.

G is the logcontrast isotropic form of **Γ**. The solutions are then as follows:

1. for principal components yielding d solutions,
$$(\mathbf{U'U} - \lambda\mathbf{I})a = \boldsymbol{0}; \qquad [5.7]$$

2. for principal coordinates yielding N solutions,
$$(\mathbf{UU'} - \mu\mathbf{I})c = \boldsymbol{0}; \qquad [5.8]$$

3. the scaling that unites solutions [5.7] and [5.8] is
$$\mathbf{c} = \mathbf{U}a.$$

5.4 Q-MODE FACTOR ANALYSIS

Imbrie and Purdy (1962) showed how "inverted" factor analysis could be applied to a common geological problem, namely, that of delineating lithofacies. Imbrie (1963) expanded the method and published computer programs for its execution. Imbrie and Van Andel (1964) success-fully used the method in several sedimentological problems.

Geological interpretation

The usual starting point is the N-by-p data matrix. The rows represent objects and the columns some properties of the objects. It will be convenient to consider the variables as constituents of the objects, for example, mineral species that make up a rock or a sediment. Often, the total amount of these constituents will sum to a constant value along any row of the matrix because they are usually expressed as propor-tions. For present purposes, this is not a problem. An exactly tailored procedure by Aitchison (1986) is now available – see the foregoing section (Section 5.3).

One, but by all means not the only, way to view the objects of the data matrix is to consider them as being combinations of some number of end-member compositions. That is, we may conceive of a situation in which the compositions of the objects in the data matrix could be achieved by mixing, in various proportions, several hypothetical, or real, objects of specific composition. We could then describe each object in terms of the proportions of the end-members, rather than by the amounts of the various constituents.

A simple sedimentological example may serve to illustrate this con-cept. Suppose that a sedimentary basin is being supplied with sediment by three rivers, as depicted in Fig. 5.3. Suppose, further, that each river system contributes a distinct mineral assemblage to the sediment. Upon entering the basin, the three assemblages are mixed in varying propor-tions by the physicosedimentary agencies. A sample of sediment from the basin can therefore be thought of as a mixture of three end-member objects, owing its composition to the end-members and the proportions in which they have been combined.

To pursue the example further, we shall assume that the basin is fossilized and that the resulting sedimentary rock unit is available for sampling. The usual procedure would be to collect numerous rock specimens and to analyze them for the minerals in relation to the total mineral composition. An objective of the study would be to determine the ultimate sources of the rocks and the sedimentary agencies involved in their accumulation. Unless the rock unit was exceptionally well exposed, it would normally be difficult to determine the sediment sources by direct mapping; however, by employing the concept of end-members as described previously, the analysis of rock composition offers a way of realizing this. We could, in terms of the Q-mode concept, try to express each rock specimen as proportions of contribut-ing assemblages. In most instances, however, we should not be in a

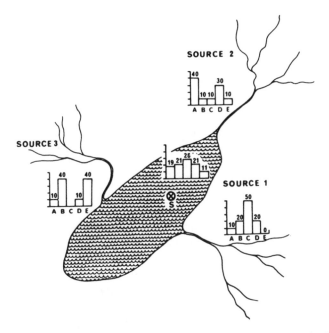

Figure 5.3. Schematical representation of a sedimentary basin being supplied with sediment from three different sources. The composition of the sediment supplied in terms of five mineral species (A–E) is given for each river. A sample, S, from within the basin can be considered as a mixture of the three "end-member" sediments in the proportions $0.5(1) + 0.3(2) + 0.2(3)$. For example, for mineral species A, we have $S_A = 0.5(10) + 0.3(40) + 0.2(10) = 19$.

position to know neither the number of distinct end-member assemblages nor their compositions.

For this example, as also for the general case, we may therefore specify the aims of the Q-mode analysis as follows:

1. to find the minimum number, k, of "end-member" assemblages of which the observed objects may be considered combinations;
2. to specify the compositions of the end-members in relation to the p constituents;
3. to describe each object in terms of the end-members; that is, to "unmix" the objects into their end-member components.

For the first of these aims, the minimum number of end-members is determined by approximating the data matrix by a matrix of lesser rank. This will define a number of linearly independent row vectors as well as the dimensionality of the system (see Section 2.13).

The second of these aims, specification of the compositions of end-members, necessitates further definition for, in fact, there are an infinite number of k compositions that will serve equally well. The method is designed so as to seek those end-members which, as compared to the observed objects, have the most divergent compositions. These may be defined as theoretical end-members or the most divergent observed objects in the data matrix may be used.

The third aim, the determination of the contribution of each end-member to each object, is achieved from the eigenanalysis, using the factor-loading matrix.

Similarity coefficient

In many geological problems involving compositional data, it is the proportions of the constituents that are of major importance. In the previous example, the proportions of mineral species in an end-member presumably reflects the proportions in the source terrain. The proportionality of minerals will hopefully be preserved while a unit of end-member composition is spread throughout the sedimentary basin. A sample of sediment taken from within the basin, the compositional elements of which are in the same proportions as that in one of the end-members, may be considered to be wholly derived from that end-member.

Imbrie and Purdy (1962) expanded on this concept and defined an "index of proportional similarity." The degree of similarity between two objects may be evaluated in relation to the proportions of their constituents. For any two objects, n and m (which are row vectors of the data matrix), the coefficient of proportional similarity, "cosine theta," is determined from

$$\cos \theta_{nm} = \frac{\sum_{j=1}^{p} x_{nj} x_{mj}}{\sqrt{\sum_{j=1}^{p} x_{nj}^2 \sum_{j=1}^{p} x_{mj}^2}}. \qquad [5.9]$$

It will be observed that [5.9] merely computes the cosine of the angle between the two row vectors as situated in p-dimensional space (cf. Section 2.7).

The value of $\cos \theta$ ranges from 0 for two collinear vectors to $+1.0$ for two vectors $90°$ apart (assuming that all values in the data matrix are positive).

In dealing with a suite of N objects, $\cos \theta$ must be computed for each possible pair of objects, after which the coefficients may be conveniently arranged in an $N \times N$ matrix of associations, **H**. Consideration of the matrix form of [5.9] gives additional insight into the nature of the coefficient.

Instead of working out $\cos\theta_{nm}$ in one operation, we may do this in two steps. First, we define

$$w_{nj} = \frac{x_{nj}}{\sqrt{\sum_{j=1}^{p} x_{nj}^2}} \qquad (j = 1, \ldots, p; \, n = 1, \ldots, N). \qquad [5.10]$$

That is, dividing every element in a row by the square root of the row sum of squares normalizes the data matrix so that

$$\sum_{j=1}^{p} w_{nj}^2 = 1, \quad \text{for } n = 1, \ldots, N.$$

Then

$$\cos\theta_{nm} = \sum_{j=1}^{p} w_{nj} w_{mj}. \qquad [5.11]$$

It is, however, more convenient to use matrix notation:

$$\mathbf{W}_{(N \times p)} = \mathbf{D}^{-1/2}\mathbf{X}, \qquad [5.12]$$

where $\mathbf{D}$ is an $N \times N$ diagonal matrix of the row sum of squares of $\mathbf{X}$. This operation ensures that every row vector of $\mathbf{W}$ is of unit length. The association matrix becomes

$$\mathbf{H} = \mathbf{W}\mathbf{W}' = \mathbf{D}^{-1/2}\mathbf{X}\mathbf{X}'\mathbf{D}^{-1/2}. \qquad [5.13]$$

This matrix is, consequently, the major product moment of the row-normalized data matrix.

Row-normalization, that is, multiplication of each row element by a constant, does not affect the proportionality relationships between variables. It does, however, remove the effects of size differences between objects. $\mathbf{H}$ contains the angular separations between all objects (row vectors) as they are situated in p-dimensional space. It now remains to determine if these angles can be contained in a space of lesser dimensionality. This is ascertained by finding a rank of the matrix by means of the eigenvalues.

Mathematical derivation

The following derivation is merely the application of the development of Section 2.13 to the matrix $\mathbf{W}$. The matrix to be factored is the N-by-N association matrix $\mathbf{H}$. The eigenvectors of this matrix provide a set of linearly independent reference vectors to which the object vectors may be referred. The steps are as follows.

The raw-data matrix $\mathbf{X}$ is row-normalized:

$$\mathbf{W} = \mathbf{D}^{-1/2}\mathbf{X},$$

where $\mathbf{D} = \text{diag}(\mathbf{XX}')$.

The association matrix is defined as the major product moment of the row-normalized data matrix:

$$\mathbf{H} = \mathbf{WW}'.$$

The row-normalized data matrix can be expressed approximately as the product of a factor-loadings matrix and a factor-score matrix:

$$\mathbf{W} \approx \mathbf{AF}'. \qquad [5.14]$$

Here, $\mathbf{A}$ is an N-by-k matrix and $\mathbf{F}$ is p by k, where k is the approximate rank of $\mathbf{W}$.

The relationships between $\mathbf{W}$, $\mathbf{H}$, $\mathbf{A}$, and $\mathbf{F}$ are given by

$$\mathbf{H} = \mathbf{WW}' = \mathbf{AF}'\mathbf{FA}'. \qquad [5.15]$$

The constraint that the $\mathbf{F}$ matrix be columnwise orthonormal is expressed by

$$\mathbf{F}'\mathbf{F} = \mathbf{I},$$

which leads to the following equation for [5.15]:

$$\mathbf{H} = \mathbf{AA}'.$$

The square, symmetric matrix $\mathbf{H}$ may be factored according to

$$\mathbf{H} = \mathbf{U}\Lambda\mathbf{U}', \qquad [5.16]$$

where $\mathbf{U}$ is the matrix of eigenvectors and Λ the diagonal matrix of associated eigenvalues. Thus,

$$\mathbf{H} = \mathbf{U}\Lambda\mathbf{U}'$$

and

$$\hat{\mathbf{A}} = \mathbf{U}_k\Lambda_k^{1/2}. \qquad [5.17]$$

The matrix of factor loadings is then the matrix of eigenvectors, scaled by the square roots of the eigenvectors.

Factor scores

Analogously to the R-mode case of Chapter 4, the matrix of factor scores may be defined in the following manner:

$$\mathbf{W} \approx \hat{\mathbf{A}}\mathbf{F}'.$$

Premultiplication by $\hat{\mathbf{A}}'$ gives

$$\hat{\mathbf{A}}'\mathbf{W} \approx \hat{\mathbf{A}}'\hat{\mathbf{A}}\mathbf{F}',$$

but $\hat{\mathbf{A}}'\hat{\mathbf{A}} = \mathbf{\Lambda}$, so that $\hat{\mathbf{F}}' = \mathbf{\Lambda}^{-1}\hat{\mathbf{A}}'\mathbf{W}$, or

$$\hat{\mathbf{F}} = \mathbf{W}'\hat{\mathbf{A}}\mathbf{\Lambda}^{-1}. \tag{5.18}$$

In the preceding derivation of the Q-mode method, we have emphasized the conceptual ideas. In practice, when the number of objects, N, is much greater than the number of variables, p, it is computationally more efficient to use the following procedure.

Step 1. Compute the minor product moment $\mathbf{W}'\mathbf{W}$ instead of $\mathbf{H} = \mathbf{W}\mathbf{W}'$, which is of order $p \times p$ rather than $N \times N$.

Step 2. Compute the eigenvalues $\mathbf{\Lambda}$ and eigenvectors $\mathbf{V}$ of $\mathbf{W}'\mathbf{W}$, that is,

$$\mathbf{W}'\mathbf{W} = \mathbf{V}\mathbf{\Lambda}\mathbf{V}'.$$

As noted in Section 2.12, the positive eigenvalues of $\mathbf{W}'\mathbf{W}$ are the same as those of $\mathbf{H}$.

Step 3. The factor scores matrix $\hat{\mathbf{F}}$ is now identical to $\mathbf{V}_k$, that is,

$$\hat{\mathbf{F}} = \mathbf{V}_k,$$

with orthonormal columns as before.

Step 4. The factor loading matrix $\hat{\mathbf{A}}$ can now be computed from

$$\hat{\mathbf{A}} = \mathbf{W}\hat{\mathbf{F}}.$$

It should be noted that this procedure is computationally simpler, not only because p is smaller than N, but also because no scaling of the columns of $\mathbf{A}$ and $\mathbf{F}$ is needed.

We hasten to point out that R-mode principal components and Q-mode factor analysis are not duals. Principal coordinate analysis is the dual of principal component analysis only when the Pythagorean distance is used (Gower, 1966a). This is a consequence of the singular value properties of a data matrix.

Number of end-members

The first objective of the analysis is to determine k, the number of independent end-member objects. This is equivalent to finding the rank of the normalized data matrix $\mathbf{W}_{(N \times p)}$ or that of its major product moment $\mathbf{H}_{(N \times N)}$. Because the rank of a product moment matrix cannot

Table 5.II. *The raw data matrix* X

Rock specimen	Variables			Row sum of squares
	x_1	x_2	x_3	
1	70	30	0	5800
2	0	10	90	8200
3	50	30	20	3800
4	40	20	40	3600
5	20	10	70	5400
Centroid	36	20	44	3632

exceed that of the matrix whose product it is, it is evident that the rank of H cannot be greater than p, the lesser order of W. Therefore, $N - p$ eigenvalues of H must be zero and, if the rank of W is k ($k < p$), then $N - k$ eigenvalues will be zero.

With real data, the exact rank of W will usually be p. Hence, in order to reproduce exactly the data matrix by [5.14], the number of factors will usually equal the number of variables. Nevertheless, it is often possible to arrive at a close approximation to W with significantly fewer factors. We shall therefore use the "approximate" rank of W, or H, to determine the minimum number of end-members. To do this, we use the fact that the sum of the eigenvalues of H must equal the trace of H.

The diagonal elements of H contain the row sums of squares of W, which, because of row normalization, are equal to 1. Geometrically, each object vector is of unit length. The trace of H thus corresponds to the sum of vector lengths and may be referred to as the total information content of H. This value will be N.

As we shall explain soon, an eigenvalue of H represents the sum of the squared projections of the object vectors onto the associated eigenvector. Therefore, the ratio of an eigenvalue to $\text{tr}\, H$ is a convenient measure of the information content of the associated factor. If this ratio indicates that a factor contributes a trivial amount of information to the solution, that factor may be taken as being insignificant. The number of useful eigenvalues thus serves as a first approximation to k.

Geometrical representation

Table 5.II contains a 5-by-3 data matrix of hypothetical compositional data. Note that the sum of each row is 100, a common enough situation for this kind of data. In Fig. 5.4 a plot of the object vectors on a triangular compositional diagram is shown; it is evident that the objects

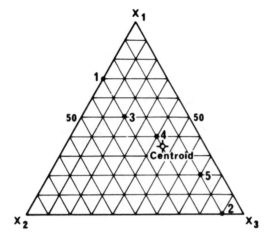

Figure 5.4. Plot of the hypothetical compositional data for Q-mode analysis.

Table 5.III. *The row-normalized data matrix $W = D^{-1/2}X$*

Rock specimen number	Variables		
	x_1	x_2	x_3
1	0.919	0.394	0
2	0	0.110	0.994
3	0.811	0.487	0.324
4	0.667	0.333	0.667
5	0.272	0.136	0.953
Centroid	0.597	0.332	0.730

can, to all intents and purposes, be considered as various mixtures of end-members 1 and 2.

The row-normalized data matrix W is listed in Table 5.III. Because the row vectors of this matrix are of unit length, elements in one of its rows may be looked upon as direction cosines of the row vector with respect to the axes w_1, w_2, and w_3. The positions of the normalized row vectors are shown in Fig. 5.5 on a right spherical triangle. Each dot represents the intersection of a row vector with a sphere of unit radius.

For further reference, the position of the centroid of the three variables is also marked on these diagrams.

As is intuitively evident, the array of row vectors almost defines a great circle in Fig. 5.5, thus indicating that the array of vectors is almost coplanar.

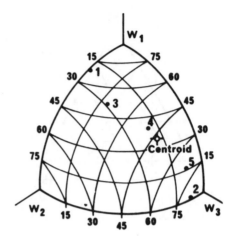

Figure 5.5. Plot of the normalized row vectors on a right spherical triangle.

Table 5.IV. *Association matrix for cos θ, H = WW'*

Rock specimens	1	2	3	4	5
1	1.000	0.044	0.937	0.744	0.304
2		1.000	0.375	0.699	0.962
3			1.000	0.919	0.596
4				1.000	0.862
5					1.000

The matrix of cosines, **H**, is given in Table 5.IV. Here, all possible values are listed. Table 5.V presents the eigenvalues of **H** along with the percentage of information explained by the associated factor and the cumulated sum of this percentage. The interpretation of Table 5.V would be that two end-members, two linearly independent vectors, are necessary in order to account for the approximate composition of the five rock specimens.

Table 5.VI is the factor loadings matrix **A**. The columns are the orthogonal eigenvectors of **H**, multiplied by the square roots of the associated eigenvalues, in accordance with [5.17]. The sum of the squared elements of any column will be equal to the eigenvalue associated with that column.

The elements of any row of **A** yield the coordinates of the row vector on the three mutually perpendicular factor axes. The element a_{nj} for the nth row and jth factor axis can be regarded as the cosine of the angle between the nth row vector of **W** and the jth factor axis of **A**.

Table 5.V. *The eigenvalues of **H***

Number	Eigenvalue	Percentage information	Cumulative percentage of information $= \lambda_j/N$
1	3.636	72.72	72.72
2	1.352	27.05	99.77
3	0.011	0.23	100.00
4	0		
5	0		

Table 5.VI. *The unrotated factor loadings matrix for k = 3*

Rock specimen	Communality	Factor 1	Factor 2	Factor 3
1	1.000	0.718	0.696	−0.037
2	1.000	0.727	−0.685	0.040
3	1.000	0.906	0.417	0.073
4	1.000	0.999	0.038	−0.024
5	1.000	0.879	−0.473	−0.051
Percentage of information explained		72.72	27.05	0.23

Alternatively, a_{nj} can be considered as the projection of the nth row vector onto the jth factor axis. Using the fact that a row vector of **W** is of unit length, then the sum of a_{nj}^2 $(j = 1, \ldots, k)$ will also equal unity, provided as many factors as variables are used; that is, if $k = p$. When $k < p$, this sum of squares is called the communality and is a measure of the proportion of the length of a row vector explained by the k factors (cf. R-mode analog discussed in Chapter 3).

The sum of the squared elements in a column of **A** represents the sum of projected vector lengths onto that particular factor axis. This, as noted previously, is equal to the associated eigenvalue. In Table 5.VI, each row vector has only a small component of its length projected along the third factor. We would conclude that the third factor is trivial; therefore, only two factors would be considered. Table 5.VII lists the loadings for the rock specimens on just two factors, along with their communalities. The departure of the communalities from unity reflects the degree to which the row vectors deviate from the plane formed by the axes of factors 1 and 2 in Fig. 5.6. The essential features of Fig. 5.4 are more concisely illustrated in Fig. 5.6.

The factor scores matrix associated with Table 5.VI is given in Table 5.VIII. The product **AF'** exactly reproduces **W**, the row-normalized data matrix of Table 5.III. Table 5.IX contains the scores for two factors

Table 5.VII. *The unrotated factor loadings matrix for k = 2*

Rock specimen	Communality	Factor 1	Factor 2
1	0.999	0.718	0.696
2	0.998	0.727	−0.685
3	0.995	0.906	0.417
4	0.999	0.999	0.038
5	0.997	0.879	−0.473
Percentage of information explained		72.72	27.05

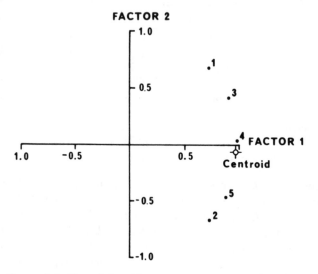

Figure 5.6. Plot of the objects on two unrotated factors.

only. Denoting this matrix as $\mathbf{F}^*$, and the matrix of Table 5.VII as $\mathbf{A}^*$, then

$$\hat{\mathbf{W}} = \mathbf{A}^* \mathbf{F}^{*\prime}$$

is a least-squares estimate of $\mathbf{W}$.

The positions of the objects on factor 1 are almost equally determined by the first and third variables. Along factor 2, the objects are ordered according to their content of variable 1, in the positive direction, and to their content of variable 3, in the negative direction.

An important use of the matrix $\mathbf{F}$ is to locate objects, which were not part of the original data matrix, in the factor space. To illustrate this we

Table 5.VIII. *The unrotated factor scores matrix for k = 3*

Variable	Factor 1	Factor 2	Factor 3
1	0.632	0.646	−0.427
2	0.345	0.258	0.902
3	0.693	−0.718	−0.606

Table 5.IX. *The unrotated factor scores matrix for k = 2*

Variable	Factor 1	Factor 2
1	0.632	0.646
2	0.345	0.258
3	0.693	−0.718

can take the vector of mean values of X as being a new object vector. The row-normalized mean vector, denoted "centroid" in Table 5.III, can now be located in the factor space by means of the matrix F. Equation [5.14] states that $W = AF'$ and, since $F'F = I$, $A = WF$. Inasmuch as the centroid vector is in the present example considered as a row vector of W, the corresponding row vector of A^* may be found from

$$a^{*\prime} = w'F^*;$$

that is, the row vector of normalized mean values only needs to be postmultiplied by the matrix of factor scores. The resulting product yields the coordinates of the centroid vector with respect to the two factor axes. In this case, [0.9986, −0.0527]. The plotted position of the centroid on Fig. 5.6 indicates that factor 1 almost coincides with the center of gravity of all the object vectors. This example shows that the F matrix may be viewed as a transformational operator that maps the row vectors of W onto the factor axes of A.

Rotation of Q-mode factor axes

The number of end-members is determined, to a first order of approximation, by the number of "significant" eigenvalues. It remains to find a set of compositionally distinct end-members.

The factors determined by the principal-axes method do not meet this requirement. As illustrated, factor 1 has a composition, in terms of normalized constituents, very near the "average" of all objects; clearly, this does not approach the concept of an end-member as defined

Table 5.X. *Varimax factor loadings matrix* **B** *for k* = 2

Rock specimen	Communality	Factor 1	Factor 2
1	0.999	0.999	−0.016
2	0.998	0.030	−0.999
3	0.995	0.935	−0.346
4	0.999	0.733	−0.680
5	0.997	0.287	−0.957
Percentage of information explained		49.89	49.88

earlier. From Fig. 5.5, it is apparent, for our simple example, that object vectors 1 and 2 conform more closely to a meaningful pair of end-members; this notwithstanding, the selection of end-members in a more complex, multidimensional problem may be obscured and an analytical method for their location is necessary.

The varimax method of factor rotation is one way of approaching the problem. This method rotates rigidly the factor axes so that they will coincide with the most divergent vectors in the space (cf. Chapter 7). This is done by maximizing the variance of the factor loadings on each factor, subject to the constraint that the factors retain their orthogonality.

For the example of Table 5.VII, the necessary transformation matrix, **T**, was computed to be

$$\mathbf{T} = \begin{bmatrix} 0.7071 & -0.7071 \\ 0.7071 & 0.7071 \end{bmatrix}.$$

This matrix, premultiplied by **A** of Table 5.VII, results in the varimax factor loadings matrix **B** of Table 5.X. The factor axes have been rigidly rotated through an angle of 45° in a counterclockwise direction. Similarly, the matrix of varimax factor scores of Table 5.XI was obtained by postmultiplying **F** of Table 5.IX by **T**. The negative signs assigned to the loadings in factor 2 simply reflect the arbitrary quadrant in which the positive end of the factor is located. Reversing the signs in the corresponding columns of **B** and **F** has no effect on the solution.

Interpretation

The comments made for the unrotated factor-loadings and factor-scores matrices apply equally to their rotated equivalents. The sum of the row-squared elements of **B**, the communalities of the objects, are, of course, the same values as in the unrotated case; the lengths of the

Table 5.XI. *The varimax factors scores matrix for k = 2*

Factor 1	Factor 2
0.904	0.010
0.427	− 0.062
− 0.018	− 0.998

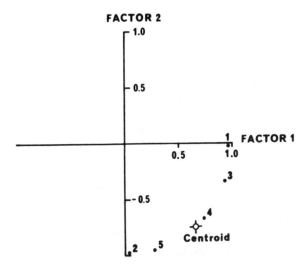

Figure 5.7. Plot of the objects on two varimax factors.

object vectors do not change due to rigid rotation of the reference axes. The sum of column-squared elements is the sum of the projected vector lengths onto the rotated factor axes. Divided by the total length of all vectors, N, this figure may be used as a measure of the information content explained by a factor. Comparing these figures with those in Table 5.VII, it can be seen that the rotated factors are more equally distributed in terms of "explanatory power." Finally, we may once again plot the positions of the object vectors in the varimax factor space, as shown in Fig. 5.7. The computed position of the centroid [0.6686, 0.7433] is marked, as before. Table 5.X clearly identifies row vectors 1 and 2 as the most compositionally different objects of the data.

Factors 1 and 2 may themselves be considered as hypothetical objects that, because of their orthogonality, are compositionally completely dissimilar. They are thus hypothetical, linearly independent end-members. We shall now turn our attention to the method of oblique rotation, which permits us to select the most different object vectors as end-members and to express all other object vectors in relation to them.

Table 5.XII. *The oblique projection matrix* $C = BV^{-1}$

Rock specimen	Factor 1	Factor 2
1	1.0	0.0
2	0.0	1.0
3	0.926	0.332
4	0.714	0.669
5	0.259	0.954

Oblique projection

It is apparent in Table 5.X and Fig. 5.6, as well as the original data matrix, that objects 1 and 2, being the compositionally most different objects, would constitute end-members. Imbrie (1963) used a procedure, *oblique projection*, that rotates the orthogonal varimax axes so that they coincide with the most divergent object vectors. The resulting factors are no longer orthogonal, but they do correspond to real objects, an advantage in some situations. The steps involved are as follows:

1. Scan the columns of **B** for the highest element and note the row index. In Table 5.X, row 1 has the highest loading in column 1 and row 2 in column 2.
2. Form matrix $V_{(k \times k)}$, consisting of the row vectors isolated in the first step. In the present example,

$$V = \begin{bmatrix} 0.9992 & -0.0155 \\ 0.0296 & -0.9988 \end{bmatrix}.$$

These elements are the cosines of the angles between the varimax axes and what appear to be the most divergent object vectors.

3. Form the inverse of **V**. In our example, this is

$$V^{-1} = \begin{bmatrix} 1.0013 & -0.0155 \\ 0.0297 & -1.0017 \end{bmatrix}.$$

The matrix V^{-1} is analogous to the **T** matrix used for rotating the initial, principal factor axes to the varimax axes, but it is not orthonormal.

4. Compute the oblique projection matrix $C = BV^{-1}$. The results are listed in Table 5.XII. The loadings for the first row [1.0, 0.0] and the second row [0.0, 1.0] indicate that these row vectors are now the reference axes. The other loadings are the projections of the row vectors onto these oblique reference axes.

Table 5.XIII. *The "denormalized" oblique projection matrix*

Rock specimen	Factor 1	Factor 2
1	100.0	0.0
2	0.0	100.0
3	70.11	29.89
4	47.30	52.70
5	18.59	61.49

5. Scan the column vectors of **C**, noting the row indices of any loadings greater than 1.0. Replace the rows of **V** with the rows of **B** corresponding to rows having the highest absolute loadings on **C** and repeat the calculations of V^{-1}, **C**, and the scanning of **C**. A loading greater than 1 means that some row vector is more divergent than the ones chosen in the first step.

This iterative procedure ultimately finds the most divergent row vectors in the data matrix, although, in some cases, a unique solution cannot be obtained.

6. The row vectors of the matrix **C** give the proportional contributions of the normalized end-members to the normalized object vectors. To recast the values in terms of the raw data, it is necessary to divide each column vector of **C** by the corresponding vector length of the end-member objects they represent; in effect, this "denormalizes" the column vectors of **C**.

In the present example, the length of the first object vector is 76.15 and that of the second object vector, 90.55. Table 5.XIII gives the adjusted values of **C**, further adjusted so that the row sums equal 100, as in the raw data. Consequently, all the object vectors are given as percentages of the two end-members. The compositions of the end-members can be taken directly from the rows of **X**.

5.5 THE ANALYSIS OF ASYMMETRY

The subject we include in this section is not often recognized by natural scientists, although problems involving asymmetric distributions occur frequently in field ecology, map analysis, and physical geography, to quote a few cases. It seems also that the method outlined here could prove useful in studying data occurring in molecular biology. The usual way of trying to analyze asymmetric data is to use multidimensional scaling, a topic not considered in the present book [cf. Jackson (1991) for a recent appraisal of this field], but this loses an essential part of the information contained in the asymmetry.

The following presentation is based on Reyment and Banfield (1981) and Gower (1977). The procedure depends on the singular value decomposition and is not a typical *Q*-mode one. We have included it in this section because of the manner in which it can be contrasted with principal coordinates. The same set of data were reassessed in Chapter 7 of Reyment (1991).

Introduction

The relationship between two geological specimens i and j can be quantified in many ways. Such quantities, a_{ij} say, are calculated from measurements made on the specimens, and these can be many and varied, depending on what aspects of the specimens are of interest. If the relationship between the two specimens is symmetric, then the quantities a_{ij} and a_{ji} will be equal. Geologists are familiar with a variety of multivariate statistical methods, including principal component analysis and multidimensional scaling, which can be used to analyze square symmetric matrices holding such values. Because distance is usually metric in nature and symmetric by definition, many of these statistical techniques aim at representing the values a_{ij}, or some simple transformation of them, as distances on a plot in which the distance d_{ij} $(= d_{ji})$ represents the relationship of specimen i to specimen j.

There are, however, many problems in science where the relationship between two specimens is nonsymmetric, and $a_{ij} \neq a_{ji}$. Asymmetric data are commonly generated in the quantitative interpretation of spatial relationships of all kinds of geological maps: for example, counts on the number of times formation i is the nearest unlike neighbor of formation j, or the number of times sediment i is surrounded by sediment j. Here, the sites have a known geographical relationship, and the significance therefore lies in the asymmetry of the observations. Another example is provided by thin sections of rocks, which also form a kind of map in which relationships between neighboring mineral species may be of interest, as in *Gefüge*-analysis. Other examples can be found in problems of spatial paleoecology, microfacies analysis, the distribution of glacial drift, and plant ecology. All of these have the common factor that they can be represented as some kind of map.

A square nonsymmetric matrix **A**, whose rows and columns are classified by the same specimens (see, for example, Table 5.XIV), cannot be treated by the same methods as used in the standard multivariate analysis of square symmetric matrices. Attempts at analyzing asymmetric data have been made by using *nonmetric multidimen-*

Table 5.XIV. *Matrix[a] of the number of nearest unlike neighbors for depositional environments in the Mississippi Delta*

Category	(1) Natural levee	(2) Point bar	(3) Swamp	(4) Marsh	(5) Beach	(6) Lacustrine	(7) Bay sound
(1)	—	117	286	148	0	2	0
(2)	38	—	5	2	0	0	1
(3)	301	10	—	175	1	138	12
(4)	538	3	168	—	29	320	281
(5)	0	0	0	9	—	0	8
(6)	2	0	168	292	0	—	20
(7)	0	1	147	617	161	25	—

[a]The asymmetric entries reflect the widely differing spatial relationships between the environments, hence order in the geological pattern. Symmetric entries would be indicative of geological disorder.
Source: McCammon (1972, page 424).

sional scaling for analyzing the symmetric matrix $\frac{1}{2}(\mathbf{A} + \mathbf{A}')$. The asymmetry may sometimes be caused by so-called noise, but in geological applications it may have an intrinsic meaning (cf. Jackson, 1991).

Gower (1977) has given attention to methods of analyzing the asymmetry in square nonsymmetric matrices whereby the differences between a_{ij} and a_{ji} can be represented graphically. Among the solutions examined by him, we have chosen that of the canonical analysis of asymmetry as being the most applicable to geological-map interpretation in its broadest sense. This technique should not be confused with spatial modeling, to which it bears no relation. We present here an application of Gower's method to a sedimentary environmental pattern (Reyment and Banfield, 1981).

Description of the method

Given an $n \times n$ nonsymmetric matrix $\mathbf{A}$, whose rows and columns classify the same n specimens, then $\mathbf{M}$ $[= \frac{1}{2}(\mathbf{A} + \mathbf{A}')]$ is a symmetric matrix, $\mathbf{N}$ $[= \frac{1}{2}(\mathbf{A} - \mathbf{A}')]$ is a skew-symmetric matrix, and $\mathbf{A} = \mathbf{M} + \mathbf{N}$. Gower (1977) shows that matrices $\mathbf{M}$ and $\mathbf{N}$ can be analyzed separately and their results considered together to form an overall representation of the asymmetry values in $\mathbf{A}$. However, if $\mathbf{M}$ is of little importance or is unknown, the analysis of $\mathbf{N}$ alone can still be considered (Section 2.14).

The canonical form of $\mathbf{N}$ can be obtained by using the singular value decomposition of $\mathbf{N}$,

$$\mathbf{N} = \mathbf{USV}',$$

where $\mathbf{S}$ is a diagonal matrix containing the singular values s_i ($i = 1, 2, \ldots, n$), $\mathbf{U}$ is an orthogonal matrix, and $\mathbf{V}$ is the product $\mathbf{UJ}$, where $\mathbf{J}$ is the elementary block-diagonal skew-symmetric matrix made up of 2×2 diagonal blocks $\begin{pmatrix} 0 & 1 \\ -1 & 0 \end{pmatrix}$. If n is odd, then the final singular value, s_n, will equal zero, and the final diagonal element of $\mathbf{J}$ will equal 1. Because $\mathbf{N}$ is skew-symmetric, the singular values will be in pairs and can be arranged in descending order of magnitude: $(s_1 = s_2) > (s_3 = s_4) > (s_5 = s_6) \ldots$.

Gower (1977) shows that the pair of columns of $\mathbf{U}$ corresponding to each pair of equal singular values, when scaled by the square root of the corresponding singular value, holds the coordinates of the n specimens in a two-dimensional space – a plane – that approximates the skew symmetry. The proportion of the total skew symmetry represented by this plane is given by the size of the corresponding singular value. Hence, as the first singular value, s_1 ($= s_2$), is the largest, the first two columns of $\mathbf{U}$ give the coordinates of the n specimens in the plane accounting for the largest proportion of skew symmetry. If s_1 is large compared to the other s_i ($i > 2$), the plot of the first plane will give a good approximation to the values in **N**. If s_1 is not comparatively large, then other planes are also required to give a good approximation. Because of the nonmetric nature of skew symmetry, the values of **N** are represented in the planes by areas of triangles made with the origin. Hence, the area of triangle $(i, j, 0)$ defined by points i, j, and the origin is proportional to the ijth value in the matrix **N**, and

$$\text{area}(i, j, 0) = -\text{area}(j, i, 0).$$

The canonical analysis of asymmetry can, in some respects, be regarded as an *analog of the method of principal component analysis*. In principal components, distance is explained by differences in transformed values, relative to principal axes (one axis for each eigenvalue), whereas the *skew symmetry* here is explained by areas of triangles relative to principal planes (one plane for each pair of singular values).

The plot or plots of the n specimens formed from analysis of the symmetric matrix **M** may also be considered with the plane or planes obtained by the method described previously, as together they give a complete graphical representation of the asymmetric values in **A**. The plots for **M** can be obtained by many methods, including multidimensional scaling. This will provide a set of coordinates for the n specimens that hopefully represent the symmetry values of **M** with good approximation in two or three dimensions. If two dimensions are sufficient, then these coordinates can be plotted to produce a graphical representation in which the distances between the specimens represent the values of **M**. If two dimensions are insufficient, a third dimension, or even more, will be necessary.

The data: an environmental study

The data used for illustrating the application of Gower's method to a typical ecological problem based on a map were taken from McCammon (1972). The material consists of observations on the nearest-neighbor relationships between the following seven environmental categories in the Mississippi Delta region of southeastern Louisiana: (1) natural levee; (2) point bar; (3) swamp; (4) marsh; (5) beach; (6) lacustrine; (7) bay sound. Further information on these variables can be obtained from McCammon (1972).

The observations on the spatial relationships between these depositional environments were extracted from a map of the delta (McCammon, 1972, page 423). Each environment is represented by a distinct variety of sediment. The matrix of nearest unlike neighbors is shown in Table 5.XIV.

Table 5.XV. *Asymmetric matrix of Table 5.XIV transformed to proportions*

Category	(1) Natural levee	(2) Point bar	(3) Swamp	(4) Marsh	(5) Beach	(6) Lacustrine	(7) Bay sound
(1)	1.0000	0.2116	0.5172	0.2676	0.0000	0.0036	0.0000
(2)	0.8261	1.0000	0.1087	0.0435	0.0000	0.0000	0.0217
(3)	0.4725	0.0157	1.0000	0.2747	0.0016	0.2166	0.0188
(4)	0.4018	0.0022	0.1255	1.0000	0.0217	0.2390	0.2099
(5)	0.0000	0.0000	0.0000	0.5294	1.0000	0.0000	0.4706
(6)	0.0041	0.0000	0.3485	0.6058	0.0000	1.0000	0.0415
(7)	0.0000	0.0011	0.1546	0.6488	0.1693	0.0263	1.0000

Table 5.XVI. *Skew-symmetric matrix formed from matrix of Table 5.XV*

Category	(1) Natural levee	(2) Point bar	(3) Swamp	(4) Marsh	(5) Beach	(6) Lacustrine	(7) Bay sound
(1)	0.0000	− 0.3073	0.0223	− 0.0671	0.0000	− 0.0003	0.0000
(2)	0.3073	0.0000	0.0465	0.0206	0.0000	0.0000	0.0103
(3)	− 0.0223	− 0.0465	0.0000	0.0746	0.0008	− 0.0660	− 0.0679
(4)	0.0671	− 0.0206	− 0.0746	0.0000	− 0.2539	− 0.1834	− 0.2195
(5)	0.0000	0.0000	− 0.0008	0.2539	0.0000	0.0000	0.1506
(6)	0.0003	0.0000	0.0660	0.1834	0.0000	0.0000	0.0076
(7)	0.0000	− 0.0103	0.0679	0.2195	− 0.1506	− 0.0076	0.0000

Asymmetry analysis of the data

The 7×7 asymmetric matrix listed in Table 5.XIV was first transformed to a matrix $\mathbf{A}$ of proportions (Table 5.XV) by dividing each value by its row total. Note that this transformation would not usually be necessary, but McCammon's matrix had no diagonal elements so the transformation was used to ensure constant known diagonal elements for the later analysis of matrix $\mathbf{M}$. From the matrix $\mathbf{A}$, the skew-symmetric matrix $\mathbf{N}$, $\mathbf{N} = \frac{1}{2}(\mathbf{A} - \mathbf{A'})$, shown in Table 5.XVI was computed.

The pairs of singular values from the singular value decomposition of $\mathbf{N}$, and the corresponding coordinate vectors, are given in Table 5.XVII. It will be seen from the relative sizes of the pairs of singular values that representation of the skew symmetry needs at least four dimensions – that is, two planes corresponding to singular values 0.4241 and 0.3078 – to represent it with good approximation. The distributional properties of the singular values are as yet unknown, but an *ad hoc* approach might be to require values to sum to more than 80% of the total for an acceptable approximation. Here, a plot of the first two

Table 5.XVII. *Singular values and coordinates in the corresponding planes, obtained from the skew-symmetric matrix of Table 5.XVI*

	Singular values and corresponding planes[a]						
	(1) 0.4241	(2) 0.4241	(3) 0.3078	(4) 0.3078	(5) 0.1050	(6) 0.1050	(7) 0.0000
	First plane		Second plane		Third plane		
(1)	−0.0891	−0.1489	0.4995	0.1779	−0.0276	−0.0116	0.0000
(2)	0.1314	−0.0237	0.1993	−0.5010	0.0205	0.0120	
(3)	0.1205	−0.0351	0.0429	0.1057	0.2455	0.0391	
(4)	0.3788	−0.4673	−0.0821	0.0376	0.0053	−0.0859	
(5)	0.2354	0.3610	0.0771	0.0467	0.0194	−0.2361	
(6)	0.2054	0.1893	0.0476	0.0382	0.1161	0.1510	
(7)	0.3785	0.1248	0.0432	0.0947	−0.1722	0.1315	

[a]Categories (1)–(7) are as described in previous tables.

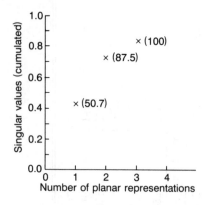

Figure 5.8. Plot of cumulative singular values (one from each pair) against the number of corresponding planar representations. The percentage of the total of the singular values is shown in brackets for the inclusion of each plane.

dimensions (that is, the first plane) alone would not show all the skew symmetry in the material – only 50.7%, in fact – and the second plane, corresponding to the second pair of singular values, has to be considered as well to improve the overall fit of **N** to 87.5% (see Fig. 5.8).

The plot of the first plane, corresponding to the first pair of singular values, is shown in Fig. 5.9. The largest skew-symmetric value of **N** is n_{12}, which equals 0.3073. The corresponding triangle formed by joining environments (1) and (2) to the origin, and here denoted $(1, 2, 0)$, has a relatively small area and, in fact, there are several triangles that greatly exceed it in area. The triangle with the greatest area is $(4, 5, 0)$, corresponding to the second-largest skew-symmetric value (0.2539). This shows that the first planar representation alone is not sufficient to give a good approximation to the values of **N**. The second plane (Fig. 5.10), however, discloses that $(1, 2, 0)$ is the triangle with the largest area, and that all other triangles have relatively small areas. So this plane, as expected from the singular values, accounts for the area not explained by the first planar representation.

The plot of the first two dimensions (Fig. 5.9) shows that all points have a large asymmetry with point **4** [that is, environment (4) marsh], because all triangles including point **4** have large areas. There seems to be a collinear relationship between points **1**, **3**, **2**, and **7**, in that order (that is, natural levee, swamp, point bar, bay sound). Gower (1977) shows that collinear points will have equal skew-symmetry values with a point on a line through the origin parallel to the collinearity, because these points will form triangles having the same base and with equal height, hence equal area. Unfortunately the line through the origin and

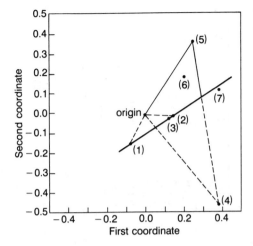

Figure 5.9. Plot of the coordinates for the seven environments in the plane corresponding to the first pair of singular values. Category of environments: (1) natural levee; (2) point bar; (3) swamp; (4) marsh; (5) beach; (6) lacustrine; (7) bay sound.

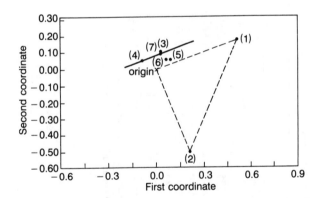

Figure 5.10. Plot of the coordinates for the seven environments in the plane corresponding to the second pair of singular values. Category of environments: (1) natural levee; (2) point bar; (3) swamp; (4) marsh; (5) beach; (6) lacustrine; (7) bay sound.

parallel to the collinearity passes through no other environment, so the theory cannot be exemplified in this planar representation (it can be seen, however, in Fig. 5.10). Points **2** and **3** (point bar, swamp) lie close together in this plane and thus appear to have similar skew-symmetry values with the other environments because triangles formed by other environments with the origin and these two points will have very similar

Table 5.XVIII. Principal coordinates obtained for the symmetric matrix derived from the data of McCammon (1972)

	Latent roots[a]						
	(1) 1.5372	(2) 1.1909	(3) 0.8086	(4) 0.7107	(5) 0.5897	(6) 0.1629	(7) 0.0000
(1)	0.6275	−0.1240	−0.2031	−0.1038	−0.3053	−0.2434	—
(2)	0.5829	−0.4592	0.4623	−0.0332	0.1939	0.1452	—
(3)	0.3256	0.3957	−0.5505	0.1937	0.2273	0.1518	—
(4)	−0.3191	0.1951	0.0526	−0.3383	−0.4770	0.1843	—
(5)	−0.5247	−0.4387	−0.0402	0.5819	−0.1464	−0.0166	—
(6)	−0.1451	0.7078	0.4614	0.1355	0.1325	−0.1300	—
(7)	−0.5472	−0.2767	−0.1825	−0.4357	0.3751	−0.0913	—

[a]Categories (1)–(7) are as described in previous tables.

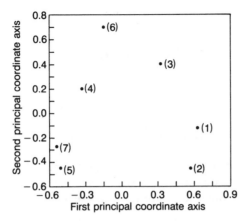

Figure 5.11. Plot of the principal coordinates for the seven environments, accounting for 54.5% of the total symmetry. Category of environments: (1) natural levee; (2) point bar; (3) swamp; (4) marsh; (5) beach; (6) lacustrine; (7) bay sound.

areas. However, in the second plane this misrepresentation is corrected, as points **2** and **3** are not close.

In Fig. 5.10, the supplementary information of the second plane is dominated by the triangle $(1, 2, 0)$. Environments (1) and (2) share the largest asymmetry, suggesting a possible genetic relationship. Inspection of McCammon (1972, Fig. 1) indicates this to be identifiable with the immediate depositional environment of the Mississippi River. The points for environments (3), (4), and (7) lie on a line parallel to that joining the origin to environment (1), indicating that in this planar representation, and this one only, these environments show equal skew symmetry with the natural levee environment. The close proximity of swamp with bay sound, and beach with lacustrine shows that these pairs have similar skew symmetries with the other environments.

In order to reify further the relationships outlined by the graphical representations of the skew symmetry, we analyzed the symmetrical part, **M**, of the asymmetric matrix **A** by principal coordinate analysis. The results of the calculations are given in Table 5.XVIII. The plot of the first pair of coordinates, illustrated in Fig. 5.11, represents 54.5% of the total symmetry between the depositional environments. Because these coordinates are plotted in a Euclidean space, we can give a metric interpretation to the interpoint distances, which represent the similarities between the environments. As might be expected, bay sound and beach are comparatively close together, showing that the degree of organization of their spatial relationships is similar; natural levee and point bar are also similar. Lacustrine is seen to be very different from beach and point bar, and most similar to swamp and marsh.

6 Q-R-mode methods

6.1 CORRESPONDENCE ANALYSIS

A method of factor analysis that combines some of the advantages of R-mode and Q-mode techniques was developed by Bénzécri (1973) and his associates under the name of "analyse factorielle des correspondances."

Bénzécri (1973) gave a detailed account of the theory of his method in a two-volume text, along with numerous examples that were drawn mainly from the social sciences. The Q-R-mode analysis of a table of contingencies has, however, a very long history in statistics and this has been excellently reviewed by Hill (1974). Several geological applications have appeared in the literature, for example, David, Campiglio, and Darling (1974), Melguen (1971) and Mahé (1974). In geology and in other subjects, clandestine applications of the method are common in that the calculations are being applied to continuously distributed variables, whereas the method was designed for scored observations ranged in a contingency table. This subject has been debated by Hill (1974) in a paper that should be studied by anybody wishing to apply correspondence analysis to continuously distributed variables.

There is a useful expository article on correspondence analysis in *News and Notes of the Royal Statistical Society* (1991, vol. 18, p. 13); it is based on a lecture to the society given by M. Greenacre. The book by Lefebvre (1976) contains an instructive introduction to the technique, especially useful because it gives the francophone philosophy of the subject along with the terminology. Other references are Greenacre (1984) and Jongman, ter Braak, and Van Tongeren (1987).

Briefly, the aim of correspondence analysis is to obtain simultaneously, and on equivalent scales, what we have termed R-mode factor loadings and Q-mode factor loadings that represent the same principal components of the data matrix.

The goal has been reached by a method of scaling the data matrix, a definition of an appropriate index of similarity, and by making use of the singular-value decomposition as discussed in Chapter 2, Section 12. The scaling procedure and analysis are, up to a point, algebraically equivalent to Fisher's (1940) contingency table (Hill, 1974).

One of the distinct advantages of the method is its ability to portray graphically relationships between variables and objects simultaneously.

Scaling procedure

The method, primarily designed for discrete variates, may be applied to continuous distributions by the stratagem of dividing the ranges of the variables into discrete pieces (Hill, 1974). In the present connection, we shall describe the technique with reference to a data matrix $\mathbf{X}_{(N \times p)}$, where the rows represent N values on each of p discretely distributed variables. A natural restriction on the elements of the data matrix is that they cannot be negative. The elements of the data matrix are first divided by the sum of all the elements in the matrix; that is,

$$b_{ij} = x_{ij} \bigg/ \sum_{i=1}^{N} \sum_{j=1}^{p} x_{ij}.$$

This transformation does not destroy the proportional relations among the rows and the columns. This and ensuing transformations are motivated by a different interpretation of the data matrix from that of the methods described earlier in this book. The transformation from $\mathbf{X}$ to $\mathbf{B}$, in this view, converts the elements to proportions that may be interpreted as probability values and matrix $\mathbf{B}$ is then an N-by-p contingency table in the sense of Fisher (1940). The vector of row sums of $\mathbf{B}$ contains the marginal probabilities for objects; the vector of its column sums contains the marginal probabilities for variables.

An important consequence of this transformation is that the elements of a row or column vector, being proportions, can be thought of in terms of multinomial probability distributions (cf. Mosimann, 1962).

For convenience later, we define a diagonal matrix $\mathbf{D}_{c(p \times p)}$, containing the column sums of $\mathbf{B}$ in the principal diagonal, and $\mathbf{D}_r$ as an N-by-N diagonal matrix, containing row sums of $\mathbf{B}$.

Finally, the row and column sums are used to transform $\mathbf{B}$ as follows:

$$\mathbf{W} = \mathbf{D}_r^{-1/2} \mathbf{B} \mathbf{D}_c^{-1/2}. \qquad [6.1]$$

The effect of this transformation is to stretch, differentially, the column vectors by the reciprocal of the square root of their column sum and stretch, differentially, the row vectors by the reciprocal of the square root of their row sum. Variables measured on disparate scales are thus differentially weighted and, as it were, equalized. Similarly for the objects.

The similarity measure

The rationale for the transformation of [6.1] is based on the development of a measure of similarity between objects and between variables. The concept of Euclidean distances does not apply for contingency data of the kind here involved. In order to produce both *R*-mode and *Q*-mode loadings in the same metric, the similarity must be defined jointly and symmetrically. Bénzécri (1973) developed the subject in relation to a chi-squared distance interpretation. Hill (1974) derives the procedure in terms of multidimensional scaling.

Similarity between objects is determined as follows:

1. Each element of **B** is scaled according to $b_{ij}/b_{i\cdot}$, where $b_{i\cdot}$ is the row sum of the *i*th row. Note that $\sum_{j=1}^{p} b_{ij}/b_{i\cdot} = 1$, thus all rows sum to unity and the dimensionality of the problem is therefore reduced by 1.

2. The contingency table distance between objects *q* and *r* is given by

$$d_{qr}^2 = \sum_{j=1}^{p} \frac{1}{b_{\cdot j}} \left(\frac{b_{qj}}{b_{q\cdot}} - \frac{b_{rj}}{b_{r\cdot}} \right)^2.$$

This expression is similar to the usual distance measures we have seen earlier, except that the term $1/b_{\cdot j}$, the column sum of the *j*th variable, is introduced as a weighting factor, which takes into account the scales for the variables.

Written in another way, we have that

$$d_{qr}^2 = \sum_{j=1}^{p} \left(\frac{b_{qj}}{b_{q\cdot}\sqrt{b_{\cdot j}}} - \frac{b_{rj}}{b_{r\cdot}\sqrt{b_{\cdot j}}} \right)^2, \qquad [6.2]$$

which shows that each of the axes of variables is stretched by a factor of $1/\sqrt{b_j}$ and the distances between objects are obtained from object points having as coordinates $b_{ij}/b_{i\cdot}\sqrt{b_{\cdot j}}$.

Distances between variables are defined, following an analogous development, as

$$d_{ij}^2 = \sum_{s=1}^{N} \left(\frac{b_{si}}{b_{\cdot}\sqrt{b_{s\cdot}}} - \frac{b_{sj}}{b_{\cdot j}\sqrt{b_{s\cdot}}} \right)^2. \qquad [6.3]$$

Again, dimensionality is reduced by 1 due to closure, and a weighting factor of $1/\sqrt{b_{s\cdot}}$ is employed.

The next step is to produce a simple cross-products matrix from which principal components may be extracted. The distances will remain unchanged when objects or variables are referred to the coordinate axes defined by the principal components. The covariance matrix for multinomial variables provides the key to the solution. Lebart and

Fénelon (1971) demonstrate that the covariance for variables i and j can be written

$$c_{ij} = \sum_{s=1}^{N} b_{s\cdot} \left(\frac{b_{si}}{b_{s\cdot}\sqrt{b_{\cdot i}}} - \sqrt{b_{\cdot i}} \right) \left(\frac{b_{sj}}{b_{s\cdot}\sqrt{b_{\cdot j}}} - \sqrt{b_{\cdot j}} \right). \qquad [6.4]$$

The terms $b_{\cdot i}$ and $b_{\cdot j}$ in [6.4] are the centers of gravity of variables i and j, respectively, for it will be observed that

$$\sum_{s=1}^{N} b_{s\cdot} \frac{b_{si}}{b_{s\cdot}\sqrt{b_{\cdot i}}} = \sqrt{b_{\cdot i}} \, ,$$

which is the multinomial mean of variable i.

As a consequence, [6.4] may be manipulated algebraically to yield a kind of modified covariance matrix, $\mathbf{C}$, of the form

$$\mathbf{C} = (c_{ij}) = \sum_{s=1}^{N} \frac{b_{si}b_{sj}}{b_{s\cdot}\sqrt{b_{\cdot i}b_{\cdot j}}} .$$

This is equivalent to the minor product moment of $\mathbf{W}$, defined in [6.1], or

$$\mathbf{C} = \mathbf{W}'\mathbf{W} = \left(\mathbf{D}_c^{-1/2}\mathbf{B}'\mathbf{D}_r^{-1/2} \right)\left(\mathbf{D}_r^{-1/2}\mathbf{B}\mathbf{D}_c^{-1/2} \right).$$

The eigenvectors of $\mathbf{C}$, used as coordinate axes, will preserve the distances between objects. As a consequence of the closure of the data matrix and then the centering of the variables to their mean values, one eigenvalue will be zero and its eigenvectors contain elements that are all equal to unity; it is therefore rejected.

To complete this discussion, we note that the major product moment of $\mathbf{W}$ is

$$\mathbf{G} = \mathbf{W}\mathbf{W}' = \left(\mathbf{D}_r^{-1/2}\mathbf{B}\mathbf{D}_c^{-1/2} \right)\left(\mathbf{D}_c^{-1/2}\mathbf{B}'\mathbf{D}_r^{-1/2} \right)$$

and

$$g_{qr} = \sum_{j=1}^{p} \frac{b_{qj}b_{rj}}{b_{\cdot j}\sqrt{b_{q\cdot}b_{r\cdot}}} .$$

Hill (1974) showed that correspondence analysis is, formally, a variety of principal components.

Derivation of factor-loading matrices

In the usual R-mode principal components analysis, the factor-loadings matrix $\mathbf{A}$ of a minor product moment matrix is obtained from its scaled eigenvectors. In the same way, we may obtain the factor loadings for the *R*-mode part of correspondence analysis as follows:

$$C = W'W \quad \text{and} \quad W'W = C = U\Lambda U',$$

where U is the matrix of eigenvectors of C, and Λ is a diagonal matrix of eigenvalues.

Then

$$AA' = U\Lambda U' \quad \text{and} \quad A = U\Lambda^{1/2}.$$

For the *Q*-mode part of the analysis, we can develop the derivation in the ensuing manner.

Let $G = WW'$ and $WW' = G = V\Lambda V'$, where V is the matrix of eigenvectors of G. (As illustrated in Section 2.12, the nonzero eigenvalues of the major product moment equal those of the minor product moment so that Λ in both of the preceding equations contains the same non-zero eigenvalues.)

If we let A^* be the desired matrix of Q-mode factor loadings, then

$$A^*A^{*\prime} = G \quad \text{and } A^*A^{*\prime} = V\Lambda V'$$

so

$$A^* = V\Lambda^{1/2} \tag{6.5}$$

This procedure would entail computing the eigenvalues of the N-by-N matrix G, a time-consuming operation on even a large computer for big samples.

We turn instead to the duality between V and U, mentioned in Section 2.12. For the major product moment we have that

$$WW' = V\Lambda V'.$$

Since V is orthonormal, this may be written as

$$WW'V = V\Lambda.$$

Premultiplying both sides of this equation by W' gives

$$W'W(W'V) = W'V\Lambda. \tag{6.6}$$

For the minor product moment we have that

$$W'W = U\Lambda U' \tag{6.7}$$

and $W'WU = U\Lambda$.

Comparing [6.6] and [6.7] shows that both U and $W'V$ contain eigenvectors of $W'W$. That is,

$$U = W'V.$$

Similarly, we can express V in terms of U to find that

$$V = WU\Lambda^{-1}. \tag{6.8}$$

By substituting [6.8] into [6.7] we obtain

$$\mathbf{A}^* = \mathbf{V}\mathbf{\Lambda}^{1/2} = \mathbf{W}\mathbf{U}\mathbf{\Lambda}^{-1/2} = \mathbf{W}\mathbf{A}\mathbf{\Lambda}^{-1}.$$

Thus, both R- and Q-mode loadings are obtained from factoring the minor product moment of the transformed data matrix.

The importance of the individual factors can be gauged by computing the ratio of each eigenvalue to the trace of $\mathbf{W}'\mathbf{W}$ as in the previous methods treated.

In order to express $\mathbf{A}$ and $\mathbf{A}^*$ in the same metric, the differential scaling of row and column vectors in $\mathbf{B}$ must be taken into account. We adjust the rows of $\mathbf{A}$ according to

$$\mathbf{A}_a = \mathbf{D}_c^{1/2}\mathbf{A} \quad \text{and} \quad \mathbf{A}_a^* = \mathbf{D}_r^{1/2}\mathbf{A}^*.$$

$\mathbf{A}_a$ and $\mathbf{A}_a^*$ are now in the common metric of the data matrix $\mathbf{B}$. Distances between objects in $\mathbf{A}_a^*$ are the same as those expressed previously. Moreover, it can be shown that the plot of each variable of $\mathbf{A}_a$ is located at its centroid. These features permit the simultaneous presentation of objects and variables as points on the same set of coordinate axes, namely, the factors of $\mathbf{A}_a$ and $\mathbf{A}_a^*$.

The usual interpretation of factor plots is, of course, possible. Objects situated near each other represent groups of similar individuals; variables that plot close together are interpreted to react similarly. However, because the positions of objects and variables are to some degree interrelated, the mutual dependencies can also be interpreted from the plots. Objects containing "large amounts of a variable" will be found to be clustered near the point representing that variable.

Example. The 10×10 matrix of artificial data (p. 142) employed previously for illustrative purposes, is now used to demonstrate the main steps in calculating the method described in this section.

Bearing in mind our observation concerning the need for using contingency-type data in correspondence analysis, we shall now suppose the artificial data to be scores on various properties of a sedimentary rock; for example, scores such as arise in point-counting work. We have therefore 10 such scored variables, observed on 10 samples. In order to obtain whole numbers, we would multiply the data matrix by 10. From the point of view of the calculations of the scaling procedure, this would, however, be immaterial, as any such initial scaling number cancels out.

One computes first the row sums and the column sums and notes these. These, together with the total sum of all the elements of the contingency table, as we now refer to our data matrix, are used to compute the similarity matrix of [6.6]. This is listed here (only the upper triangle is given).

	x_1	x_2	x_3	x_4	x_5	x_6	x_7	x_8	x_9	x_{10}
x_1	0.0677	0.1242	0.0964	0.0873	0.0746	0.0872	0.0676	0.0538	0.0411	0.0335
x_2		0.2350	0.1805	0.1542	0.1300	0.1685	0.1269	0.0990	0.0794	0.0693
x_3			0.1442	0.1271	0.1212	0.1412	0.1103	0.0859	0.0693	0.0578
x_4				0.1235	0.1220	0.1198	0.0989	0.0791	0.0595	0.0437
x_5					0.1554	0.1340	0.1171	0.0915	0.0730	0.0520
x_6						0.1543	0.1219	0.0924	0.0796	0.0680
x_7							0.0997	0.0763	0.0643	0.0519
x_8								0.0589	0.0485	0.0381
x_9									0.0423	0.0355
x_{10}										0.0328

Extraction of the roots of this matrix gives for λ_1 a value of 0.07881, and for λ_2 a value of 0.03398. There is actually a third root, but it is trivial, owing to the scaling procedure (cf. p. 173).

By using the scaling procedure outlined in the text, the following projections for the 10 variables were obtained:

	Factor 1	Factor 2		Factor 1	Factor 2
x_1	−0.329	−0.156	x_6	0.101	0.213
x_2	−0.366	0.029	x_7	0.246	0.090
x_3	−0.154	0.000	x_8	0.223	−0.008
x_4	−0.044	−0.299	x_9	0.281	0.239
x_5	0.478	−0.202	x_{10}	0.128	0.559

The projections for the 10 samples were calculated to be

	Factor 1	Factor 2		Factor 1	Factor 2
Rock 1	−0.117	0.355	Rock 6	0.270	0.057
Rock 2	−0.377	−0.273	Rock 7	−0.234	−0.134
Rock 3	0.528	−0.142	Rock 8	−0.066	0.067
Rock 4	−0.247	0.041	Rock 9	−0.144	−0.121
Rock 5	0.012	0.256	Rock 10	0.373	−0.105

The scaling procedure of correspondence analysis ensures that the lengths of the vectors representing variables and the lengths of the object vectors are on the same scale. It is therefore a simple matter to plot both of these sets of coordinates on the same figure, as in Fig. 6.1. The lengths of the variable vectors reflect their relative loadings and the nearness of two vectors is an expression of the correlation between them. Inspection of the plots for the variables indicates that x_1, x_2, and x_3 are more intimately correlated with each other than, say, they are

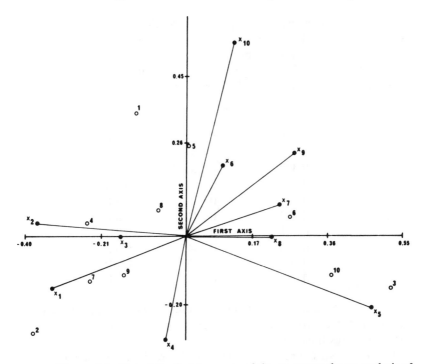

Figure 6.1. Plot of the first two axes of the correspondence analysis of the artificial data, regarded as scores on 10 characters of 10 rock specimens. The plot does not "distort the data." If you graphically rescale lines joining end-members (objects 1, 2, and 3) to form a triangular diagram, the input data of Table 5.I can be read off exactly.

with x_5. These are just examples of possible interpretations and many more are required to give a complete picture of the interrelationships in the material. One should also inspect the plots for clustering of the objects, and their ordination. Objects with high values on certain variables tend to be concentrated to the space occupied by those variables; this is perhaps the basic message conveyed by a correspondence analysis.

6.2 COMPARISON OF Q-MODE METHODS AND CORRESPONDENCE ANALYSIS

There is often difficulty associated with deciding which of several available methods is most suitable for a particular data set. We shall now compare the results of the Q-mode factor analysis, principal coordinate analysis, and correspondence analysis of the same data. We consider first relevant parts of the factor analysis. The trace of the matrix of cosine associations is 10 and there are three eigenvalues $\lambda_1 = 9.05$, $\lambda_2 = 0.74$, and $\lambda_3 = 0.21$. These eigenvalues show that the exact rank of the matrix is 3. The matrix of factor scores is

Factor 1	Factor 2	Factor 3
0.166	−0.140	−0.198
0.577	−0.614	−0.022
0.392	−0.126	−0.025
0.312	0.068	−0.598
0.337	0.666	−0.302
0.410	0.184	0.605
0.261	0.264	0.210
0.156	0.154	0.025
0.104	0.110	0.185
0.070	0.025	0.256

The three end-members are not obvious from an inspection of this matrix; however, rotation of the factor axes by means of the varimax method brings them out.

The exact structure is revealed by the oblique projection method. The oblique projection matrix, shown next, identifies objects 1, 2, and 3 as being the most divergent in composition, thus identifying them as the end-members sought (cf. Table 5.I).

	Factor 1	Factor 2	Factor 3
Object 1	1.000	0.000	0.000
Object 2	0.000	1.000	0.000
Object 3	0.000	0.000	1.000
Object 4	0.497	0.537	0.001
Object 5	0.847	0.000	0.207
Object 6	0.441	0.000	0.644
Object 7	0.200	0.753	0.098
Object 8	0.530	0.343	0.207
Object 9	0.206	0.668	0.201
Object 10	0.108	0.116	0.839

Taking into account the differing vector lengths and forcing the rows to sum to unity gives the "correct answer" exactly. This we leave as an exercise for you.

For this particular set of data, only the *Q*-mode method gives a true representation of what was entered. Correspondence analysis also gives a "correct answer" but in a form that requires considerable adjustment to obtain the real picture. Principal coordinate analysis yields a distorted answer owing to the use of coefficient [5.3]; we leave it to you as an exercise to prove that the Pythagorean distance in principal coordinates also gives an "exact plot." (The difference in locations of points in Figs. 5.2 and 6.1 is a result of the computer programs for eigenvalues and eigenvectors used, one giving the mirror image of the other.) The main strength of principal coordinates lies with the fact that this method allows one to mix quantitative, qualitative, and dichotomous data, using Gower's coefficient (1971). The attractive feature of correspondence analysis is that it permits a graphical display containing both the objects and the variables. A somewhat similar result can also be obtained (on unscaled axes) for principal coordinate analysis of the Pythagorean distances between objects and between variables.

Horseshoe plots

It is appropriate here to mention briefly a feature of some factor plots, namely, the "horseshoe effect," which can be rather frightening when first run into. In principal components, principal coordinates, Q-mode factor analysis, and correspondence analysis, scores on one of the axes may show up the existence of a quadratic relationship with scores of the other axes (usually, axis 2 is quadratically related to axis 1). This tends to be the case when correlations are high and approximately equal. The plot of the scores will form a quadratic cloud even although the points may be scattered. Hill (1974) proves that this situation must exist for equal correlations between variables and he gives a discussion of the literature on the subject. Other polynomial relations can exist between scores of other axes. The fact that scores based on eigenvectors can be interrelated in this manner is, perhaps, unexpected, as the statistical theory of factor analysis is based on the eigenvectors not being correlated. It is, however, possible to show algebraically, as does Hill (1974), that under certain conditions, identifiable interactions occur. This has been observed many times empirically, to which witness the terms applied, for example, Kendall's horseshoe, the Guttman effect, Goodall plot, and so on.

The occurrence of horseshoe plots and other geometrical figures tend to frighten people more than they should. There is nothing at all "wrong" about their turning up in an analysis, and they are quite interpretable. This has, however, led to a desire to detrend such data or in some other manner attempt to make them look like normal scatter plots. Reyment (1991) provides comments and references to this topic.

6.3 THE GABRIEL BIPLOT METHOD

In the First Edition, Gabriel's (1971) method of the biplot was not given much room. Since then it has proven to be far more important than could have been suspected at first. A succinct account of the procedure was given by Gordon (1981) on which the present section is based. The biplot gives a graphical description of

1. the relationships among p-dimensional observations
$$x_1, x_2, \ldots, x_N,$$

2. the relationships among the p variables.

The biplot utilizes the well-known result that any matrix $\mathbf{X}_{N \times p} = (x_{ij})$ of rank r can be nonuniquely decomposed as

$$\mathbf{X} = \mathbf{GH}', \qquad [6.9]$$

where matrices $\mathbf{G}$ and $\mathbf{H}$ are $N \times r$ and $p \times r$, respectively, of rank r.

Thus, a typical element x_{ij} is expressed as

$$x_{ij} = g_i' h_j \qquad (i = 1, \ldots, N; j = 1, \ldots, p).$$

Here, g_i' denotes the ith row of $\mathbf{G}$ and h_j is the jth column of $\mathbf{H}'$.

In the derivation of the biplot, matrix X is standardized so that the mean on each variable is 0. This maneuver produces deviation matrix $\mathbf{Y}$ (see Chapter 2 for a discussion).

The vectors g_i $(i = 1, \ldots, N)$ are assigned one to each row of $\mathbf{X}$ and the vectors h_j $(j = 1, \ldots, p)$ are assigned one to each column of $\mathbf{X}$. The set

$$\left\{ g_i, h_j \qquad (i = 1, \ldots, N; j = 1, \ldots, p) \right\}$$

gives a representation of $\mathbf{X}$ as $N + p$ vectors in r-dimensional space.

The visual impact of the biplot is greatest when $r = 2$. If $r > 2$, the singular value decomposition can be invoked so as to yield a matrix $\mathbf{Y}_2$ of rank 2, which is the best approximation to $\mathbf{Y}$ from the point of view that the sum of the squares of the elements of the matrix of differences $(\mathbf{X} - \mathbf{X}_2)$ is a minimum.

Let the singular value decomposition of $\mathbf{X}$ be, by [2.43],

$$\mathbf{X} = \mathbf{V\Gamma U}'$$
$$= \sum_{j=1}^{r} \gamma_{jj} v_j u_j'. \qquad [6.10]$$

Here $\mathbf{\Gamma}$ is a diagonal matrix with r nonzero elements, the positive singular value of $\mathbf{X}$.

Reverting to the situation for $\mathbf{X}_2$, a least-squares approximation is

$$\mathbf{X}_2 = \mathbf{V}_2 \boldsymbol{\Gamma}_2 \mathbf{U}_2'. \qquad [6.11]$$

The adequacy of the rank 2 approximation can be illustrated by formula [6.12] applied to the first two columns of the orthogonal matrices $\mathbf{V}$ and $\mathbf{U}$:

$$\sum_{j=1}^{2} \frac{\gamma_{jj}^2}{\Sigma_{j=1}^r \gamma_{jj}^2}. \qquad [6.12]$$

Hence,

$$\mathbf{V}_2' \mathbf{V}_2 = \mathbf{U}_2' \mathbf{U}_2 = \mathbf{I}_2. \qquad [6.13]$$

The indeterminacy in $\mathbf{G}$ and $\mathbf{H}$ (if, say, $\gamma_{11} = \gamma_{22}$) can be eliminated by the simple trick of replacing each $\mathbf{g}_i$ by a point located at the tip of the vector. A useful computational step is to make the distance between the ith and jth points in the plot equal to the "distance" between the objects located in the ith and jth rows of $\mathbf{X}_2$. This is done by defining $\mathbf{G}$ as follows:

$$\mathbf{G}\mathbf{G}' = \mathbf{X}_2 \mathbf{M} \mathbf{X}_2',$$

where $\mathbf{M}$ denotes the metric used on the rows of $\mathbf{X}_2$. Two special cases of $\mathbf{M}$ are

1. $\mathbf{M} = \mathbf{I}_2$, the identity matrix,
2. $\mathbf{M} = \mathbf{S}_2^{-1}$, where $\mathbf{S}_2$ is the sample covariance matrix for the reduced data matrix $\mathbf{X}_2$.

These stipulations ensure that the Euclidean distances between the points at the ends of each pair of $\mathbf{g}$-vectors corresponds to (1) Euclidean distance and (2) Mahalanobis distance between the corresponding objects in $\mathbf{X}_2$. The approximate biplot of $\mathbf{X}$ is then yielded by the *exact* biplot of $\mathbf{X}_2$.

From [6.11] and [6.13]

$$\mathbf{X}_2 \mathbf{X}_2' = \mathbf{V}_2 \boldsymbol{\Gamma}_2 \mathbf{U}_2' \mathbf{U}_2 \boldsymbol{\Gamma}_2' \mathbf{V}_2'$$
$$= (\mathbf{V}_2 \boldsymbol{\Gamma}_2)(\mathbf{V}_2 \boldsymbol{\Gamma}_2)'. \qquad [6.14]$$

By choosing

$$\mathbf{G} = \mathbf{V}_2 \boldsymbol{\Gamma}_2$$

and

$$\mathbf{H} = \mathbf{U}_2$$

it is ensured that

$$\mathbf{G}\mathbf{G}' = \mathbf{X}_2 \mathbf{X}_2'.$$

Table 6.I. *Data matrix for female diabetic mortality*[a]

Year	Mortality per million	Fiber (g/100 g)	Fat (lb/cap)	Sugar (lb/cap)	Protein (lb/cap)	Carbohydrate (lb/cap)	Calories (kcal/day)
1939	152	10	47	106	79	414	30.5
1940	153	10	43	80	80	390	28.9
1941	141	15	42	72	83	409	29
1942	130	68	41	72	87	403	29.3
1943	129	55	39	72	85	411	29.2
1944	122	45	41	77	86	429	30.6
1945	122	20	38	74	90	427	30.1
1946	111	61	37	82	88	421	29.4
1947	107	51	36	87	89	433	30
1948	98	48	41	88	87	438	31.2
1949	104	44	47	98	86	446	31.2
1950	109	42	48	90	87	419	31.2
1951	109	32	50	99	83	429	30.8
1952	98	28	45	94	82	424	30.3
1953	93	30	46	105	82	425	31
1954	86	18	49	112	82	431	31.9
1955	96	12	48	115	82	425	31.7
1956	92	12	48	116	84	422	31.7
1957	91	10	49	118	83	422	31.8

[a] Mortality (measured as deaths per million) due *directly* to diabetes; fiber as grams per 100 or crude wheat flour; lb/cap = pounds per capita per year.
Source: Trowell (1975, p. 231).

At this point the recommended procedure is to use the singular value decomposition to specify the plotting axes. The coordinates of the *g*-points will then display the projections of the original configuration on the best-fitting least-squares plane. By the same token, the configuration of the *g*-points would be yielded by principal component analysis of the covariance matrix, $\mathbf{X'X}/(N-1)$ [note that Gabriel (1971) divided by $\sqrt{N}$].

Example. Female diabetes mortality in England and Wales

The data analyzed here come from the reports of the Registrar General of the United Kingdom, Ministry of Agriculture, Fisheries and Food.

Diabetes mellitus is virtually absent in human communities living under natural conditions. Once such populations have become urbanized and learn to each junk food, diabetes shows up. The example presented here is for female diabetes mortality in England and Wales over the period 1939–57, an interval encompassing food rationing

Table 6.II. *Biplot results for the female diabetic mortality data*

Matrix H
The matrix for the variables

Mortality per million	− 0.4830	− 0.2273
Fiber	− 0.1985	0.4657
Fat	0.0762	− 0.0518
Sugar	0.3790	− 0.1336
Protein	− 0.0156	0.0620
Carbohydrate	0.2150	0.1778
Kcalories per day	0.0242	0.0000

Matrix G
The matrix for the units

1939	− 0.2305	− 0.7075
1940	− 0.5658	− 0.7306
1941	− 0.4463	− 0.4304
1942	− 0.5875	0.3836
1943	− 0.4882	0.2472
1944	− 0.2473	0.2295
1945	− 0.1814	− 0.1155
1946	− 0.2058	0.4819
1947	− 0.0286	0.4151
1948	0.1171	0.4465
1949	0.1998	0.3383
1950	− 0.0303	0.1599
1951	0.1350	0.0233
1952	0.1938	0.0422
1953	0.3322	0.0645
1054	0.5444	− 0.0595
1955	0.4619	− 0.2592
1956	0.4967	− 0.2479
1957	0.5336	− 0.2814

during the Second World War. Organized rationing began in 1940 and had the effect that it was associated with the greatest reduction in diabetic mortality since 1918.

Medical science has not been able to provide a rational explanation for the marked fall in diabetic death rates in 1941, since the decrease in energy input was only 5%, the fall in fat-intake 12%, and the drop in the consumption of sugar 25%. Trowell (1975) concluded that the obligatory increase in fiber-enriched foods can have been the cause. He suggested that fiber-depleted starchy foods, such as low-extraction white flour, are diabetogenic in susceptible genotypes, whereas fiber-rich foods tend to be protective.

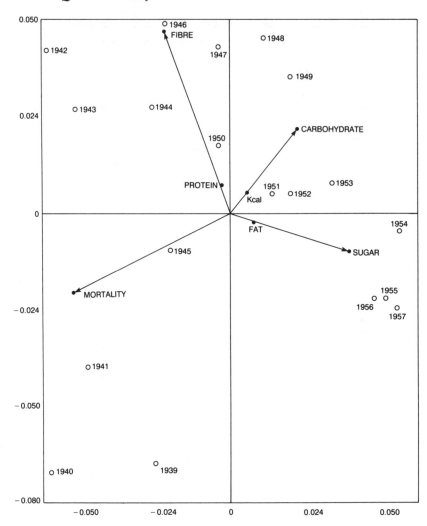

Figure 6.2. Biplot for the female diabetics. Note the clearly demarcated circular trend, beginning in the lower left quadrant with the first years of observation then continuing progressively through adjacent quadrants. There is a strong secular trend in the plot. Protein and fiber lie on the same vector, as do calorific level and carbohydrate, and fat and sugar on respective vectors.

The data comprise observations made over 19 years on mortality, fiber-rich flour, fat, sugar, protein, carbohydrates, and energy (as kilocalories).

The biplot method was applied to the data matrix listed in Table 6.I to yield the decomposition summarized in Table 6.II. The plot of these two sets of results is given in Fig. 6.2. Several interesting features are apparent. Variables located close to each other in a biplot are interpreted as being similar in behavior. In Fig. 6.2 we see that "kcal" and "fat" are comparable in behavior. The lengths of the arrows reflect the variance associated with a particular character. In Fig. 6.2, the greatest length is connected with "fiber," then "mortality," followed by "sugar," then "carbohydrates." The early periods of higher mortality fall near the location "mort," but there is a circular path into the next quadrant of the graph.

"Kcal" and "carbohydrate" lie on the same vector and "fat" is on the same vector as "sugar." "Protein" and "fibre" are in turn located on the same vector. These relationships are significant. They indicate that certain effects coincide and that some traits, for example, fiber and protein, operate along the same path.

6.4 CANONICAL CORRELATION ANALYSIS

Introduction

C. ter Braak (1986, 1987) has expanded the method of correspondence analysis by amalgamating it with canonical correlation analysis. The original name of his technique was canonical correspondence analysis but, in order to emphasize the fundamentally synecological (= community ecology) aspect of the synthesis, it is now named CANonical community analysis (for which the acronym is CANOCO).

In the past, various multivariate statistical techniques have been applied to synecological data, including principal components, principal coordinates, Q-mode factor analysis, correspondence analysis, and canonical correlation, but with varying levels of success. Canonical correlation was not included among the techniques reviewed in the First Edition, despite its undeniable connection to factor analysis. One reason for this omission was the difficulty attaching to the reification of the results.

Quantitative community (paleo)ecology has attained such importance over the last two decades that we deemed it to be of value to the users of this book to provide an introduction to the ideas involved. A useful reference for data analysis in community and landscape ecology is the text edited by Jongman et al. (1987). In the following brief review, the notation used by ter Braak in his article in that volume is retained.

Mathematical model

Consider n sites at each of which occurrences of m species (presence scored as "1", absence scored as "0") are observed and the values of q environmental variables determined ($q < n$).

Let y_{ik} be the abundance (presence or absence) of species k ($y_{ik} = 0$) and z_{ij} the value of environmental variable j measured at site i.

By the method known to ecologists as *indirect gradient analysis*, the main variation in the data for the species frequencies is summarized by *ordination*. The usually adopted procedure is "Gaussian ordination," which is done by constructing an axis such that the data optimally fit Gaussian response curves along that axis. The appropriate response model for the species is the bell-shaped (hence Gaussian) function

$$\mathscr{E}(y_{ik}) = c_k \exp\left(\frac{-\frac{1}{2}(x_i - u_k)^2}{t_k^2}\right) \qquad [6.15]$$

where $\mathscr{E}(y_{ik})$ denotes the expected value of y_{ik} at site i that has score x_i on the ordination axis, c_k is the maximum of the response curve for species k, u_k is the mode, or optimal value, of x, and t_k is the tolerance, an indicator of ecological amplitude.

It can be demonstrated that correspondence analysis approximates the maximum likelihood solution of Gaussian ordination if the sampling distribution of the species abundance is Poissonian under certain conditions pertaining to the parameters in [6.15] (e.g., ter Braak, 1986, p. 61).

The second step undertaken in indirect gradient analysis is to relate the ordination axis to the environmental traits. This is usually done by graphical means, by computing simple correlation coefficients, or by multiple regression of the site scores on the environmental variables as in Equation [6.16]:

$$x_i = b_0 \sum_{j=1}^{q} b_j z_{ij}, \qquad [6.16]$$

where b_0 is the intercept and b_j the regression coefficient for environmental variable j.

The technique of *canonical correlation analysis* simultaneously estimates the species optima, the regression coefficients, and, hence, the site scores using the model of Equations [6.15] and [6.16]. The transition formulas are as given in Equations [6.17]–[6.20]:

$$\lambda u_k = \sum_{i=1}^{n} \frac{y_{ij}x_i}{y_{+k}} \qquad [6.17]$$

$$x_i^* = \sum_{k=1}^{m} \frac{y_{ik}u_k}{y_{i+}}, \qquad [6.18]$$

$$b = (Z'RZ)^{-1}ZRx^*, \qquad [6.19]$$

$$x = Zb, \qquad [6.20]$$

where y_{+k} and y_{i+} are totals for species and sites, respectively; $\mathbf{R}_{(n \times n)}$ is a diagonal matrix with y_{i+} as the iith element; $\mathbf{Z}_{(n \times [q+1])}$ is a matrix holding environmental entries and a column of 1s; $b_{(1 \times q)}$ and $x_{(1 \times n)}$ are column vectors.

The eigenproblem resembles that of canonical correlation analysis (note that λ in [6.17] is an eigenvalue).

Just as in correspondence analysis, there is always one trivial solution in which all scores for sites and species are equal and $\lambda = 1$. This can be disregarded or excluded by centering the scores for sites to have mean zero.

The transition formulas [6.17]–[6.20] are solved by an iterative algorithm of reciprocal averaging and multiple regression. This is done in the following order:

Step 1. Begin with arbitrary, but *unequal*, initial site scores.

Step 2. Calculate scores for species by weighted averaging of the scores for sites (this is done by Equation [6.17] with $\lambda = 1$).

Step 3. Compute new site scores by weighted averaging of the scores for species (this is done by Equation [6.18]).

Step 4. Use Equation [6.19] to obtain regression coefficients by weighted multiple regression of the site scores on the environmental variables. The weights are the totals for sites, y_{i+}.

Step 5. Compute new site scores by Equation [6.20] or, equivalently, by Equation [6.17]. These new scores are, in effect, the fitted values of the regression of Step 4.

Step 6. Center and standardize the scores for sites such that

$$\sum_i y_i + x_i = 0,$$
$$\sum_i y_i + x_i^2 = 1. \qquad [6.24]$$

Step 7. Stop on reaching convergence, that is, when the new scores for sites are close enough to the site scores of the previous iteration. Otherwise, go back to Step 2 and repeat the sequence.

Ecologists will recognize the general format of this procedure as being like the reciprocal averaging technique of standard correspondence analysis.

Ter Braak classifies his synthesis as being a correspondence analysis with the restriction imposed by Steps 4 and 5 on the scores for sites. The final regression coefficients are called *canonical coefficients* and the multiple correlation coefficient of the final regression is called the *species–environment correlation* this correlation is a measure of how well the extracted variation in community composition can be explained by environmental variables. It equals the correlation between the scores for sites $\{x_i\}$, which are the weighted mean species scores of Equation [6.18], and the site scores $\{x_i\}$, each of which is a linear combination of the environmental variables (yielded by Equations [6.19] and [6.20]).

Ordination diagram. The primary goal of CANOCO is to produce ecologically instructive ordinations of the results. Sites and species are indicated by points on the diagrams and environmental variables are represented by arrows. The diagrams are to be understood in the following manner.

The plot of species and sites jointly depicts the overriding patterns in community composition to the extent that these can be explained by the environmental variables. The plot of species points and environmental arrows together reflects the distribution of each species along each of the environmental variables, just as in Gabriel's biplot. For example, an arrow pointing toward "silica content" groups about it the points for those species that occur preferentially at sandy sites. The length of an arrow for an environmental trait equals the rate of change in the weighted average as indicated by the biplot: It is therefore a measure of how much the species distributions differ along that environmental variable. Hence, an important environmental component will have a long arrow, whereas one of minor significance will have a short arrow.

Example The analysis of vegetation–environment relationships: the dune meadow environment

The data for the dune meadow environment of the Dutch Friesian Islands have been used in several instances to illustrate CANOCO (ter Braak, 1986; Jongman et al., 1987). Reyment (1991) applied CANOCO to oceanological data for planktonic foraminifers. The data were obtained from a project on the Friesian island of Terschelling. They are described in detail in Jongman et al. (1987, p. xvii). The object of the Terschelling project was to try to find a possible connection between vegetation and management in dune meadows where both farming and national park type milieus occur.

The environmental variables determined are the following:

1. thickness of the A1 soil horizon (a pedological convention);
2. moisture content of the soil (an ordinal variable);
3. grassland management type (four categories);
4. agricultural grassland use (expressed in ranking order);
5. quantity of manure applied (five classes, an ordinal variable).

Table 6.III. *Biplot scores of environmental variables*

N	Name	AX1	AX2	AX3	AX4
	R(SPEC.ENV)	0.9580	0.9018	0.8554	0.8888
1	A1	0.5629	−0.1732	0.5894	−0.1094
2	Moisture	0.9221	−0.1702	−0.1402	0.1696
3	Manure	−0.3092	−0.7646	−0.1972	−0.1803
4	Hayfield	−0.0756	0.6046	−0.2523	0.2821
5	Haypasture	−0.1719	−0.5535	−0.1320	−0.0864
6	Pasture	0.2795	−0.0312	0.4279	−0.2110
7	SF	0.1484	−0.6956	−0.4210	−0.0864
8	BF	−0.3645	0.1750	−0.0298	−0.5845
9	HF	−0.3611	−0.1161	0.4394	0.5224
10	NM	0.5704	0.7381	0.0008	0.0426

Thirty species of plants were studied; they are listed in Table 01 of Jongman et al. (1987).

Results. Table 6.III contains the biplot scores for the environmental variables, that is, the species environmental correlations. These results indicate that the measured environmental variables account for the main variation in the association of species in all canonical axes (the top row of the table). These are the intra-set correlations of the environmental variables with the first four axes. Table 6.IV lists the weighted correlation matrix for axes and environmental categories.

The information in Table 6.III is suggestive of a moisture gradient, as indicated by the correlations. The second axis is concerned with the effects of manuring; it separates the meadows included in a nature reserve from the farmed meadows. This relationship is clearly illustrated in Fig. 6.3, the ordination diagram for species and environmental variables (arrows). This is a biplot of the weighted averages of each species in relation to each of the environmental variables.

The ranking of the species in relation to, say, the arrow for manuring is obtained by projecting the points for each species onto that axis. Thus, the species *Cirsium arvense* (**Cir arv** in Fig. 6.3) occurs in highly manured meadows, whereas *Anthoxanthum odoratum* (**anto odo** in Fig. 6.3) is found in slightly manured meadows.

Generally, species that do better in a wetter environment group in the two right quadrants. Those that thrive under drier conditions are located in the two left quadrants.

The inferred weighted average is higher than average if a point lies on the same side of the origin as the head of the environmental arrow.

Table 6.IV. *Weighted correlation matrix (weight = sample total)*

	SPEC AX1	SPEC AX2	SPEC AX3	SPEC AX4	ENVI AX1	ENVI AX2	ENVI AX3	ENVI AX4	A1	Moisture	Manure	Hayfield	Haypasture	Pasture	SF	BF	HF	NM
SPEC AX1	1.0000																	
SPEC AX2	−0.0387	1.0000																
SPEC AX3	0.0773	−0.0400	1.0000															
SPEC AX4	−0.0506	0.1273	−0.1110	1.0000														
ENVI AX1	0.9580	0.0000	0.0000	0.0000	1.0000													
ENVI AX2	0.0000	0.9018	0.0000	0.0000	0.0000	1.0000												
ENVI AX3	0.0000	0.0000	0.8554	0.0000	0.0000	0.0000	1.0000											
ENVI AX4	0.0000	0.0000	0.0000	0.8888	0.0000	0.0000	0.0000	1.0000										
A1	0.5392	−0.1562	0.5042	−0.0972	0.5629	−0.1732	0.5894	−0.1094	1.0000									
Moisture	0.8833	−0.1535	−0.1199	0.1507	0.9221	−0.1702	−0.1402	0.1696	0.4154	1.0000								
Manure	−0.2962	−0.6895	−0.1687	−0.1603	−0.3092	−0.7646	−0.1972	−0.1803	−0.2283	−0.2204	1.0000							
Hayfield	−0.0724	0.5453	−0.2158	0.2508	−0.0756	0.6046	−0.2523	−0.1845	0.0251	−0.6118		1.0000						
Haypasture	−0.1647	−0.4992	−0.1129	−0.0768	−0.1719	−0.5535	−0.1320	−0.0864	0.1588	−0.1671	−0.6023		1.0000					
Pasture	0.2677	−.0281	0.3660	−0.1875	0.2795	−0.0312	0.4279	−0.2110	0.0210	0.1634	0.1231	−0.4096	−0.4816	1.0000				
SF	0.1421	−0.6273	−0.3601	−0.0768	0.1484	−0.6956	−0.4210	−0.0864	0.0768	0.1595	−0.4661	−0.1283	0.5600	−0.1083	1.0000			
BF	−0.3491	0.1578	−0.0255	−0.5195	−0.3645	0.1750	−0.0298	−0.5845	−0.3069	−0.3759	−0.1809	0.0277	−0.0512	0.0282	−0.2956	1.0000		
HF	−0.3459	−0.1047	0.3758	0.4643	−0.3611	−0.1161	0.4394	0.5224	−0.1444	0.1361	0.0857	−0.2581	0.2008	−0.4375	−0.3049		1.0000	
NM	0.5464	0.6656	0.0007	0.0379	0.5704	0.7381	0.0008	0.0426	0.3551	0.3641	−0.7422	0.3933	−0.2831	−0.1083	−0.3463	−0.2413	−0.3572	1.0000

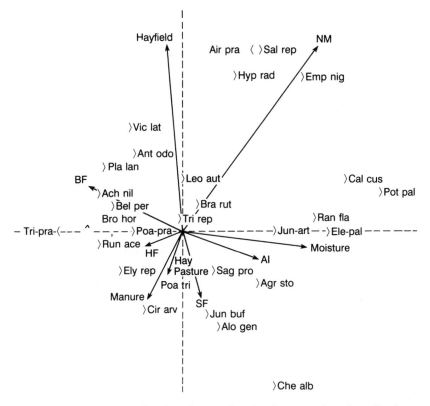

Figure 6.3. Ordination diagram for the dune meadow data. Environmental variables are indicated by arrows, species by dots.

It is lower if the origin lies between a point and the head of the arrow. The two species just referred to are an example of this.

Environmental variables with long arrows are more strongly correlated with ordination axes than are those with short arrows. Consequently, they are more closely bound to the pattern of variation in species composition exhibited in the ordination.

A complete canonical community analysis can comprise plots of species against sites, sites against environmental categories, and a triplot of species, environment, and sites on the same graph. All of these modes of presentation can be easily produced from the results yielded by the main analysis.

The computations were made using ter Braak's latest version of his program for CANOCO, now available on diskette for the PC. This diskette is distributed by Microcomputer Power, 111 Clover Lane, Ithaca, NY 14850.

7 Steps in the analysis

7.1 INTRODUCTION

The ever increasing availability of computer programs, personal computers, and computer facilities leaves no question that you can perform a factor analysis with little difficulty. Because of this ease, there is often a tendency to apply methods without fully understanding either the logic of the method or the more subtle question of what actually happens to the data at various stages during computation. These questions, although inextricably tied to the mathematics, can be more closely examined if one goes through the various stages in the analysis.

Few available texts on factor analysis focus their attention on these practical aspects. An exception is the book by Rummel (1970), in which factor analysis is presented as a problem in research design. We shall in this chapter outline points brought up by Rummel that are important to investigations in the natural sciences, seasoned with comments from our own experience.

7.2 OBJECTIVES

The goals of any research problem must be as precisely stated as possible, for it is from these specifications that both the characteristics of the problem as well as the methodology to be employed in its resolution emanate. The analysis may involve a simple summarization or delineation of properties, or, interrelationships between properties may need to be evaluated. Another type of problem may require the study of dependencies between the variables. Still another may be concerned with associations among objects and the existence of groups in the data, thence heterogeneity.

Finally, the objectives must specify whether the results are to be used for work outside the specific study. The following list of criteria suggests the general specifications that would justify the use of a factor-analytical approach.

1. Properties reflecting the underlying phenomena of interest can be expressed as quantities; they are observable and measurable.

194

2. Interrelationships between the variables or the objects is the principal aim in that such interrelationships may lead to an expression of inherent structure.
3. There is no prior partitioning of the data into sets of variables or objects. No particular variable, or set of them, can be isolated as being of any special importance in the analysis and there is, therefore, no question of predictability or dependency. Similarly, objects cannot be subdivided into mutually exclusive sets and there is, at this stage of study, no question of assessing similarities or differences between sets of objects.
4. The nature and number of underlying causal influences is unknown, poorly known, or known. In the first two cases, the objective is to gain some appreciation of this number by determining the basic dimensionality of the variables in question. In the third case, the problem may involve the testing of the hypothesis that this particular number of influences are governing the data under analysis.

7.3 CATEGORIES OF DATA

Cattell (1952) was among the first to analyze the specific nature of psychological data. He pointed out that any phenomenon can be described in terms of three fundamental properties or dimensions. The first of these defines the objects under study, the second, the variables, and the third, the time or occasion at which the phenomenon took place. We shall illustrate Cattell's ideas by recourse to a psychometrical example. The objects may be defined as male, first-year university students, the variables, their scores on various tests of aptitude, and the occasions, three or four testing periods, say, at the beginning, the middle, and at the end of the year. He illustrated this concept by means of a "data cube," such as we have depicted in Fig. 7.1.

With some minor modifications, this idea can be transferred to general studies. The object and variable dimensions of the data cube are largely defined by the nature of the scientific investigation. In studies of modern sedimentary processes, it may be possible to sample repeatedly at one particular site (the object) and measure its properties at different times, thus adding the dimension of occasion to the data.

In many scientific situations, especially in stratigraphy, the dimensions of object, or entity, and occasion may become blurred. If an entity is defined as a rock specimen, and specimens are collected at several stratigraphical levels, then each specimen can be thought of as representing a certain moment in time. We cannot, in this case, observe one specific entity over time. This distinction is important as it may well

DATA CUBE

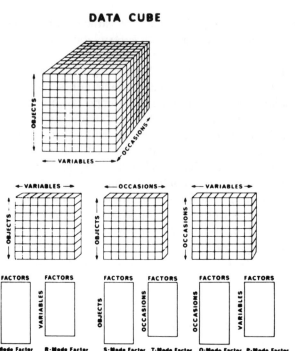

Figure 7.1. Cattell's data cube.

affect decisions on the choice of correlation indices. The two modes of factor analysis can now be more logically defined.

An interesting extension of factor analysis to three-way data has been made by Tucker (1966). In this book, however, we shall only consider analysis of one "slab" of the data cube of Fig. 7.1 at a time. Thus, a data matrix corresponds to any one slab of the data cube. There are three possibilities and, within each of these, two alternate modes of analysis.

1. The object-by-variable slab: In this situation, we direct our attention toward studying objects and their properties at a given point in time. The analysis will be confined to one of the slabs perpendicular to the axis for occasions. Now, there are two possible views of the data. An *R*-mode analysis unravels the interrelations between the variables. A *Q*-mode analysis resolves relations between objects.

2. The object-by-occasion slab: In this slice of the data cube, interest centers on one variable, measured on a number of objects located at various points in time. Again, one may

analyze the matrix in two manners. A "*T*-mode analysis," as it were, would reveal relationships between occasions, considering the varying amounts of one characteristic in the objects. An "*S*-mode analysis" would, conversely, attempt to identify those objects that behaved in a similar fashion during the period of study.

3. The variable-by-occasion slab: The data matrix here shows the variation of characteristics of one object over a period of time. An "*O*-mode analysis" would concentrate on finding relationships between occasions, whereas a "*P*-mode analysis" would illustrate the relations between the variables of one entity as they varied through time.

7.4 THE DATA MATRIX

The data matrix forms the basis of applied factor analysis. In it are quantified representations of the phenomenon of interest. It must contain sufficient information to portray adequately the problem under study.

Although it is true that the scientist is often severely constrained in the ability to organize and assemble data, owing to the difficulty of finding adequate material, and so forth, it is likewise true that some initial forethought will greatly increase the chances of obtaining representative data.

Choice of variables

The specifications of the problem should determine what properties are chosen to reflect the underlying processes. This is clearly a matter for professional judgment. Some guidelines exist, however. One may consider a universe of properties that completely describes the objects. With lithological samples, a complete description would comprise mineralogy, chemistry, texture, and other properties. Each of these categories consists of a great number of possible variables that might be measured. You cannot measure them all and so, in practice, some means of selection must be resorted to. Cattell (1952), supported by Rummel (1970), lists the following criteria for helping achieve full selection.

1. Random selection: If all the variables in the universe are known, random selection can be employed. Many textbooks in statistics can tell you how to do this.
2. Variables that reflect known or theoretical processes should be included. Thus, in a study of the chemical variation of soil,

elements known to be sensitive to pH, Eh changes should be included.

3. In some work, variables relevant to a hypothesis of interest may be possible to include. In a study of the chemistry of subsurface brines, Hitchon, Billings and Klovan (1971) chose variables that were relevant to the hypothesis that there was a relation between certain elements and osmotic diffusion through shale membranes.

4. The set of variables should embrace those that represent the entire range of variation of the phenomenon under study. In morphometrical studies of organisms, it is obvious that features relating geometrical aspects of form must be integrated with those reflecting function.

5. It is common practice in geology and chemistry to create compound variables from other variables by some arithmetic operation. For example, ratios of elements or compounds are often utilized as chemical indices. Combining these variables with those from which they were derived can lead to artificial factors that reflect nothing more than the manipulations on the data. Aitchison (1986) is an essential reference for this kind of problem.

Mixed populations

A topic that does not always appear to have been given full appreciation in the past concerns what we here call "mixed populations." This is particularly relevant in *R*-mode studies. Suppose that the object of the analysis is simply determining the pattern of covariation in the elements of a suite of subsurface brines. To establish a true picture, it is essential that the suite consist of waters that are homogeneous with respect to the processes that have affected them. If the set of water samples (objects) is an inadvertent mixture of two or more water types, then a confused pattern of covariation between the variables may emerge. It is therefore advisable to adhere to the following course of action to avoid running into a complication of this kind.

1. Analyze the frequency distributions of the variables to disclose the eventual presence of multimodality deriving from mixed populations. Dissection of cumulative frequency curves will supply information on the means and variances for each component population.

2. Preliminary *Q*-mode or cluster analysis will indicate heterogeneity in the data. If well-defined clusters are found, the data should be partitioned into these groupings prior to *R*-mode analysis.

Dimensions of the data cube

In assembling a data matrix for analysis, the question always arises as to the number of rows and columns of the data slice under consideration. In the case of the *R*-mode analysis of an object-by-variable slab, for example, the question would entail how many objects (= rows) are needed for a given number of variables (= columns). There are several reasons for wanting to know this.

1. The rank of a matrix cannot exceed the minimum dimension of the matrix; therefore, the number of factors that can be extracted cannot exceed this number. Thus, if fewer objects than variables are included, there will be fewer factors than variables.
2. In *R*-mode analysis, the number of objects must be sufficiently great to bring about stability in the variances and covariances (or correlations). Likewise, in *Q*-mode analysis, the number of variables must be great enough to provide stability in the distances or other measures of similarities between objects.
3. It is sometimes claimed (e.g., Rummel, 1970) that factor analysis will yield a description of variability in the data, regardless of the number of objects in relation to the number of variables. This is a truth requiring much modification. Consider the simple case of two variables. To describe any relationship between them would require at least two data points. A correlation computed from these two objects would be, naturally, 1. Although a result has emerged from the calculation, it has no practical meaning. The situation is more complicated for more variables, but the explanation remains essentially the same.

Data collection

Once the variables, objects, and/or occasions to be included in the analysis have been decided upon, the actual process of accumulating the observations commences.

The *scale* on which the variables are measured will have a profound effect on the resulting analysis. Siegel (1956) listed four scales of measurement, to wit:

1. The nominal scale, which is used to describe the presence or absence of some property, for example, the presence or absence of a posterior spine on an ostracod.
2. The ordinal scale, which is employed when various levels of abundance can be established for a variable. A nominal scale

can be used to record the direction or the increase. Thus, a certain rock component may be recorded as being rare, common, or abundant. Arbitrary numerical tags of 1, 2, and 3, respectively, describing these three attribute states would represent an ordinal scale. We point out that only the order is specified, not the magnitude.

3. Interval scale; when both the order and the magnitude of an attribute state can be determined relative to some arbitrary zero-value, the variable states can be expressed on an interval scale. Temperature scales are well-known examples, counts of some repeated organ are another (e.g., vertebrae).

4. A ratio scale is used if a variable state can be fixed with respect to order and magnitude to a fixed zero-value. This scale has the property that the ratios between scale values are meaningful as well as their sums and differences. Variables for length and weight, for example, are measured on ratio scales.

In factor analysis, one should keep well in mind that the scale on which a variable is measured determines the mathematical operations to which it can be submitted. The coefficient used to express the association between variables (R-mode) and objects (Q-mode) may be sensitive to the scale used, such as is the case for the usual correlation coefficient. The choice of scales will, clearly, affect the interpretation of factor results. Mixing of scales in a data matrix is usually to be avoided.

The *metric* used to measure variables is the units of measurement employed. On an ordinal scale, 1, 2, 3, or 1, 3, 5, or 25, 50, 100 might be arbitrarily utilized to denote the ordered position of three objects. Inasmuch as the ordinal scale is only concerned with the rank order, any of these three schemes will do. The interval variable, temperature, will give different factor results depending on whether Celsius or Fahrenheit was used.

Gordon (1981, p. 15) has given an enlightening account of the various categories of variables.

If the correlation coefficient is used, the variables will be re-metricized to zero mean and the units of measurement become the respective standard deviations.

Error in the observations can arise in several ways. Clerical errors and keyboarding mistakes are common, and you should make it a general part of your working routine to scrutinize carefully your data listings. Systematic errors can be generated by instruments or by systematic bias on the part of the data collector.

Atypical observations are common. Special methods for treating atypicalities are now available (see Section 4.5).

Transformations of data

Most statistical methods of data analysis require that the observations conform to some particular statistical distribution; this is usually the normal distribution. The discussion of this topic here is based on Rummel (1970) and Koch and Link (1970).

When the subject of transforming data is raised, an understandable reaction is not uncommonly that some trickery is being undertaken in order to make the data fit in with some preconceived notion. Let us look at some of the reasons why observations sometimes must be transformed. The unit of measure can seldom be justified on a-priori grounds. In the analysis of trace elements, to give a well-known example, amounts quoted in parts per million are no more reasonable than the logarithms of these amounts. Diameters of sedimentary particles can be expressed either in millimeters or as their logarithms. Experience has proved that the logarithmic values more closely reflect the distributions of the variables.

The correlation coefficient, which is at the heart of R-mode factor analysis, is adversely affected if the variables have highly skewed distributions. Another source of aberration is provided by nonlinearity in the relationships between variables. Tests for linearity should be used wherever there is the shadow of a doubt and graphical analyses of pairs of variables should form a routine first step in any analysis, as nonlinearity can usually be picked up by inspection of the plots.

Another problem is posed by "outliers," that is, rogue points that lie well away from the principal grouping and that may, with good reason, be suspected of being in error in some manner or other. Gnanadesikan (1974) has given this subject close attention. A single outlier can cause severe effects in a correlation matrix.

In Q-mode analysis, careful consideration must be given to the effects of the data metric and the distributions, as in unfortunate cases, the similarity measures can be adversely affected.

The choice of transformation

The correct choice of a transformation tends to be a rather hit-and-miss procedure. Kendall and Stuart (1958) provide several helpful tips. These tips may be summarized as follows: (1) examine the empirical frequency distribution of each variable separately; (2) depending on the shape of the curve, choose a transformation that will bring about a more symmetrical pattern; and (3) examine the results closely.

Two transformations are commonly applied to skewed distributions, all of which tend to pull in the tail of the distribution by reducing the arithmetic size of the intervals as the values increase. The logarithmic (natural or Briggsian) transformation is in wide use. The second transformation is the $x^{1/N}$-transformation, especially, the square-root member of the family. You need to keep in mind that in order to use the logarithmic transformations, all observations must be greater than 0, otherwise, a constant has to be added to all values.

Where a set of variables is to be transformed, the aim is usually to make all of them as nearly conformable as possible. In addition to the procedures briefly noted in the foregoing paragraph, some situations may warrant one of the following methods.

Normalization. This involves changing the values of a vector of observations so that the sum of its squared elements equals 1. For data matrices, one may normalize columns or rows. By rescaling each variable, a column of the data matrix, to a sum of squares of 1, one is, geometrically, "stretching" or "compressing" the vectors so that all have unit length. For rows, the normalization makes the sum of squares for each object equal to 1. For the variables, the ratios of numbers within any one vector are not changed, although the proportional relationships along rows are altered. For the objects, the proportionality of variables within an object is unaffected, but the differences in magnitude between objects are destroyed.

Range restriction. When the range of values is scaled to lie between 0 and 1, the proportionality within rows and columns is destroyed. Variables with an originally small range of values will have a similar influence in determining the correlation, or other similarity coefficient, as variables with a large range. For example, in the mineralogical analysis of a sediment, garnet may display a range of from 0 to 3%, whereas quartz may range from 10 to 80%. The effect of the range transformation is to equate very small changes in garnet content to large shifts in the quartz content. In fact, one unit of garnet content will, in this case, be equated to 23 units of quartz content.

Centering. This transformation, as its name implies, consists of shifting the origin of the observations, usually to the mean. It is often difficult to justify row centering. If variables are expressed in different scales and metrics, what does a mean of an object signify? One instance where such a transformation might apply is for a matrix of grain-size analyses where the variables represent the amount of sediment in each size fraction. If row-centered, the mean differences in grain size would be removed. Double centering refers to an operation in which rows and columns are centered.

Standardizing. This is the most commonly employed transformation in *R*-mode analysis. For a variable x_i, the transformation is

$$z_i^* = \frac{x_i - \bar{x}_i}{s_i},$$

where $\bar{x}_i$ and s_i are the mean and standard deviation of variable i. This transformation combines the effects of column normalization and column centering, and each variable will have a mean of zero and it will be expressed in units of standard deviation. The transformation is appropriate in *R*-mode studies when variables are measured on different scales and in different metrics. For example, if one variable is the strontium content of a rock expressed in parts per million and another is percent calcium carbonate. This transformation causes variables with a comparatively small variance to be preferentially weighted in relation to those with greater variances.

Column standardization is invoked automatically when the correlation coefficient is used in *R*-mode analysis. It can be applied in *Q*-mode analysis as a preliminary step to remove differences in means and standard deviations. Double standardization obviously involves rows as well as columns. The only variance remaining is within objects and it can be used in order to study interrelationships within, rather than among, objects or variables.

Multivariate normality

As theory indicates and demonstrated empirically by Reyment (1971), variables which, when adjusted to be univariate normal comply with the requirements of this distribution, when tested by Mardia's (1970) procedures, often turn out to be multivariate non-normal. The problem is very complex and cannot be treated in this book. We refer you to Gnanadesikan (1977), Mardia (1970), and Reyment (1971). Another important reference is Campbell (1979).

7.5 SELECTION OF THE MEASURE OF ASSOCIATION

A key step in factor analysis is the selection of a mathematical description of the "relationship" between variables in the *R*-mode and objects in the *Q*-mode. Coefficients may be selected to emphasize certain aspects of the variability within the data. It is well to remember that most of these measures are essentially major or minor products of the data matrix or a transformation of it. Moreover, many of them have some of the transformations of the preceding section baked into them. Lamont and Grant (1979) reviewed the field of similarity coefficients.

R-mode coefficients

We mention here three *R*-mode coefficients. The first of them is the *raw cross product* (Section 2.8). This is seldom used but might conceivably be of interest if all the variables were expressed in compatible units and had a common origin. The second coefficient is the *covariance* (Section 2.8). This is appropriate when the units of measure are compatible and one wishes to retain the information deriving from the deviations from the means; this measure is often useful in morphometrical studies. The third, and most popular, measure is the *correlation coefficient* (Section 2.8). As already mentioned in Chapter 2, the correlation coefficient measures the degree of linear relationship between two variables with the mean differences subtracted and the metric of the variables expressed in units of standard deviation, thereby ensuring their equal weighting. An alternative interpretation of the correlation coefficient is that it is the cosine of the angle between two variables, expressed as vectors in object space. In this interpretation, the column standardization brings about a shift of the origin of the vectors to the centroid of the object space and a scaling of each vector to unit length.

Q-mode coefficients

A great number of similarity coefficients have been proposed, thus attesting to the difficulty in defining this concept in suitable mathematical terms. Very little work has been done by professional mathematicians on the problem. In this book, we restrict ourselves to a few of these measures. A full account is given in Sneath and Sokal (1973).

Major product moment.   The matrix of cross products between specimens may be used to express similarities between objects. It contains the effects of overall size differences between objects as well as differences due to varying proportions of the constituent variables. The factors deriving from this coefficient are difficult to interpret.

The cosine association measure.   This coefficient was introduced into geological work by Imbrie and Purdy (1962). It assesses the similarity between objects solely on the proportions of the variables measured. In terms of the matrix operations of Section 2.6, the coefficient is computed as the major product moment of the row-normalized data matrix. The formula is given in Sections 2.4 and 2.6.

Euclidean distance measures.   The idea of using the distance between two objects situated in *p*-dimensional space has long been of interest in *Q*-mode analysis. Such distances are used in principal coordinates and correspondence analysis (see Sections 5.2 and 6.1).

Gower's similarity measure. Gower (1971) proposed a general coefficient of similarity for similarity between two sampling units. This matrix is positive semidefinite and was designed to cope with a hierarchy of characters. It is therefore possible to combine in the one analysis quantitative, dichotomous, and qualitative characters.

7.6 CHOICE OF THE FACTOR METHOD

The various *R*-mode factor methods available that we believe are suitable for scientific problems were outlined in Chapter 4. We shall review here some of the features of these so that an initiated choice between them can be made. Some of these methods differ mainly with respect to the method used for scaling the observations.

Most applications referred to as being factor analysis have made use of the principal components (Solution 1 of Chapter 4) of a correlation matrix, subsequently rotated by a criterion such as varimax. In these applications, only the first few factors, those explaining a major part of the total variance, are retained as meaningful by most workers; the remaining factors are often taken to represent error and other noninterpretable variability. Following on results by mathematical statisticians, a few geostatisticians also study the factor associated with the smallest variance, that is, the loadings that do not vary from specimen to specimen, as this factor can yield useful information on constant relationships between the variables. This interpretation is relevant in taxonomical and geochemical work.

In *Q*-mode studies, different criteria come to the fore. If your interest lies mainly with the graphical appraisal of your data by means of a distance-preserving method, you would be advised to choose principal coordinate analysis. This method has the added advantage that it can be related directly to discriminant analysis by the Mahalanobis generalized distance (Gower, 1966b) and hence to the method of canonical variates. It can therefore form a natural link in a chain of multivariate analyses. *Q*-mode factor analysis is indicated if you wish to test various hypotheses about your data, such as is possible in the Imbrie method of analysis. If you are interested in the graphical relationships between data points in various projections and the variables measured on them, and if your data can be said to form a real contingency table, then correspondence analysis may be the correct avenue of approach. Further aspects of what the methods do are taken up in Chapter 8.

7.7 SELECTION OF THE NUMBER OF FACTORS

Two important concerns in the practical application of factor analysis are, as shown in Chapter 3, the two "unknowns," the communality of the variables, and the "correct" number of factors. As we have shown in Chapters 3 and 4, the first of these problems can be dealt with by the suitable selection of the factor method, whereas the second one is not so easily dispensed with. As you have seen, these two unknowns are closely bound to each other in that the one affects the other.

There is at least one situation in which neither of these concerns is relevant. If the objective of the analysis is merely to create new, uncorrelated variables, a principal component analysis with as many factors as there are variables will do the job, provided communalities consist of total variances (or 1s, in the case of the correlation matrix).

Theory tells us that the rank of the data matrix will supply us with the minimum number of factors necessary to describe the data. However, in practice, the rank of the data matrix is almost invariably equal to its smallest dimension. So we must be content either to accept no simplification whatsoever, or a certain loss of information by an approximation.

Experience in interpreting the results of factor analysis has led to many suggestions for choosing the optimal number of factors. We discussed the question in Chapters 3 and 4. Next we summarize some of the suggestions of Chapter 4 plus ideas gleaned from textbooks on factor analysis. Some of these may turn out to be useful to you.

1. The "amount" of the cumulative variance explained: Some workers suggest that this be set at 95 or 99%, particularly in components analysis.
2. The variance explained by a factor: The sum of squared loadings for a given factor represents the information content of that factor. The ratio of the sum of squares to the trace of the correlation matrix, or whatever association matrix is being used, gives the proportion of total information residing in the factor.
3. The sizes of factor loadings: The factor loadings squared, for orthogonal factors, indicate the variance accounted for of any variable (in R-mode) by a particular factor. Factors not contributing much may be dropped. As an empirical rule, we propose that if you do not have at least three significant loadings in your factors, you have probably gone too far in taking out factors.
4. The "profile" of the eigenvalues, that is, a plot of the total variance explained against factor number, can be used as an indication of the correct number of factors by taking the point at which the graph flattens out to indicate the diagnostic number of factors.

5. The significance of the residuals: A residual correlation matrix may be calculated after each factor has been extracted. The point at which no further factors need be extracted is reached when the residual matrix consists of correlations solely due to random error. The standard error of the residual correlations, estimated roughly as $1/\sqrt{N-1}$, can be used to decide the cutoff for factoring, this being when the standard deviation of the residuals becomes less than the standard error.

6. Some people prefer to stop factoring a correlation matrix when the eigenvalues become less than 1, a procedure due to Guttman (1954). This criterion is not regarded as very sound by some workers.

7. Mention has already been made of the role of significance testing in factor analysis, particularly as regards the number of significant eigenvalues. Suffice it here to say that these tests are of little practical value unless all the assumptions upon which they have been based are fulfilled. They are important, however, from the point of view of the completeness of the statistical theory of factor analysis. Further methods are taken up in the section on cross-validation (Section 4.5).

Most of these criteria apply to the unrotated factor matrix. It is recommended that whatever criterion, or combination of criteria, is selected, more, rather than fewer, factors are chosen initially. Most factor studies pass through to the rotational stage and if too few factors are picked at the start, severe distortion of the data will result. Assuming that you have been generous in choosing factors at the beginning of the analysis, you can discard the last of them if, after rotation, it does not meet the requirements for retention. This procedure can be repeated until a logically acceptable result has been obtained.

Therefore, in summary, we recommend that you follow the following course of action when embarking upon the selection of the optimal number of factors.

> Extract as many factors as necessary to account for a minimum of 95 to 99% of the total variance. Usually this decision will roughly coincide with the "knee" formed by the plot of the size of the eigenvalues against their order.
>
> Rotate this number of factors by means of an orthogonal rotation procedure such as the varimax method.
>
> Use the various rules of thumb previously set out on this matrix. If the last factor does not meet the requirements for inclusion, delete the last unrotated factor and reenter the rotational procedure.
>
> By iterating the first two steps, you will be able to compare the results of repeated rotation with a decreasing number of

factors and so select the matrix that has no trivial factors remaining and yet has not undergone any major distortion due to your having used too few factors.

7.8 ROTATION OF FACTOR AXES

One of the aims of factor analysis is to display the configuration of variables or objects in as simple a manner as possible. In Chapters 3 and 4, models were developed that essentially determine what aspects of the data are to be considered. The eigenvalue technique was then used to find a set of axes to which variables and objects could be referred. Two points should be brought to mind at this juncture.

1. The number of reference axes necessary to display the data will be determined by the approximate rank of the data matrix.
2. The axes are located according to some mathematical criterion. In the principal components method, the first factor is located in a manner such that the variance connected to it is a maximum, the second factor accounts for the maximum remaining variance at right angles to the first, and so on. Thus, the initial solution defines the dimensionality of the data and provides a uniquely positioned set of factors.

Although mathematically unique, they are only one of an infinite number of sets of factors that will describe the configuration of the data just as well. In Fig. 7.2, eight hypothetical variables are referred to two factors derived by the extraction of the eigenvectors of a matrix. The factor loadings are the projections of the variables onto the factors. In Fig. 7.3, the factors have been rotated rigidly through 45°. The configuration of variables has, of course, not changed and the new factor loadings, the projections of the variables onto the new axes, describe their positions as well as do those of the unrotated case. It is obvious that the factors could be rotated to an infinite number of positions without negating the preceding statement.

This indeterminacy of a "correct" factor solution has been much debated in the past. Some claim that the solution by eigenanalysis, giving a mathematically well-defined set of axes as it does, should be used as the final reference scheme; others choose to rotate the axes to positions they hope will give a meaningful perspective to the data. There are several compelling reasons for proceeding to a rotated solution.

By the spectral character of the eigenstructure of a matrix, the variance of the variables is spread across all factors. The first factor will be a weighted average of all the variables present. In Fig. 7.2, Factor 1

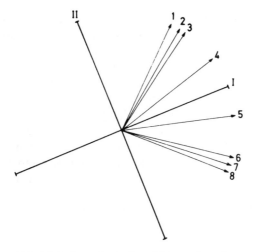

Figure 7.2. Eight hypothetical variables referred to two factors.

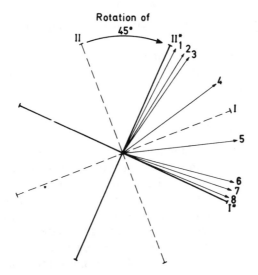

Figure 7.3. Rotation of the factors of Fig. 7.2 through 45°.

lies near the center of gravity of the vectors of the variables. The second and succeeding factors will tend to contain about equal loadings of opposite sign. The factors thus do not separate out independent clusters of variables (or objects in the Q-mode). By rotating the axes, we try to place the factors so that each contains only a few highly loaded variables, whereby the factors become combinations of interdependent variables only, as in Fig. 7.3. It is difficult to find an objective definition of where the exact position of the factors will be in order that they will satisfy this requirement. However, the concept of simple structure put forward by Thurstone (1935) goes a long way toward doing this.

The positions of the unrotated factors are strongly dependent on the particular set of variables used in the analysis. Should even one variable be left out, the factor positions will almost certainly change. In order to facilitate comparison between factor studies, and to find factors that may be related to important underlying influences, it is desirable to produce "invariant" factors. In other words, the factorial description of a variable must remain the same when it is a part of one set or another that involves the same factors (Harman, 1960). Rotated factors, whose positions are determined by clusters of variables, approximate factorial invariance to a high degree (Kaiser, 1958).

The geometrical implications of rotation have already been reviewed in Chapter 2. This treatment showed that the rotation of the factor axes can be thought of as a problem in rigid rotation, that is, we may consider

$$\mathbf{B} = \mathbf{AT},$$

where $\mathbf{A}$ is the matrix of unrotated factor loadings, $\mathbf{B}$ is the matrix of rotated factor loadings, and $\mathbf{T}$ is the transformation.

The objective of the rotational procedures to be discussed in this section is to find a matrix $\mathbf{T}$ that yields some matrix $\mathbf{B}$ with certain desirable properties.

Simple structure

Mathematically, it is a simple matter to rotate one set of axes to another set, provided that the angle of rotation is given. This is, however, not so when the problem is to better the position occupied by the factor axes and the major task is then to determine the angle through which the axes must be rotated. In order to do this, the aims of the rotation must be clearly stated and formulated in mathematical terms. Historically, Thurstone stated the objectives in 1935 and 1947, but the analytical means for the task were not firmly established until the mid-1950s.

Thurstone's criteria for establishing a simple structure solution are as follows:

1. There should be at least one zero in each row of the factor-loadings matrix.
2. There should be at least k zeros in each column of the factor matrix, where k is the number of factors extracted.
3. For every pair of factors, some variables (in the R-mode) should have high loadings on one and near-zero loadings on the other.

4. For every pair of factors, several variables should have small loadings on both factors.
5. For every pair of factors, only a few variables should have nonvanishing loadings on both.

These criteria were worked out in order to achieve a simple solution, "simple" here meaning that as many near-zero and high loadings as possible are produced. This implies, that with a simple-structure solution, a variable will be loaded heavily on one or, at least, a small number of factors only. At the same time, each factor will have only a few variables highly loaded on it and the factors are therefore specifically related to clusters of interdependent variables. This is in stark contrast with the unrotated solution, where factors are spread across all the variables and the first factor is usually a general factor containing significant contributions from all variables.

Thurstone's criteria were primarily developed on the premise that the factors should correspond to underlying causal influences and as such, these influences should not necessarily affect all the variables of the analysis. Thurstone's principle is also valid for many situations, although it was originally developed for psychometrical work. Given a set of variables, the real meaning of which is in doubt, it is surely important to isolate those subsets of variables that react similarly to the same causal influence. In a sedimentological study, for example, certain attributes of the sediment will be interrelated owing to the fact that they are dominated by physical agencies. Other attributes may reflect the chemistry of the environment, whereas others may be partially dependent on both. The simple-structure solution attempts to portray these relationships, whereas an unrotated solution does not.

Analytical rotation

Translation of Thurstone's qualitative criteria into precise mathematical terms took several years and the efforts of many workers. As a consequence, numerous solutions were developed; these are explained in detail in Harman (1960). We shall restrict ourselves to two methods that appear to fulfil Thurstone's requirements and are widely used in most computer programs for doing factor analysis.

The first of these methods, the varimax procedure, was designed to bring about orthogonal, or rigid, rotation of the unrotated factor axes. As with the unrotated axes, the varimax axes are consequently orthogonal to each other. The second method, the promax procedure, gives an oblique solution. The factors are positioned in such a manner as to yield simple structure but they are no longer orthogonal; the factors themselves will therefore be correlated.

The varimax procedure

Kaiser (1958) developed what is perhaps the most popular of all the analytical rotational procedures. His method entails the simplification of the columns of the unrotated factor matrix. A "simple" factor would be one having a few high loadings and many zero, or near-zero, loadings. This notion can be interpreted in relation to the variance of the factor loadings. Clearly, when the elements of the vector of loadings for a given factor approach ones and zeros, the variance of the factor will be at a maximum.

To avoid complications due to signs of the factor loadings, the variance of the squared factor loadings is used. Hence, the variance of the squared factor loadings for the jth factor is given by

$$s_j^2 = \frac{1}{p} \sum_{i=1}^{p} \left(b_{ij}^2\right)^2 - \frac{1}{p^2} \left(\sum_{i=1}^{p} b_{ij}^2\right)^2. \qquad [7.1]$$

When the variance is at a maximum, the factor is said to have achieved simplicity. For the entire matrix of factor loadings, simplicity is attained when the sum of each individual factor variance, s_T^2, is at a maximum

$$s_T^2 = \sum_{j=1}^{k} s_j^2 = \frac{1}{p} \sum_{j=1}^{k} \sum_{i=1}^{p} b_{ij}^4 - \frac{1}{p^2} \sum_{j=1}^{k} \left(\sum_{i=1}^{p} b_{ij}^2\right)^2. \qquad [7.2]$$

Kaiser (1958) suggested that each row of the matrix should be normalized to unit length before this variance is computed. Since the sum of squared elements of a row of the factor matrix is equal to the communality of the variable, the normalization can be easily carried out by dividing each element in a row by the square root of the associated communality (cf. [3.11]). After rotation, the rows are brought back to their original lengths.

The final quantity to be maximized for producing simple structure via the varimax procedure is then

$$s_v^2 = p^2 s_T^2 = p \sum_{j=1}^{k} \sum_{i=1}^{p} \left(\frac{b_{ij}}{h_i}\right)^4 - \sum_{j=1}^{k} \left(\sum_{i=1}^{p} \frac{b_{ij}^2}{h_i^2}\right)^2, \qquad [7.3]$$

where we write, for reasons of clarity, h_i^2 for the communality.

For any pair of factors, j and l, the quantity to be maximized is

$$s_{(v_{jl})}^2 = p \left[\sum_{i=1}^{p} \left(\frac{b_{ij}}{h_i}\right)^4 + \sum_{i=1}^{p} \left(\frac{b_{il}}{h_i}\right)^4 \right] - \left[\sum_{i=1}^{p} \left(\frac{b_{ij}^2}{h_i^2}\right)^2 - \sum_{i=1}^{p} \left(\frac{b_{il}^2}{h_i^2}\right)^2 \right]. \qquad [7.4]$$

To maximize [7.4], we rotate factor axes j and l through some angle θ_{jl} chosen such that [7.4] is a maximum and leave all other factor axes unchanged. By repeating this procedure on pairs for all possible combinations of factors, [7.3] will be maximized.

Finding the required angle θ_{jl} that will maximize [7.4] involves the following steps.

We remind you first of all that rigid rotation of axes is done by the following process (cf. Section 2.9):

$$b_{ij} = a_{ij} \cos \theta_{jl} + a_{il} \sin \theta_{jl},$$
$$b_{il} = a_{ij} \sin \theta_{jl} + a_{il} \cos \theta_{jl}, \qquad [7.5]$$

where a_{ij} and a_{il} are the loadings of variable i on unrotated factors j and l, and b_{ij} and b_{il} are the respective loadings on the factors, rotated through θ_{jl} degrees.

To maximize [7.4] we substitute these expressions for the b's into [7.4] and we differentiate with respect to θ_{jl}. Setting the derivative to zero and solving for θ_{jl} yields the angle through which factors j and l must be rotated so as to maximize [7.4]. The actual expression used to determine θ_{jl} is given by Harman (1960). Kaiser (1959) has given all the steps required for programming the procedure.

Determination of θ_{jl} for each of the possible pairs of j and l factors is called a cycle. The matrix of rotated loadings after a cycle can be obtained from

$$\mathbf{B} = \mathbf{AT}_{12}\mathbf{T}_{13} \cdots \mathbf{T}_{jl} \cdots,$$

where $\mathbf{T}_{jl}$ represents the transformation matrix of the form [7.5], derived from the rotation of the factors j and l. The cycle is repeated and a new value s_v is computed that will be as large or larger than that obtained in the previous cycle. This process is continued until s_v converges to a stable value. The final transformation matrix can be viewed as an operator that transforms the unrotated factor matrix $\mathbf{A}$ into the varimax factor matrix $\mathbf{B}$.

Example of varimax rotation

In Table 7.I, the unrotated factor-loadings matrix for a principal components solution of four variables onto the principal factors are listed. Note that the sum of the squared elements of any row is the communality of the variable representing that row. In addition, the sum of the squared elements in a column is the variance along the factor representing that column. This latter sum of squares, divided by p, the number of variables, is the percentage of the total variance accounted for by the factor, provided that the variables are standardized. It is

Table 7.1. *Principal component factor matrix for the varimax example*

Variables	Communality	Factors		
		1	2	3
1	0.9974	0.4320	0.8129	0.3872
2	0.9911	0.7950	−0.5416	0.2565
3	0.9950	0.5944	0.7234	−0.3441
4	0.9986	0.8945	−0.3921	−0.1863
Variance		49.302	40.780	9.222
Cum. var.		49.302	90.082	99.304

Table 7.II. *Transformation matrix for the varimax example*

0.8417	0.3299	−0.4273
−0.5301	0.6552	−0.5382
0.1024	0.6796	0.7264

Table 7.III. *Varimax factor loadings matrix*

Variables	Communality	Factors		
		1	2	3
1	0.9974	−0.0276	0.9383	−0.3409
2	0.9911	0.9825	0.0818	0.1381
3	0.9950	0.0816	0.4363	−0.8933
4	0.9887	0.9418	−0.0884	−0.3066
Variance		46.491	27.132	25.681
Cum. var.		46.491	73.622	99.304

evident that the first factor is tied to most of the variance and the remaining factors account for steadily diminishing amounts thereof.

The transformation matrix derived from maximizing [7.3] is given in Table 7.II and the rotated varimax factor loadings matrix is presented in Table 7.III. Several properties of this matrix should be noted:

1. The communalities of the variables are identical to those in the matrix of unrotated loadings; communalities are invariant under rigid rotation.
2. The major product moment of the varimax factor-loadings matrix is the same as the major product moment of the unrotated factor-loadings matrix. This implies that the transformed factors are mutually orthogonal. If $\mathbf{B} = \mathbf{AT}$ with $\mathbf{T}$ orthonormal, then $\mathbf{BB'} = \mathbf{ATT'A'} = \mathbf{AA'}$.

3. The sum of squares of a column, divided by p, still represents the percentage of total variance accounted for by a factor. Observe, however, that the variance has been more evenly distributed among the factors than was the case for the unrotated factors. The total variance has been divided into orthogonal components.
4. The distribution of the variable loadings approaches the ideals of simple structure.

Oblique rotation

Oblique rotation refers to the relocation of factor axes in order to arrive at simple structure, but without the constraint that the factors be orthogonal. Thurstone (1947) strongly advocated the use of oblique factors because in most practical situations, orthogonal factors did not seem to him to meet the criteria of simple structure as well as did oblique factors.

A more fundamental reason for using oblique factors is that it is unlikely that the underlying causal influences, which the factors may represent, would be entirely independent of one another in nature. In most natural processes, it will be found that the factors directly influencing the measurable variables are themselves linked by higher-order effects. In an ecological study, for example, the abundance of a particular organism may be directly governed by such influences as salinity, temperature, competition between species, and so on. Some of these, such as temperature and salinity, may in turn depend on depth. Hierarchies of causality have been the subject of many publications. By letting the factors be oblique, the correlations between them can be examined and the so-called higher-order factors established by factoring the matrix of correlations between factors.

We develop the subject for the R-mode in the following text; what we say applies also to the Q-mode.

Geometrical considerations of oblique rotation

In Fig. 7.4, we illustrate a configuration of eight hypothetical variables whose coordinates are given with reference to two orthogonal factors, I and II. A simple-structure configuration will be obtained if the axes are rotated to the positions shown in Fig. 7.5, whereby each factor is clearly related to one group of interdependent variables. Here, the factors are no longer orthogonal to each other, a state of affairs that complicates concepts, terminology, and the interpretation of resulting matrices.

Ignoring for the moment the way in which the factor positions are located, let us examine some of the properties of the oblique solution.

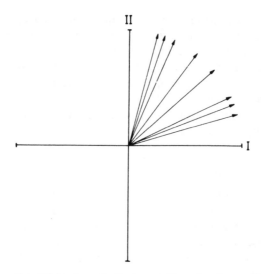

Figure 7.4. Eight hypothetical variables showing orthogonal simple structure.

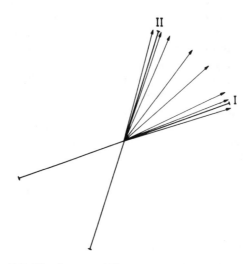

Figure 7.5. The factors of Fig. 7.4 rotated to oblique simple structure.

The general situation is depicted in Fig. 7.6A; we see variable x_j situated in relation to two oblique factor axes. For reasons to be explained later, these are termed *primary factor axes*.

The coordinates of x_j may be specified in two ways. Fig. 7.6B shows the coordinates of x_j specified by projections parallel to the primary factor axes. The coordinates are known as *primary factor loadings* and the p-by-k matrix for all variables is called the *primary factor pattern*

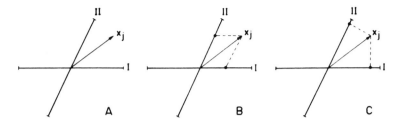

Figure 7.6. (A) Vector of the variable x_j, located with respect to two oblique primary factor axes. (B) Projection of variable vector x_j parallel to the primary factor axis yields the primary pattern loadings. (C) Projection of variable vector x_j perpendicular to the primary factor axes yields the primary structure loadings.

matrix. A different view is presented in Fig. 7.6C. Here, the projection of x_j onto each axis is by lines perpendicular to the axes. Coordinates determined in this fashion are termed *primary structure loadings* and the corresponding matrix of coefficients is the *primary factor structure matrix*. In using oblique rotation, both the primary pattern and the primary structure must be specified to give a complete solution.

The primary pattern loadings are interpretable as measures of the contribution of each factor to the variances of the variables. Pattern loadings are regression coefficients of the variables on the factors (see Chapter 3). The matrix of pattern loadings is useful in the interpretation of the factors themselves, for variables highly loaded on one factor will have low loadings on other factors. Clusters of variables can be seen as a result of a similarity of pattern-loading coefficients. Pattern loadings can be greater than 1, even if the variables are standardized.

The primary factor structure loadings, on the other hand, do not illustrate the configuration of data points in a manner conforming to simple structure. The structure loadings are, in fact, the correlation coefficients of the variables with the oblique factors if the variables are standardized. This will become apparent if you look at Fig. 7.6C. If $\theta_{j\mathrm{I}}$ is the angle between oblique axis I and vector x_j, then the structure loading on I is

$$r_{j\mathrm{I}} = h_j h_{\mathrm{I}} \cos \theta_{j\mathrm{I}},$$

where we write, as in the previous section, h_j for the length of vector x_j, which is the square root of the communality, and h_{I} is the length of the oblique axes, which is 1.

Primary factor structure loadings, although not useful for interpreting factors, are important in the determination of factor scores (Section 7.8).

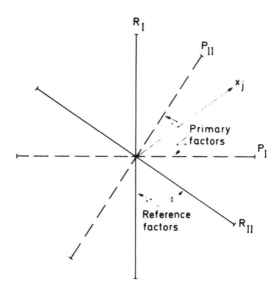

Figure 7.7. Relationship between primary factors and reference factors.

One other matrix is important in the oblique solution. It presents the correlations between the primary factors. These represent the cosines of the angles between the primary axes as well as the correlations between the factors. The coefficients are used to assess the obliqueness of the factor solution. Furthermore, they are necessary for the computation of the factor scores. Lastly, the *primary factor correlation matrix* may itself be factored to yield higher-order factors.

In orthogonal factor solutions, the factor correlation matrix is an identity matrix and need not be considered. Moreover, insofar as the factors are at right angles, there is only one way in which to project a data point onto them. Hence, the pattern loadings and structure loadings are one and the same for orthogonal factors.

An alternative set of axes, which we mention here solely because they occur in the *promax* method, may be employed in place of the primary factor axes just discussed. These are the *reference factor axes*, which are defined as axes normal to planes, or hyperplanes, demarcated by the primary factor axes. That is, for three oblique factors, reference factor I would be normal to the plane defined by primary factors II and III; reference factor II is normal to the plane delineated by primary factors I and III, and so on. The situation for two factors is illustrated in Fig. 7.7.

As with primary axes, data points may be projected onto the reference axes to yield both *reference pattern loadings* and *reference structure*

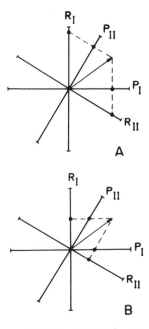

Figure 7.8. Diagram showing that (A) primary structure loadings are proportional to reference pattern loadings and (B) primary pattern loadings are proportional to reference structure loadings.

loadings. The relationship between the pattern and the structure in the two schemes is indicated in Figs. 7.8A, 7.8B. These figures show that the reference pattern loadings are, in fact, correlations between the variables and the reference factors and are proportional to the primary pattern loadings on the corresponding primary factors.

A matrix of correlation coefficients between the reference factors may also be produced.

The reference pattern and structure are, to an extent, mirror images of the primary structure and pattern, respectively. The reference matrices are seldom utilized in interpretation, but the reference pattern matrix is usually the one employed to determine the positions of the oblique axes. Thurstone (1947) defined the reference factor scheme because he found that the reference structure loadings reflect a simple structure solution to a higher degree than did the matrices in the primary scheme. The reference structure loadings are geometrically constrained to lie between plus and minus 1. In addition, variables lying in the hyperplane normal to a reference factor will have near-zero loadings on that factor. Hence, there will be many loadings near zero on the reference factor, one of the criteria for simple structure. We shall now take up the question of how the positions of the oblique axes may be objectively determined.

The promax method

Of the many methods put forward for obtaining an oblique simple-structure solution, the so-called promax method of Hendrickson and White (1964), is treated here. The name "promax" derives from "oblique Procrustean transformation" (Mulaik, 1972, p. 293, Chapter 12). The calculations are fairly simple and quickly done on a computer. Results seem to be in good agreement with more complicated methods and they are consonant with the criteria of simple structure. Promax rotation has been applied to geological problems by Cameron (1968). The technique involves the following steps:

1. Development of a "target matrix": The procedure begins with a matrix of factor loadings that has been rotated to orthogonal simple structure. The varimax factor-loadings matrix B is the usual starting point. This matrix is normalized by columns and rows so that the vector lengths of both variables and factors are unity. Next, the elements of this matrix are raised to some power, generally 2 or 4. All loadings are thereby decreased, the high loadings much less than the low loadings. This results, hopefully, in an "ideal pattern matrix," here denoted B^*. The concept behind this procedure is that an ideal oblique factor pattern should have its loadings as near to 0 or 1 as possible. Powering the varimax matrix tends to do this.
2. The least-squares fit of the varimax matrix to the target matrix: As was the case for orthogonal rotation, some transformation matrix T_r is required to rotate the original factor axes to new positions. The original factors are those specified by the varimax factor matrix B, and the new positions are specified by the reference structure matrix $S_r = BT_r$.

One would wish to determine T_r in such a way that S_r is as close to B^* as possible in the least-squares sense. As shown by Mosier (1939) and Hurley and Cattell (1962), the least-squares solution for T_r is obtained as

$$T_r = (B'B)^{-1}B'B^*.$$

The columns of T_r are then normalized, which is necessary in order for the new oblique factors to have unit variances.

Table 7.IV. *The promax oblique primary pattern matrix derived from Table 7.III* ($k_{min} = 4$)

Variables	Factors		
	1	2	3
1	0.008	0.912	−0.160
2	1.023	0.162	0.254
3	−0.012	0.168	−0.907
4	0.899	−0.182	−0.300

The matrix T_r represents the transformation that converts the orthogonal varimax matrix B to the oblique reference structure matrix S_r. The elements of T_r are the direction cosines between the orthogonal axes and the oblique axes. The following equations (from Harman, 1967) show how the remaining oblique rotation matrices may be derived.

The relationship between T_r, the reference structure transformation matrix, and T_p, the primary structure transformation matrix, is given by $T_p' = T_r^{-1}$, and T_p' is thereafter row-normalized.

Correlation between primary factors is expressed by

$$\Phi_p = T_p' T_p,$$

where Φ is defined as in Chapters 3 and 4. The primary factor structure matrix S_p is given by

$$S_p = BT_p$$

and the primary factor pattern matrix P_p by

$$P_p = S_p \Phi_p^{-1} \quad \text{or} \quad P_p = B(T_p')^{-1}.$$

The reference pattern matrix P_r is yielded by

$$P_r = S_r \Phi_r^{-1} \quad \text{or} \quad P_r = B(T_r')^{-1},$$

and the correlations between the references axes are

$$\Phi_r = T_r' T_r.$$

Example of the promax method

To illustrate oblique rotation, we shall find the promax solution to the example given earlier on in this section. Table 7.IV lists the primary pattern loading matrix, and Table 7.V the primary structure loadings. It

Table 7.V. *The promax oblique primary structure matrix* $(k_{min} = 4)$

	Factors		
Variables	1	2	3
1	-0.005	0.989	-0.601
2	0.971	-0.007	-0.005
3	0.141	0.606	-0.986
4	0.960	-0.079	-0.371

will be observed that the primary pattern loading matrix conforms to the tenets of simple structure to a greater degree than does the varimax factor matrix, especially on the second factor. The correlations between the primary factors indicate that factors 1 and 2 are still almost orthogonal (92.6°), whereas factors 2 and 3 are oblique, being separated by an angle of 119°.

Figures 7.9–7.11 summarize the various solutions we have taken up in this section. For illustrative purposes, we plot the positions of the four variables in the space of the three factors: first for the unrotated, principal components factors; second, for the varimax factors; and, third, for the primary, oblique, promax factors. All combinations of the factors are shown.

Oblique-projection method

Imbrie (1963) devised a method of oblique rotation that is essentially different from promax and similar techniques. In promax rotation, the factor axes are located so as to minimize some function and they are usually found to be associated with some cluster of variables. The factor axes represent composite variables and are therefore abstract. Imbrie thought that a suitable frame of reference could be obtained by forcing the factors to be collinear with the most divergent, actual variable vectors in the factor space.

We refer to the discussion in Chapter 5 for further details.

7.9 FACTOR SCORES

In the previous chapters, we have dealt almost exclusively with one part of the factor equation, to wit:

$$\mathbf{Z}_{(N \times p)} = \mathbf{FA'}.$$

The matrix of factor loadings **A** (or its rotated counterparts) provides a description of the relationship between a set of variables (in the *R*-mode) and a set of orthogonal or oblique factors. In the solution for **A**, various mathematical constraints were imposed so that **F** dropped out of the equations.

In *R*-mode analysis, a factor may be looked on as a function of the original variables. It is a hypothetical construction, a sort of "new" variable created from the old ones. It may be of interest to determine the "amount" of this new variable in each object. The elements of the factor-score matrix provide this information.

The usefulness of factor scores can be appreciated by bearing in mind two points. The first of these is that there will, in general, be fewer factors than original variables. Hence, if the variables are mappable quantities, for example, mapping the derived factor scores will provide the same amount of information with fewer maps. The second point is that for an orthogonal solution, the "new" variables will be uncorrelated, an important feature in using other statistical procedures.

In summary, a factor-scores matrix, in the *R*-mode, is a condensed data matrix, the columns of which are linear combinations of the original variables and the rows are the original objects in the analysis.

A number of methods have been proposed for finding factor scores. We deal with two that are in wide use.

Direct factor scores

Under certain conditions, the solution for the factor-scores matrix can be relatively simple and straightforward. Starting with the basic factor equation, $\mathbf{Z} = \mathbf{FA'}$, we can derive **F** from

$$\mathbf{F} = \mathbf{Z}(\mathbf{A'})^{-1}.$$

This assumes that **A** is a square, nonsingular matrix; it contains as many factors as variables. Such a situation might arise in using principal components and retaining all the factors. When fewer factors than variables are used, the solution is slightly more complex. Assuming k factors and p variables, we have, approximately, assuming **E** small,

$$\mathbf{Z}_{(N \times p)} = \mathbf{F}_{(N \times k)}\mathbf{A'}_{(k \times p)}.$$

Postmultiplying by **A** gives

$$\mathbf{ZA} = \mathbf{FA'A}.$$

Finally, after postmultiplication by $(A'A)^{-1}$, we get

$$F = ZA(A'A)^{-1}.$$

This expression is fairly general and can be used for principal components or factor analysis. Inasmuch as there are, in both cases, fewer factors than variables, only a portion of the total variance of the variables has been taken into account. Therefore, the original observations, the elements of Z, can be considered only as approximations to the values necessary to compute the exact factor scores in terms of this partial variance. The factor scores, in this case, are approximations to the factor measurements in the reduced space delineated by the k-factors.

Regression estimation

Although the direct method yields generally favorable results for principal components, it is not always appropriate for true factor analysis.

Consider the basic equation for true factor analysis, remembering that E is not necessarily small now:

$$Z = FA' + E,$$

where F and A are matrices concerned with the common part of Z, whereas E involves the unique parts of the variables in Z.

In the general case of k common factors, the common factor-score matrix F will have k columns, whereas E will have p columns, one for each of the original variables. Therefore, there is actually a total of $p + k$ factor scores to estimate, not just k. In this situation, a regression model may be utilized to obtain a least-squares fit of F to the data.

If we let $\hat{F}$ be the matrix of estimated common factor scores, the usual regression equation becomes

$$\hat{F}_{(N \times k)} = Z_{(N \times p)} Q_{(p \times k)}, \qquad [7.6]$$

where Z is the standardized data matrix and Q is a matrix of regression coefficients. Premultiplication of both sides of [7.6] by Z' and division by N yields

$$\frac{1}{N} Z' \hat{F} = \frac{1}{N} Z'ZQ. \qquad [7.7]$$

Since the term $(1/N)Z'Z$ is the correlation matrix between the variables, R, [7.7] becomes

$$\frac{1}{N} Z' \hat{F} = RQ.$$

The term $(1/N)Z'\hat{F}$ is a p-by-k correlation matrix of variables with the

Table 7.VI. *Principal factor score matrix corresponding to Table 7.I*

Factor	1	2	3
1	− 0.3092	− 1.5966	1.1693
2	− 2.1721	0.9256	0.6446
3	0.0239	0.9382	− 1.4333
4	− 0.2382	− 1.1718	− 0.1664
5	1.2996	1.0841	1.2985
6	0.9111	0.5652	0.5533
7	− 0.2537	0.1612	− 1.0989
8	0.7386	− 0.9058	− 0.9671

Table 7.VII. *Varimax factor score matrix*

Factor	1	2	3
1	0.7058	− 0.3536	1.8409
2	− 2.2529	0.3279	0.8982
3	− 0.6240	− 0.3514	− 1.5564
4	0.4036	− 0.9595	0.6116
5	0.6523	2.0215	− 0.1955
6	0.5240	1.0469	− 0.2916
7	− 0.4116	− 0.7249	− 0.7767
8	1.0028	− 1.0070	− 0.5306

common factors. These correlations are, however, given by the factor-loadings matrix $\mathbf{A}$ or $\mathbf{B}$ for unrotated and rotated orthogonal factors, respectively, and by the primary structure matrix $\mathbf{S}_p$ for oblique factors. Thus, we may write $\mathbf{A} = \mathbf{R}\mathbf{Q}$ and

$$\mathbf{Q} = \mathbf{R}^{-1}\mathbf{A}. \qquad [7.8]$$

Substituting [7.8] into [7.6] yields the final solution:

$$\hat{\mathbf{F}} = \mathbf{Z}\mathbf{R}^{-1}\mathbf{A}. \qquad [7.9]$$

The appropriate varimax or oblique factor scores may be obtained by substituting $\mathbf{B}$ or $\mathbf{S}_p$ for $\mathbf{A}$ in [7.9].

Interpretation of factor scores

The N-by-k matrix of factor scores, $\mathbf{F}$, contains elements f_{nj} that indicate the "amount" of factor j in object n. The factor scores are interpreted in the same fashion as observations on any variable.

Table 7.VIII. *Promax factor score matrix*

Factor	1	2	3
1	0.5808	−0.7687	1.7694
2	−2.3081	0.2679	1.0087
3	−0.5179	0.0164	−1.2976
4	0.3620	−1.0877	0.8214
5	0.6630	1.9760	−0.8661
6	0.5419	1.0517	−0.6500
7	−0.3582	−0.5225	−0.4712
8	1.0365	−0.9330	−0.3144

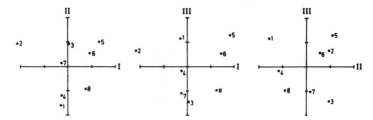

Figure 7.9. Principal factor score plots.

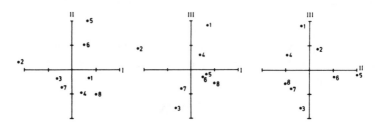

Figure 7.10. Varimax factor score plots.

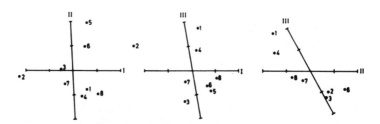

Figure 7.11. Promax factor score plots.

Earlier on, we underlined that factors are "new" variables that are linear combinations of the original variables; consequently, they are composite variables. Factor scores derived from unrotated varimax and oblique solutions are different composite variables. Factor-score matrices for the example used in Section 7.7 are listed in Tables 7.VI, 7.VII, and 7.VIII. A column of Table 7.VI denotes the distribution of the composite variable whose make-up is described by the corresponding column of the principal factor matrix of Table 7.I. A column of the F matrix is in standardized form, with a mean of zero and variance of 1. Hence, the elements measure the amount of the composite variable in each object as deviations from the mean in units of standard deviation. The actual values of the mean and variance remain, of course, unknown, meaning that the values can only be used in relative terms. In the case of unrotated scores and varimax factor scores, the composite variables (factors) are uncorrelated, a particularly useful property if they are to be used as input for further statistical treatment. Oblique factor scores are, naturally, correlated.

Plots of factor scores, such as shown in Figs. 7.9, 7.10, and 7.11, are often used for studies directed toward the ordination of objects on the basis of composite variables. Since these are fewer in number than the original variables, a more parsimonious classificatory scheme often emerges.

8 Examples and case histories

8.1 INTRODUCTION

In this chapter, we present applications of the methods we have taken up in this book. We want you to appreciate right at the outset that we have not tried to cover all facets of factor analysis in the natural sciences, our aim being to give you samples of what various specialists have done when faced with particular problems requiring some kind of factor analysis in their solution. Many more applications occur in the literature and it is expected that, armed with the knowledge imparted to you by our book, you will be able to avail yourselves of special slants devised in answer to special problems.

You will notice that we have given much space to Q-mode analysis, a reflection of its exceptional usefulness in many spheres of geological and biological research.

8.2 R-MODE PRINCIPAL COMPONENTS IN MINERAL CHEMISTRY

Statement of the problem

The chemically complex minerals of the amphibole set may consist of as many as 12 ingredients. Some of the elements occupy two or more dissimilar structural sites as, for example, Al, which may be located at the tetrahedral T_1 and T_2 sites and some of the M positions. As another example, we may take Na, which can occupy M_4 and/or A sites. These substitutional relationships may be caused by a variety of short- or long-term effects in the crystal structure and various kinds of mutual correlations are identifiable in the concentrations of the elements.

Material

Saxena and Ekström (1970) considered relationships in 639 calcic amphiboles. Finding that multiple-regression studies did not yield more than the most common set of interrelationships, they used the method

228

Table 8.I. *Principal component analysis of the chemical data on 639 calcic amphiboles*

Eigenvalues	4.301	1.614	1.556	1.254	0.847	0.780	0.636
% variation	35.8	13.4	12.9	10.4	7.05	6.49	5.30

Variables	Eigenvectors						
Si	0.423	−0.064	0.172	0.186	−0.117	−0.268	−0.063
Al^{iv}	−0.433	0.074	−0.178	−0.186	0.109	0.257	0.063
Al^{vi}	−0.168	0.306	−0.186	−0.586	−0.460	−0.084	0.045
Ti	−0.287	−0.277	−0.288	0.094	0.407	0.105	−0.522
Fe^{3+}	−0.311	−0.158	0.148	0.221	0.163	−0.033	0.775
Fe^{2+}	−0.297	0.383	0.191	0.341	−0.139	−0.311	−0.277
Mn	−0.045	−0.060	0.513	0.150	−0.405	0.701	−0.131
Mg	0.402	−0.318	−0.142	−0.176	0.086	0.194	0.045
Ca	0.064	0.188	−0.598	0.347	−0.161	0.385	0.095
Na	−0.246	−0.435	0.236	−0.372	−0.036	−0.047	−0.089
K	−0.319	−0.223	−0.043	0.282	−0.190	−0.136	−0.058
OH	0.060	0.519	0.308	−0.146	0.563	0.218	−0.004

of principal components to identify other substitutional relationships, which are less common, but of equal significance in the crystal chemistry of the amphiboles.

Methods

Saxena (1969) applied principal component analysis of the correlations between variables for analyzing chemical relationships in amphiboles. The interpretations put forward by Saxena were then based on the plots of the transformed observational vectors, $u'x$ (cf. Anderson, 1963, p. 125). Saxena and Ekström (1970) used published information on amphiboles containing more than 6% by weight of CaO for their study. The ionic formula uses 24 oxygen atoms; the following 12 ions were considered: Si, Al^{iv}, Al^{vi}, Ti, Fe^{3+}, Fe^{2+}, Mn, Mg, Ca, Na, K, and OH. The eigenvalues and eigenvectors for the seven largest roots are listed in Table 8.I. Note that these data are constrained.

Results

In the first column of Table 8.I, Si and Mg can be seen to covary in a positive relationship, whereas Al^{iv}, Al^{vi}, Ti, Fe^{3+}, Na, and K are united in a negative relationship. In other words, the one set of variables is negatively associated with the second set of variables. An increase in the concentration of Al^{iv} is accompanied by an increase in the concentration of Al^{vi}, Ti, Fe^{3+}, and Fe^{2+} in the M positions. This agrees with earlier results and the average ionization potential model of silicates, according to Saxena and Ekström. Inasmuch as the loading for Ca is small, Saxena and·Ekström thought that changes in Na and K are related to the variation in Al^{iv} and Si.

The second, third, and fourth principal components were interpreted as reflecting the complex substitutions in amphiboles. These are all roughly of equal importance (as shown by the percentages of the trace of the eigenvalues), suggesting that several substitutional relationships are equally common in amphiboles. The second component could be interpreted as representing a charge balance among the various sublattice positions. Substitutions of the following type may occur:

$$Al^{vi} \rightleftharpoons Fe^{3+}$$

$$Fe^{2+} \rightleftharpoons Mg$$

$$Al^{vi} + Ca \rightleftharpoons Ti + Na$$

The third component is marked by a high negative association between Mn and Ca; Ca is negatively associated with Fe^{2+} and with Na. The fourth component was thought to indicate that with increasing concentration of Al^{vi}, Na may begin to enter the Ca position. The variation in the concentration of Ca could not be due to its replacement by Fe^{2+}, because there is a positive association between Ca and Fe^{2+}. The fifth eigenvector was believed to indicate the substitution

$$Ti^{4+} + O_2 \rightleftharpoons Al^{3+} (M \text{ position}) + OH^-.$$

No more than the first five eigenvectors, tied to 80% of the total variation, were considered by Saxena and Ekström, who thought this to be quite sufficient for their purposes.

In Fig. 8.1, the plot of the first two sets of transformed observations is shown. The theoretical Fe and Mg end-members are joined by straight lines. Saxena and Ekström concluded that if these end-members suffice for defining the composition of the natural amphiboles of the study, the points representing compositions of the individual amphiboles should lie within the area marked by the end-members. For example, no points should lie above the Tr–Fe–Tr line or below the Mg–Ka–Ka line. Actually, about 10% of the points fall below the latter line, which would seem to be ascribable to increasing concentration of Ti and Fe^{3+} and a decrease in OH. Many of these points represent amphiboles, such as kaersutite, from volcanic rocks.

The greater variation in the graph lies along the first axis, apparently connected with variation in Si and Al^{iv} and with the substitution $Fe^{2+} \rightleftharpoons Mg$.

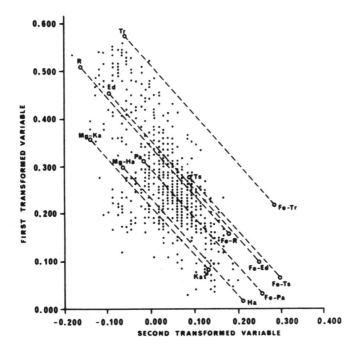

Figure 8.1. Plot of the first transformed observations against the second set of transformed observations for the data on the amphiboles. The following abbreviations for mineral names are used: tremolite (Tr), ferrotremolite (Fe-Tr), edenite (Ed), ferroedenite (Fe-Ed), pargasite (Pa), ferropargasite (Fe-Pa), tschermakite (Ts), ferrotschermakite (Fe-Ts), magnesiohastingsite (Mg-Ha), hastingsite (Ha), richterite (Ri), ferrorichterite (Fe-Ri) magnesiokataphorite (Mg-Ka), and kataphorite (Ka).

8.3 *R*-MODE FACTOR ANALYSIS IN PALEOECOLOGY

The example summarized here is based on maximum likelihood factor analysis according to Jöreskog's (1963) model and was originally published by Reyment (1963). The results were briefly noted in the First Edition but owing to the fact that it has become one of the most frequently cited applications of factor analysis in ecology, we decided to present it now in more detail. The data come from boreholes in the Maastrichtian–Paleocene sequence in western Nigeria.

Statement of the problem and materials

The first 600 randomly selected individuals of 28 borehole samples containing 17 species of Paleocene ostracods were used to construct the correlation matrix given in Table 8.II (significant correlations are printed

Table 8.II. *Correlation matrix based on collections of seventeen species of fossil ostracods*

1.0000																
-0.3176	1.0000															
0.1363	-0.1902	1.0000														
0.1804	-0.1775	-0.3495	1.0000													
-0.2547	0.1606	-0.1285	-0.1838	1.0000												
-0.0267	-0.2299	-0.1599	0.2067	-0.0948	1.0000											
-0.0038	-0.1499	-0.2642	0.0481	-0.1279	**0.4711**	1.0000										
-0.1540	-0.1756	-0.2558	-0.2026	-0.1547	0.0639	0.0332	1.0000									
-0.3659	-0.1807	-0.1862	-0.0920	-0.2624	0.0435	0.0945	0.0808	1.0000								
-0.0509	-0.0028	0.2996	-0.3229	0.2043	-0.0899	0.3722	0.2368	-0.0499	1.0000							
-0.1097	-0.2288	-0.2509	-0.2021	0.3699	-0.1206	0.0400	-0.1809	-0.0083	0.0675	1.0000						
-0.2104	-0.1734	-0.1110	0.0320	**0.5888**	-0.1685	-0.2358	-0.2040	-0.1603	-0.1889	**0.7303**	1.0000					
-0.3085	-0.3166	0.3419	0.0981	-0.2349	0.0722	-0.2635	0.0903	0.2735	-0.1878	-0.2679	0.0075	1.0000				
-0.0213	-0.2386	0.2331	-0.0183	-0.1927	0.0792	-0.1421	-0.1107	-0.0871	-0.1408	-0.1212	-0.1286	0.2957	1.0000			
0.0130	-0.2420	-0.1388	-0.0855	-0.1567	0.3008	0.0942	-0.0079	0.3098	-0.0515	-0.0136	-0.2747	0.0277	-0.0811	1.0000		
0.1513	-0.1549	**0.5050**	**0.4269**	-0.3088	0.2614	0.2442	-0.3362	0.3636	-0.2613	0.0118	-0.2447	-0.17651	-0.0099	**0.4261**	1.0000	
0.1389	-0.1714	**-0.4500**	0.1173	0.2005	0.2841	0.3086	**-0.4612**	0.0315	-0.1857	**0.4265**	0.3249	-0.3127	-0.1052	**0.4563**	-0.0165	1.0000

N.B.: These correlations are based on compositional data because the rows of the data matrix have a constant sum. For such large matrices, however, the effects of the constraint are slight and do not impinge negatively on the ordination of points.

in bold type). The data form a constrained set and should by rights be logratio correlations. The differences between the logratio correlations and usual correlations were, however, ascertained to be too small to be of any practical significance.

The ordering of the 17 species in the data matrix is as follows: *Cytherella sylvesterbradleyi, Bairdia ilaroensis, Ovocytheridea pulchra, Iorubaella ologuni, Brachycythere ogboni, Dahomeya alata, Anticythereis bopaensis, Leguminocythereis lagaghiroboensis, Trachyleberis teiskotensis, Veenia warriensis, Buntonia fortunata, Buntonia beninensis, Buntonia bopaensis, Quadribuntonia livida, Ruggieria ? tattami, Schizocythere* spp., *Xesteroleberis kekere.*

The Analysis

The model used for analyzing these data is that expounded in Section 4.3 and exemplified by a hydrogeological example. The analysis indicated five factors to be significant but owing to the "smallness" of the data matrix, it was thought advisable to look at as many as 10 factors. The factor-loadings matrix thus obtained is listed in Table 8.III.

The *first factor* indicates that species 5, 11, 12, and, possibly, 17 are affected in the same way by some environmental factor; the other elements of the vector do not differ significantly from zero. The *second factor* suggests an environmental control that mainly influences species 2 but also to a lesser degree, in the same direction, species 5. Species 1, 15, and 16 are influenced in the opposite direction. The third factor seems to represent an environmental control that affects species 3, 8, and 13 in one direction and species 4, 7, 16, and 17 in another direction. The fourth vector would appear to indicate the influence of an environmental factor that rather strongly affects species 1 in one direction and about equally as strongly species 9 in the reverse direction; other affected species are 3, 13, 15, and 16. The fifth vector suggests relatively strong influences on species 4, 7, and 10 in opposite directions, and lesser influences on species 8 and 13. These five factors seem to be the most informative. The remaining five factors included in the matrix of factor loadings are remarkable in that none of them suggests a strong reaction of any environmental agent on any of the species.

Hence, it would seem that most of the variation in frequencies of the 17 species may have been controlled by five environmental factors of some kind or other (for example, temperature, light, salinity, variation in chemical proportions of seawater, pH, redox).

If as a criterion of nonreactivity to environment we take small factor loadings, it may be suggested that species 6, 14, and 15 are euryoic and this agrees extremely well with what is to be observed qualitatively in the material. Judging from occurrences in the borehole samples one

Table 8.III. *The unrotated factor loadings (k = 10)*

Variables	Communalities	Factors									
		1	2	3	4	5	6	7	8	9	10
1	0.90	−0.15	**−0.47**	0.30	**−0.72**	−0.03	0.09	0.09	0.01	−0.04	−0.06
2	0.94	−0.25	**0.90**	0.21	−0.02	−0.06	0.03	0.00	−0.03	0.00	−0.01
3	0.80	−0.21	−0.12	**−0.69**	**−0.36**	0.02	0.13	−0.18	0.03	−0.13	−0.05
4	0.84	−0.02	−0.30	**0.38**	0.07	**−0.54**	**−0.48**	0.09	0.00	−0.09	0.20
5	0.81	**0.59**	**0.40**	−0.06	−0.10	0.10	−0.14	−0.11	**0.44**	0.02	0.05
6	0.62	−0.10	−0.31	0.21	0.28	0.12	−0.29	−0.25	0.20	0.32	−0.24
7	0.80	−0.09	−0.21	**0.35**	0.19	**0.56**	**−0.37**	−0.31	−0.11	−0.06	−0.08
8	0.83	−0.25	−0.09	**−0.52**	0.06	**0.36**	−0.32	**0.41**	−0.10	0.22	−0.03
9	0.77	−0.09	−0.18	−0.01	**0.69**	0.12	0.26	0.17	−0.07	−0.27	−0.08
10	0.74	−0.09	0.08	−0.21	−0.14	**0.66**	−0.16	−0.21	0.07	−0.31	0.27
11	0.89	**0.85**	−0.02	0.08	0.02	0.27	0.18	0.05	−0.17	0.02	0.10
12	0.93	**0.93**	0.09	−0.13	−0.02	−0.17	−0.07	0.04	0.00	−0.01	−0.04
13	0.83	−0.11	−0.26	**−0.57**	**0.38**	**−0.44**	0.01	−0.14	0.05	−0.14	−0.10
14	0.74	−0.11	−0.24	−0.23	0.01	−0.26	0.20	**−0.49**	−0.20	0.37	0.25
15	0.75	−0.16	−0.32	0.16	**0.33**	0.22	**0.35**	0.17	**0.46**	0.15	0.09
16	0.80	−0.10	**−0.36**	**0.69**	**0.34**	−0.06	0.16	0.05	0.00	−0.02	0.12
17	0.25	**0.47**	−0.18	**0.58**	0.05	0.07	0.03	−0.23	−0.02	0.01	−0.29
Variance		2.47	1.93	2.54	1.70	1.66	0.95	0.84	0.56	0.54	0.40

N.B.: Significant loadings in bold type.

gains the impression that species 1 and 9 are also euryoic. Both of these are strongly affected only by the fourth factor, which may indicate that this factor is an unimportant one with respect to actual distribution. None of the species appears to give the impression of being stenoöic.

From the point of view of the criterion of meaningful factors we have that the first five vectors contain several important loadings, whereas the remaining five are devoid of larger entries, or are dominated by one or two elements. Such factors are usually taken to be uninformative.

8.4 R-MODE FACTOR ANALYSIS IN SEDIMENTOLOGY

Statement of the problem

The use of Q-mode methods for studying grain-size distributions of sediments was reviewed in an earlier example. Mather (1972) attempted to elucidate relationships between grain-size categories using R-mode methods. A similar study was made earlier by Davis (1970).

Materials

Mather (1972) collected 100 samples of fluvioglacial sediments of Wurm age. Each sample was split into $\frac{1}{2}\phi$-size classes in the range -7ϕ to 9ϕ. Direct measurement, sieving, and pipette determinations were used to assemble the data. In all, the data matrix consisted of 100 rows and 33 columns. Each variable is the weight of sediment within one size class. Now, these data are certainly constrained in the sense of Section 4.6, but experience shows that with so many variables as in the present example, the effects of closure are slight, and hardly affect factor analysis.

Methods

Because his sediment samples were not of constant weight, Mather computed the correlation matrix between the 33 variables and an additional variable, the total sample weight. The correlation matrix was then adjusted by partialing out the effect of total weight. This was done in order to provide an estimate of the strength of association between all pairs of size classes, independent of the sample weight. The same effect could have been obtained by expressing each variable as a weight percentage rather than weight. The following steps were then gone through: (Note, this introduces the constraint.)

1. Eigenvalues of the 33-by-33 correlation matrix were extracted. The cutoff for significant factors was set at the number of eigenvalues greater than or equal to 1.

2. Squared multiple correlation coefficients were computed and placed along the diagonal of the correlation matrix as estimates of the communality.
3. Using the number of factors determined as before, and using squared multiple correlations as communality estimates, an iterative procedure was employed to derive new communality estimates (cf. Chapters 3 and 4).
4. The principal (unrotated) factor-loadings matrix, based on the final estimates of the communality, was calculated.
5. Varimax rotation was employed to achieve simple structure.
6. The promax method of oblique rotation was used to determine a factor pattern.
7. The matrix of factor intercorrelations, computed in the foregoing step, was used as input in a factor analysis in an attempt to produce so-called second-order factors. In this phase of the analysis, the steps involving the squared multiple correlations in the diagonal of **R**, the computation of the principal factor loadings, and the varimax and promax rotations were used.

Results

Six factors were judged to be significant. Unfortunately, Mather does not list the eigenvalues of his principal component analysis, nor does he indicate the percentage of total variance explained by the six factors.

The squared multiple correlations (Table 8.IV) are consistently high, indicating a high degree of common variance among the variables. The derived communality estimates, using six common factors are consistently less than the initial estimates and, in the cases of variables 1, 2, 4, 29, 30, and 32, they are very low. Mather himself questions the validity of the assumption that six factors form an adequate model (we echo this doubt), but he justifies his choice on "practical grounds."

A comparison of the varimax matrix (Table 8.V) with the oblique primary pattern matrix (Table 8.VI) shows how the oblique rotation more clearly brings out simple structures. Whereas 12 variables have loadings of 0.4 or more on 2 or more factors in the varimax matrix, only 4 variables have that pattern in the oblique solution.

The factor intercorrelations from the promax solution are listed in Table 8.VII. Squared multiple correlations are entered in the diagonal. The resulting varimax matrix is given in Table 8.VIII. The six first-order factors are seen to form two interrelated groups. First-order factors 1, 3, and 6 show a high degree of interrelationship. The main reason for this dependency becomes apparent on inspection of Table 8.V; factors 1 and 3 are related through mutual loadings on variables 3, 8, 9, 18, and 19, whereas factors 1 and 6 are connected through loadings on variables

Table 8.IV. *Communality values*

riable	SMC	Derived	Size range ϕ From	To	Classification
1	0.852	0.274	-7ϕ	-6.5ϕ	
2	0.693	0.342	-6.5	-6	Cobble
3	0.900	0.727	-6	-5.5	
4	0.867	0.312	-5.5	-5	
5	0.920	0.820	-5	-4.5	
6	0.952	0.870	-4.5	-4	
7	0.923	0.790	-4	-3.5	Pebble
8	0.972	0.905	-3.5	-3	
9	0.995	0.932	-3	-2.5	
10	0.995	0.924	-2.5	-2	
11	0.992	0.958	-2	-1.5	
12	0.988	0.957	-1.5	-1	Granule
13	0.987	0.949	-1	-0.5	
14	0.997	0.925	-0.5	0	Very coarse sandstone
15	0.998	0.872	0	0.5	
16	0.995	0.701	0.5	1	Coarse sandstone
17	0.987	0.918	1	1.5	
18	0.987	0.857	1.5	2	Medium sandstone
19	0.981	0.849	2	2.5	
20	0.981	0.850	2.5	3	Fine sandstone
21	0.983	0.842	3	3.5	
22	0.985	0.845	3.5	4	Very fine sandstone
23	0.916	0.851	4	4.5	
24	0.849	0.739	4.5	5	
25	0.941	0.806	5	5.5	
26	0.901	0.837	5.5	6	
27	0.912	0.777	6	6.5	Silt
28	0.886	0.760	6.5	7	
29	0.620	0.481	7	7.5	
30	0.558	0.321	7.5	8	
31	0.881	0.878	8	8.5	Clay
32	0.705	0.434	8.5	9	
33	0.906	0.847	>9		

8, 9, and 19. Factors 3 and 6 are related through variables 1, 5, 8, 9, 19, and 20. Similarly, factors 2 and 5 are united by mutual loadings on variables 24 to 30. A tie also exists between factors 1 and 4.

Mather made the important point that the distributions of grain size of this particular suite of samples are governed at two levels of influence. At the higher level of generalization, two second-order factors appear dominant. The first of these produces variations in the range of

Table 8.V. *Reordered varimax factor matrix*

Variable	1	2	3	4	5	6
			Factor			
12	0.964					
13	0.960					
11	0.958					
10	0.931					−0.210
14	0.909			0.257		
9	0.872					−0.393
15	0.803			0.445		
19	−0.499	0.274	−0.458		−0.426	0.349
30		−0.488			0.247	
24		−0.624			0.574	
29	0.210	−0.632				
32		−0.647				
25		−0.676			0.587	
26		−0.816			0.408	
28		−0.831			0.246	
27		−0.855			0.203	
33		−0.897				
31		−0.929				
3			0.813			−0.231
5			0.642			−0.552
2			0.581			
1			0.475			
4			0.473			−0.249
17		0.205	−0.304	0.882		
16	0.503			0.647		
18	−0.339	0.246	−0.389	0.636	−0.288	0.203
22	−0.231			−0.597	0.596	
20	−0.426		−0.377	−0.611		0.317
21	−0.311		−0.274	−0.751	0.248	0.206
23	−0.421				0.783	
8	0.639					−0.691
7	0.475					−0.720
6	0.333		0.313			−0.803

Loadings < |0.2| omitted.

Table 8.VI. *Reordered promax primary pattern matrix for k = 4*

Variable	Factor					
	1	2	3	4	5	6
13	1.062					
12	1.051					
11	1.011					
10	0.978					
14	0.968					
9	0.802		-0.265			-0.352
15	0.796			0.373		
31		-1.013				
33		-0.938				
27		-0.792				
28		-0.745			0.210	
29	0.223	-0.714			-0.212	
32		-0.673				
26		-0.657			0.408	
30		-0.386			0.239	
3			0.849			
2			0.656			
1			0.600			
5			0.540			-0.525
4			0.455			
17			-0.346	0.950		
18	-0.356		-0.352	0.699	-0.244	
16	0.398			0.637		
20			-0.300	-0.591		0.215
21			-0.248	-0.705	0.263	
23					0.872	
22			-0.202	-0.511	0.658	
25		-0.423			0.626	
24		-0.373			0.615	
19	-0.347		-0.366		-0.436	0.243
8	0.323		-0.263			-0.798
7						-0.839
6						-0.950

Loadings < |0.2| omitted.

Table 8.VII. *Correlations between promax factors for k = 4*

	1	2	3	4	5	6
1	0.43					
2	−0.03	0.25				
3	0.29	−0.09	0.30			
4	0.29	0.10	0.15	0.13		
5	0.03	−0.49	0.15	−0.18	0.28	
6	−0.63	−0.01	−0.51	−0.20	−0.00	0.52

Table 8.VIII. *Reordered second-order varimax matrix*

First-order	Second-order factor	
factor	1	2
6	0.788	
3	−0.545	
1	−0.695	
4	−0.316	−0.239
5		0.637
2		−0.591

Loadings < |0.2| omitted.

grain sizes of sand–granule–pebble–cobble. The other reflects independent variations in the fine sand–silt–clay range. At the lower level, six
factors influence the characteristics of the grain size. Factor 1 mirrors
variation in the size ranges for coarse sand–granule–pebble. Factor 2 is
essentially a silt–clay factor, whereas factor 3 expresses variation in the
pebble–cobble range. Factor 4 is bipolar, indicating that modes in the
medium sand–coarse sand range do not generally coexist with modes in
the fine and very fine sand ranges. Factors 5 and 6 influence very fine
sand and silt, and pebbles, respectively.

8.5 R- AND Q-MODE ANALYSES IN SUBSURFACE BRINE GEOCHEMISTRY

Statement of the problem

Chemical and physical changes that take place in subsurface brines are
usually difficult to explain and document, due to the complexities of
geological history and an insufficient understanding of the range of
processes operating below the surface.

A given volume of subsurface fluid sampled today may owe its
composition to many influences: composition of the connate water

trapped during deposition, mixing with percolating water, reaction with the surrounding rock, as well as modification due to compaction and other physical and chemical processes.

Hitchon et al. (1971) tried to document some of the possible chemically and physically induced reactions that affect the geochemistry of brines by applying factor-analytical techniques to compositional data of subsurface brines, but without recognizing the restrictions imposed by the constant-sum constraint.

Materials

Twenty major and minor chemical components for 78 formation waters from oil and gas fields in Alberta, Canada, provide the data. In addition, certain physical and chemical properties of the water and the surrounding environment were also measured. Independent considerations indicate that the volume-weighted, mean composition of the brines is roughly similar to the composition of seawater.

Methods

Because most of the variables included in the analysis appear to be lognormally distributed and in order to avoid curvilinear relations between variables, all variables were transformed to natural logarithms. Rare constituents were not used.

The transformed data matrix was analyzed by several factor methods.

Imbrie's Q-mode method was applied in order to evaluate the degree of heterogeneity shown by the suite of formation waters. The $\cos \theta$ measure of similarity was used, the waters being compared in terms of the relative proportions of the chemical variables. Significant factors were rotated by the varimax criterion.

R-mode principal component analysis was used to study the interrelations between 16 ions, employing the correlation matrix. Varimax rotation and oblique rotation by the "biquartimin" method of rotation were used to obtain simple structure. The factor scores associated with the varimax factors were used in a further principal component analysis, together with nine physical and chemical properties such as temperature, pressure, and pH. Correlations between the factor scores and the properties as revealed in the resulting varimax matrix were employed to interpret the significance of the factors based solely on ionic concentrations.

Results

Q-mode analysis. Three principal components account for 98% of the total information in the data matrix. Fig. 8.2 is a plot of the

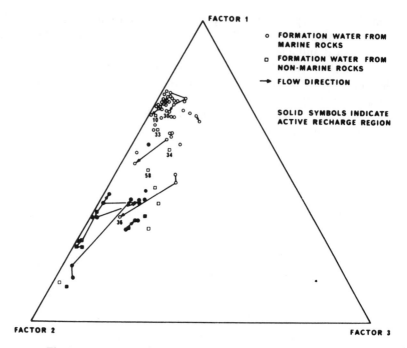

Figure 8.2. Ternary diagram of normalized varimax factor components for *Q*-mode factor analysis on 78 formation waters from oil and
gas fields of Alberta, Canada. Arrows indicate flow directions based
on hydrodynamical studies.

normalized varimax factor components. Note that the water samples
cluster along the edge of the plot, thus indicating that the set of
specimens is fairly homogeneous. Study of the associated factor scores
led the authors to conclude that Factor 1 can be identified with the
processes of membrane filtration, which tends to concentrate brines,
and Factor 2 represents dilution by freshwater recharge. Factor 3 could
not be explained. Water samples on the same flow path, as determined
by independent hydrodynamical studies, are connected to each other on
the diagram. Those arrows showing flow from Factor 2 toward Factor 1
represent fresh waters entering the basin from areas of recharge. The
arrows showing flow from Factor 1 toward Factor 2 are for formational
waters moving from the deeper parts of the basin through shales. This
results in more concentrated brines being left behind the "shale membrane" and less concentrated brines emerging from the membrane.

R-mode principal component analysis.   Table 8.IX lists the varimax factor matrix derived from the principal component analysis of the
correlation matrix for 16 ions and total dissolved solids (salinity). One

Table 8.IX. *R-mode varimax factor matrix of chemical data and salinity for 78 formation waters from oil fields and gas fields of Alberta, Canada*

Variable	Factor 1	2	3	4	5	6	7	8	Communality
Ca	0.884		−0.210			−0.260			0.956
Cu				0.993					0.996
Fe		0.954							0.996
K	0.959								0.979
Li	0.954								0.969
Mg	0.874					−0.207			0.878
Mn	0.261	0.201					0.915		0.998
Na	0.851		−0.396						0.957
Rb	0.933								0.924
Sr	0.834		−0.389			−0.240			0.954
Zn					0.951				0.999
Br	0.757		−0.570						0.950
Cl	0.847		−0.430						0.982
HCO₃	−0.387		0.238			0.847	−0.202		0.995
I	0.359		−0.893						0.960
SO₄	0.619							0.755	0.983
Total dissolved solids	0.888		−0.343						0.980

Eigenvalues

10.495	1.848	1.127	0.961	0.689	0.569	0.403	0.360

Percent of variance explained by factor

50.18	6.22	11.06	6.03	6.43	6.26	6.23	4.37

Cumulative percent of variance

50.18	56.40	67.46	73.49	79.92	86.18	92.41	96.78

Loadings < |0.2| omitted.

dominant factor accounts for more than 50% of the total variance examined. The authors interpreted this factor, with high loadings for salinity, as indicating that most of the waters owe their composition to their derivation from seawater. Accordingly, they termed Factor 1 the "seawater factor." Factor 3 is dominated by the halogen ions I and Br and was called a "halogen factor." The remaining six factors have strong loadings on only one variable and are essentially unique factors. This shows up more clearly in the oblique primary factor pattern matrix of Table 8.X.

In order to relate these factors to known physical and chemical controls, the authors entered the associated factor scores along with the physical and chemical measurements into another factor analysis. The varimax factor matrix for this is listed in Table 8.XI, it demonstrates

Table 8.X. *R-mode biquartimin factor matrix of chemical data and salinity for 78 formation waters from oil fields and gas fields of Alberta, Canada*[a]

| | Factor | | | | | | | |
Variable	1	2	3	4	5	6	7	8
Ca	0.877							
Cu		0.992						
Fe			0.925					
K	0.944							
Li	0.955							
Mg	0.907							
Mn						0.867		
Na	0.871							
Rb	0.950							
Sr	0.851							
Zn				0.912				
Br	0.796				0.363			
Cl	0.856				0.207			
HCO$_3$	−0.334						0.759	
I	0.425				0.738			
SO$_4$	0.575							0.664
Total dissolved solids	0.891							
Eigenvalues	8.639	1.020	0.940	0.870	0.843	0.810	0.642	0.514

[a] Row vectors normalized for eight factors. Loadings $< |0.2|$ omitted.

that the processes of membrane filtration, solution of halite, formation of authigenic chlorite, cation exchange on clays, and other reactions, are important in the geochemistry of subsurface brines.

8.6 IMBRIE Q-MODE FACTOR ANALYSIS IN SEDIMENTOLOGY

Statement of the problem

One of the long-sought objectives of sedimentologists has been to establish sedimentary criteria for determining environments of deposition. For clastic sediments, various criteria based on the grain-size characteristics of the sediments, or rocks, have been put forward as environmental indicators (Folk and Ward, 1957; Friedman, 1961). Most of the admittedly often questioned criteria are based on simple moments of the grain-size frequency distribution of the sedimentary samples.

Table 8.XI. *R-mode varimax factor matrix of physical properties and factor scores for 78 formation waters from oil fields and gas fields of Alberta, Canada*

Variable	Factor 1	2	3	4	5	6	7	8	9	Communality
pH	-0.585	0.506	-0.405					-0.359		0.912
δD (% SMOW)	0.842			-0.276						0.836
δO^{18} (‰ SMOW)	0.870			-0.287						0.880
Depth (feet)	0.924									0.918
Pressure (psi)	0.917		-0.303							0.956
Temperature (°C)	0.947									0.928
Fluid Potential (feet)	0.251		-0.857	0.287						0.901
P_{H_2S} (psi)	0.722				-0.303				-0.254	0.789
P_{CO_2} (psi)	0.418		0.433	0.424		-0.314		-0.307	-0.203	0.833
Factor 1 (sea water)	0.800		0.423	0.232						0.910
Factor 2 (Fe)									0.995	0.973
Factor 3 (Halogens)				0.901						0.890
Factor 4 (Cu)							0.992			0.987
Factor 5 (Zn)						0.979				0.967
Factor 6 (HCO$_3$)		0.976								0.964
Factor 7 (Mn)								0.984		0.970
Factor 8 (SO$_4$)					0.967					0.956
Eigenvalues	5.907	1.977	1.446	1.239	1.150	1.029	1.002	1.000	0.819	
Percent of variance explained by factor	34.55	7.78	8.75	7.67	6.66	6.57	6.06	7.13	6.41	
Cumulative percent of variance	34.55	42.33	51.08	58.75	65.41	71.98	78.04	85.17	91.58	

Loadings < |0.2| omitted.

Klovan (1966) and, later, Solohub and Klovan (1970) attempted to apply a Q-mode technique to the problem. The question posed was "on the basis of the entire spectrum of grain sizes contained in a suite of sediment samples, can the specimens be separated into environmentally distinct groups?"

The materials in the study dating from 1936 consisted of 69 recent sedimentary samples from Barataria Bay, Louisiana. Each sediment had been sieved into 10 size classes, as reported by Krumbein and Aberdeen (1937), from which a data matrix of 69 rows and 10 columns was constructed.

Method

Imbrie's Q-mode method was used in an attempt to establish "grain-size facies." The cos θ measure of similarity was considered appropriate because the relative proportions of the grain-size classes were judged to be a significant means of assessing similarity between samples.

Factors were rotated by the varimax method in order to make them as close as possible to the extreme sedimentary types. Because the factors turned out to be nearly orthogonal, oblique rotation was not resorted to. The squared factor loadings were made to sum to unity to permit ease of plotting on a triangular diagram.

Results

Three eigenvalues account for 97.5% of the total variability. A plot showing the distribution of the objects in three-dimensional factor space revealed a remarkable pattern of variation. In Fig. 8.3, only a few of the sediment samples are shown; with all sample points plotted, a continuous trend from Factor II to Factor I, and from Factor III to Factor I was well marked. Many samples clustered near the apex for Factor I of the triangle, but few plotted in the middle.

Analysis of results

The continuous variation exhibited among the specimens indicates that there are no distinct grain-size facies present within the suite of specimens considered. Examination of the grain-size characteristics of the end-member specimens and of specimens lying between them, as shown in Fig. 8.3, led Klovan to believe that the end-members represented types of sediment deposited under the influence of the three depositional processes. The coarsely grained, well-sorted end-member, represented by Factor III, results from a high-energy depositional setting with removal of fines. Factor II, characterized by poorly sorted,

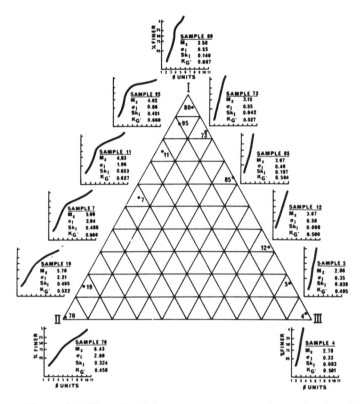

Figure 8.3. Normalized factor components and associated grain-size distributional curves for the *Q*-mode study of sediments.

fine-grained sediments, represents deposition under the dominant influence of gravitational settling. Factor I, which displays a bimodal frequency distribution, suggests that mixing of sedimentary types is occurring. Current activity was thought to constitute the dominant control of the grain-size distribution of sediments highly loaded on Factor I. Sediment samples that plot along trends between any two factors owe their grain-size distributions to the combination of two of the processes.

Finally, a plot showing the spatial distribution of the samples in Barataria Bay indicates that the foregoing interpretation is at least reasonable. Factor II–type sediments occur in sheltered, quiet parts of the Bay, whereas Factor I sediments are found in the channels. The *Q*-mode method, although failing to identify distinct environmental facies, does offer the advantage of fixing the type and the roles of certain varieties of depositional processes that have formed a particular sedimentary sample.

8.7 IMBRIE *Q*-MODE FACTOR ANALYSIS
IN FORAMINIFERAL ECOLOGY

Statement of the problem

Planktonic foraminifers have been frequently used in the reconstruction of Pleistocene environmental conditions. The relative abundances of certain species may be used as indicators of paleotemperature according to the observations of foraminiferologists. In a well-executed use of factor analysis, Imbrie and Kipp (1971) tried to establish a quantitative basis for predicting Pleistocene climates from planktonic assemblages. To this end, they first studied the relations between modern foraminiferal assemblages and certain environmental variables (temperature and salinity) and then applied these relationships, in the form of regression equations, to Pleistocene assemblages, with the aim of predicting the environmental variables. The work summarized here has become one of the classics of paleooceanography.

Imbrie and Kipp made the following assumptions:

1. Faunas in the upper few centimeters of Atlantic bottom sediment are related to the physical nature of the overlying surface water.
2. The ecosystem under study has not changed significantly in the past 450,000 years.

Materials

The study focuses attention on two separate sources of data. Firstly, sedimentary samples from the uppermost layers of 61 cores taken from geographically widespread locations in the Atlantic Ocean were analyzed in order to determine environmentally significant, modern assemblages of foraminifers. Each sample can be accepted as a record of sedimentation over the past 2000–4000 years, roughly. Foraminifera from the size fraction > 149 μm were identified, 27 species and "varieties" being recognized. The percentage abundance of each species was defined as the raw, "core-top" data.

At each of the core sites, the average winter temperature, average summer temperature, and average salinity of the surface waters was deduced by interpolation from oceanographical charts. One core, the length of which was 10.9 m, obtained in the Caribbean, was used as a record of Pleistocene depositional history. The age span of this core was ascertained to be approximately 450,000 years. The core was sampled at intervals of 10 cm and the foraminiferal species contained as for the samples for the core top. The resulting 110-by-27 data matrix constitutes the "core data."

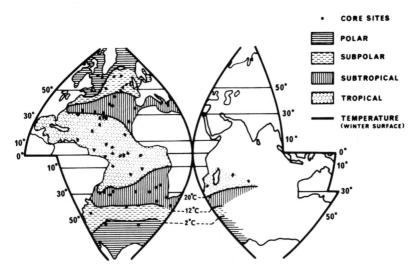

Figure 8.4. Map of the varimax factor loadings on a geographical projection for the foraminiferal frequencies.

Method

The core-top data were analyzed first in order to permit recognition of foraminiferal assemblages. Rare forms were omitted from the records, the effect of which was to produce a data matrix comprising 61 rows (the samples) and 22 variables (the species). The data matrix was transformed so as to express each variable as a percentage of its range. This was done in order to give roughly equal weight to each species. We note that the variables here, frequencies of species, are discrete variables. The matrix was then row-normalized and the major product moment, corresponding to the calculation of the $\cos \theta$ similarity measure, was computed, the principal components of which were extracted. Imbrie and Kipp selected five principal components and rotated these by the varimax method. They interpreted each of the five factors as a theoretical assemblage of foraminiferal species. These factor loadings were mapped onto a geographical projection of the regions sampled. As is illustrated in Fig. 8.4, the foraminiferal assemblages are geographically and ecologically meaningful. The results of the analysis thus far may be expressed algebraically as follows:

$$\mathbf{W}_{ct} = \mathbf{B}_{ct}\mathbf{F}', \qquad [8.1]$$

where $\mathbf{W}_{ct}$ is the row-normalized core-top data matrix, $\mathbf{B}_{ct}$ is the varimax factor-loadings matrix, and $\mathbf{F}$ the factor score matrix.

The next step was to produce a set of predictive equations relating the theoretical varimax assemblages to the measured environmental controls. Although the reasoning employed is rather involved, it may be summarized as follows. The functional relationship between an ecological stimulus, say, average winter temperature and the five assemblages, is of the form

$$T_w = k_1 b_1^2 + k_2 b_2^2 + \cdots + k_5 b_5^2 + k_6 b_1 + k_7 b_2$$
$$+ \cdots + k_{10} b_5 + k_{11} b_1 b_2$$
$$+ k_{12} b_1 b_3 + \cdots + k_{20} b_4 b_5 + k_0, \qquad [8.2]$$

where T_w is the estimated temperature, the b's are the five column vectors of the varimax factor matrix, and k_0 is a constant.

This is a nonlinear equation. The values for winter and summer temperatures as well as salinity are available, which means that the k's can be computed, using the matrix $\mathbf{B}_{ct}$. If we denote the vector of estimated ecological controls as y_{ct} [8.2] may be written in matrix form as

$$y_{ct} = \mathbf{B}_{ct}^2 k + k_0, \qquad [8.3]$$

where $\mathbf{B}_{ct}^2$ is the square of the core-top varimax matrix, now with 20 columns made up of the terms of [8.2], k is the vector of regression coefficients for a particular ecological control, and k_0 the associated constant. The adequacy of the predictive equation can be judged by a test of the multiple correlation coefficient derivable from it. The foregoing steps have attained the goals:

1. Established five environmentally meaningful modern foraminiferal assemblages.
2. Established quantitative predictive equations that relate the assemblages to their environmental stimuli.

The next phase of the study brought in the core data. These too are expressed in the form of percentage range and they are row-normalized. Rather than factor-analyzing the resulting data matrix, $\mathbf{W}_c$, Imbrie and Kipp used the factor score matrix of the core-top analysis in order to describe the core assemblages in terms of those of the core top. This step is one of the main innovations of the paper under discussion. It will be recalled that for the core-top data, the factor equation was given by [8.1]. Since matrix $\mathbf{F}$ is columnwise orthonormal ($\mathbf{F'F} = \mathbf{I}$), [8.1] can be recast as

$$\mathbf{B}_{ct} = \mathbf{W}_{ct}\mathbf{F}, \qquad [8.4]$$

from which it is again evident that $\mathbf{F}$ is a transformational matrix mapping $\mathbf{W}_{ct}$ into $\mathbf{B}_{ct}$.

The matrix $\mathbf{F}$ was then applied to the core-data matrix in order to transform it to a varimax factor matrix, $\mathbf{B}_c = \mathbf{W}_c\mathbf{F}$. Here, each core sample is resolved into contributions of the five core-top varimax assemblages. The final step of the study is to employ matrix $\mathbf{B}_c$ to estimate Pleistocene environmental factors. Firstly, the matrix $\mathbf{B}_c^2$ was

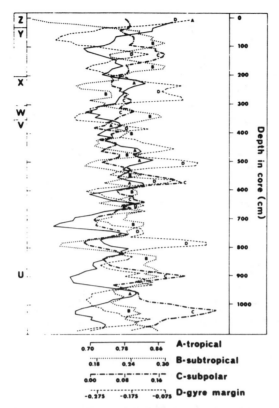

Figure 8.5. Plot of the fluctuations in proportions in the four varimax foraminiferal assemblages from the Caribbean core.

formed as in an earlier step and then [8.3] was utilized in order to estimate values of the environmental stimuli at each position of the core samples. For example, if average winter temperatures are inserted into vector y_c, [8.3] will provide a prediction of the average winter temperature at specific times during the Pleistocene, based on the composition of extant foraminiferal assemblages.

Results

Q-mode analysis of the core-top planktonic foraminifers established the existence of five recent, ecologically meaningful assemblages. Four of these were clearly related to geographically distinct zones, which could be identified as tropical, subtropical, subpolar, and polar assemblages (Fig. 8.4). The fifth assemblage is geographically restricted to the periphery of the North and South Atlantic current gyres and for this

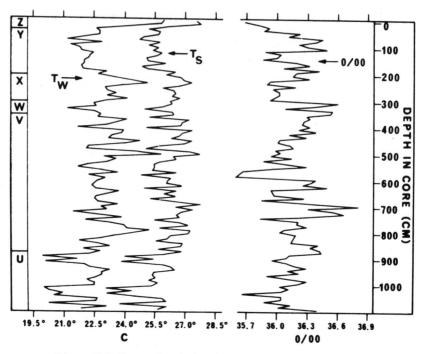

Figure 8.6. Curves for deduced temperatures and salinity of surface waters from the Caribbean during the Pleistocene.

reason, it was called the Gyre Margin Assemblage. The relative abundances of the first four assemblages are clearly related to surface-water conditions. A plot of abundances against mean winter temperatures, for example, shows that there is a close, roughly paraboloid relationship between the two, with temperature optima at 27, 17, 6, and 0°C. This result was used to develop an ecological response model in which mean winter and summer temperatures, and salinity at each core-top locality were predicted, using the abundances of the assemblages in a nonlinear regression equation. Multiple correlation coefficients of 0.99, 0.96, and 0.87, respectively, were obtained for these predictions. The standard error of estimate for mean winter temperature was 1.537 and only two of the estimated values deviated from the true values by more then 3°C.

Samples from the Caribbean core (10.9 m) were analyzed into proportions of the assemblages identified in the core tops. The polar assemblage did not occur in any of these Pleistocene samples. Fluctuations in the abundance of the remaining four assemblages within the core show variations suggestive of systematic climatic fluctuations during the Pleistocene (Fig. 8.5).

Using the ecological response model for the core-top data, in conjunction with the abundances of the assemblages in the core, estimates of mean winter temperature, mean summer temperature, and the salinity of Caribbean surface waters during the Pleistocene were made. The resulting curves are illustrated in Fig. 8.6.

Imbrie and Kipp made use of time series and the analysis of power spectra to show that the peaks and troughs in the derived temperature curves match fairly well those predicted by the Milankovitch model of astronomically induced climatic fluctuations.

We have presented this example of factor analysis in order to illustrate a principle; however, we wish to point out a statistical weakness in the analytical structure that may, or may not, have practical repercussions. Imbrie and Kipp have tended to disregard the element of random variation, which means that the tacit assumption is made that repeated sampling would yield essentially the same result. Reyment (1991) reanalyzed these data using CANOCO (= canonical community analysis).

8.8 PRINCIPAL COORDINATE ANALYSIS OF SOILS

Statement of the problem

Soil surveyors, when examining the soils of an area, often choose a "soil series" as a unit for mapping or for classification. After completion of the work, they prepare an idealized description of a soil considered to be typical of this "soil series."

Rayner (1966) was interested in improving on this subjective type of approach by attempting to identify natural groupings in the relationships between soil profiles. Descriptions of soil profiles are accounts of numerous properties of the soil exposed in the wall of a pit together with the results of laboratory analyses of samples from horizons in the pit, chosen where the appearance or texture of the soil undergoes obvious changes. Very thick layers are, however, subdivided to see whether there are differences that are not visually detectable.

Rayner computed a matrix of similarities between the various soil profiles, and the plot of the first two principal coordinates was employed to establish the groupings.

Materials

Twenty-three profile descriptions of soils in Glamorganshire and the laboratory determinations on soil samples of the 91 horizons into which they were divided by the soil surveyor provided the basic materials for the data matrix used in Rayner's study.

Method

The three classes of information about the objects to be analyzed for the calculation of similarities were as follows. First, dichotomous variables (data scored as "present or absent"; for example, the presence or absence of manganese concretions). The second category, consisting of qualitative variables, was termed "alternatives" by Rayner, by which properties such as color, types of stones (sandstone, shale, limestone) were coded arbitrarily as $0, 1, 2, \ldots$. The third category of variables, termed "scales" in Rayner's paper, embraced continuously and discontinuously varying properties such as chemical constituents and pH of the soil. Rayner considered both Q-mode factor analysis of standard type as well as Gower's Q-mode method, principal coordinates, the latter being particularly suited to data matrices constructed of mixtures of the three kinds of variables involved in the study. Two association matrices were computed, the one based on maximum similarities, the other on averaged soil-horizon similarities.

The eigenvalues and eigenvectors were extracted from the scaled matrix of associations for the 23 soils and the first two vectors of $U\Lambda^{1/2}$ plotted. These plots were compared with those obtained from the eigenvectors of the unscaled matrix of associations. The two sets of plots did not turn out to be greatly different, but Rayner found it possible to give those derived from principal coordinates a more exact pedological interpretation (Rayner, 1966, p. 89).

Results

The soil groups named Brown Earth (BE) and Acid Brown Earth (ABE) by the soil surveyor are well separated on the coordinate plot (Fig. 8.7) and from the single sample of a Rendzina soil. The Gley soils are less clearly separated, although the diagram based on average similarity gives a better separation for them. Usually, it is a difficult task to provide the axes of principal coordinates plots with a precise meaning. Rayner succeeded, however, in doing this tentatively. He suggested one of the axes to be a reflection of alkalinity–acidity, the other to represent an oxidation–reduction bipole. This interpretation was deduced from the distributions of the soil varieties on the graph for, as observed by Rayner, the elements of the eigenvectors cannot be interpreted in a manner analogous to those of principal component analysis. The analysis succeeded in separating almost completely the soil groups to which the soil profiles were allocated.

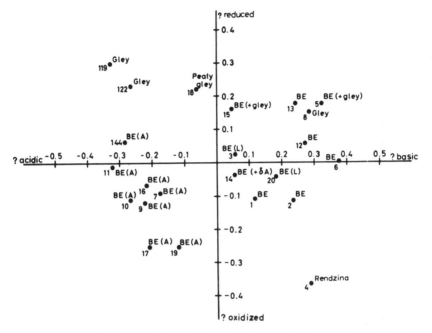

Figure 8.7. Principal coordinates plot for the Glamorganshire soils.

Rayner took his study a stage further by carrying out a numerical taxonomical analysis of the same data, using standard methods. These results gave support to the groupings produced using the method of principal coordinates.

8.9 Q-R-MODE ANALYSIS OF MATURATION OF CRUDE OILS

Statement of the problem

Experimental design, multivariate analysis, and multivariate calibration are important concepts in chemical analyses. An important source of examples is contained in the work by Kvalheim (1987). Many methods for extracting relevant information in petroleum chemistry begin with the configuration variable or object space whereby the problem is oriented toward finding useful projections in *R*- and *Q*-space.

Materials

A good illustration of petrochemical analysis occurs in an article by Telnaes et al. (1987) in which the relationship between phenanthrenes

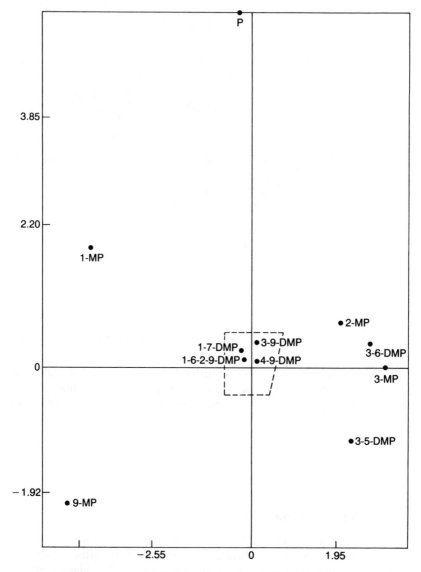

Figure 8.8. Biplot for the phenanthrene data of Telnaes et al. (1987). The points for objects lie within the convex hull in the center of the figure: P = phenanthrene; MP = monomethyl phenanthrene; DM = dimethyl phenanthrene.

in crude oil from the North Sea is studied. Distributions of phenanthrenes and alkylated phenanthrenes in 36 North Sea crude oils were determined by gas chromatography and then analyzed by principal component factor analysis. This analysis showed the occurrence and distribution of the monomethyl phenanthrenes are independent of the relative abundance of phenanthrene. The variables encompassed by the original factor analysis were phenanthrene, four monomethyl phenanthrenes, and six dimethyl phenanthrenes. The values for statistical analysis were obtained by observing the locations of the peaks made by these compounds on the gas chromatograms.

Reyment, who was chief examiner for Kvalheim's thesis in chemistry at the University of Bergen in 1987, made an analysis of the same data by the biplot method for presentation at the public examination.

Results

Using the table of values listed by Telnaes et al. (1987), the biplot shown in Fig. 8.8 was obtained. In all essential details this figure agrees with the principal component plot in the original paper. The component analysis was said to refute existing ideas regarding maturity in crude oils.

The distribution of points in Fig. 8.8 forms a tight grouping of dimethyl phenanthrenes in the center of the graph. Monomethyl phenanthrenes are distributed largely separately and no points fall in the neighborhood of phenanthrene. The tight clustering of sample points is marked by the convex hull in the middle of the figure.

8.10 CORRESPONDENCE ANALYSIS OF METAMORPHIC ROCKS

Statement of the problem

Detailed chemical study of metamorphic rocks is usually made difficult because of the complex, interactive effects of primary composition and secondary, metamorphic alteration. It is one of the goals of metamorphic petrologists to disentangle these effects and discover the operative processes that have determined the ultimate chemical composition of the rocks.

In an expositionary paper on correspondence analysis, David et al. (1974) use this method to group a suite of metamorphic rocks according to their chemical nature.

Table 8.XII. *Correspondence analysis on major and minor elements (all samples), loadings of the first five factors with percentage of the total variance explained*

	Factor 1	Factor 2	Factor 3	Factor 4	Factor 5
Ca	−0.1359	0.1631	−0.0191	−0.0413	−0.0576
Na	0.0107	−0.0485	0.0068	0.0749	−0.0263
K	0.1934	−0.3050	0.2834	−0.2078	−0.0334
Mg	−0.0348	0.2118	0.0272	0.0298	0.0401
Si	0.0115	−0.0479	−0.0054	0.0032	0.0045
Fe^{3+}	−0.0977	0.0488	−0.0742	−0.0699	0.0686
Fe^{2+}	−0.0549	0.1755	0.0985	0.0363	0.0341
Ti	−0.0429	0.1846	0.0328	0.0164	0.0726
Mn	−0.1191	0.1053	0.0268	−0.0022	0.0094
HO	0.1148	0.0923	−0.0371	−0.0582	0.0705
Al	−0.0262	0.0208	0.0014	−0.0013	−0.0242
P	−0.0772	0.0169	0.1296	−0.0327	0.0703
OC	1.1179	0.3721	−0.0401	−0.0053	−0.0597
Percentage of variance explained	46	34	5	4	3

Materials

Seventy-five rocks, analyzed for 22 major, minor, and trace elements form the data matrix. The rocks were collected from the Bourlamaque batholith, a meta-diorite body outcropping near Val d'Or, Québec. The primary igneous mineralogy has been almost entirely replaced by metamorphic assemblages consisting of chlorites, epidotes, paragonite?, leucoxene, quartz, acidic plagioclase, and erratically distributed carbonate minerals. The authors were, however, of the opinion that the batholith has largely retained its primary chemistry, despite the metamorphism.

Method

The determinations of the 22 elements were used directly in a correspondence analysis and no preliminary transformations were applied. We observe in passing that these variables are at variance with the strict theoretical requirements of correspondence analysis, as was pointed out in the foregoing chapter (they are constrained). Separate analyses were carried out on the major and minor elements, trace elements, and all elements combined.

Results

We discuss the results of the analysis made on the thirteen major and minor elements. Two factors account for 80% of the total variation.

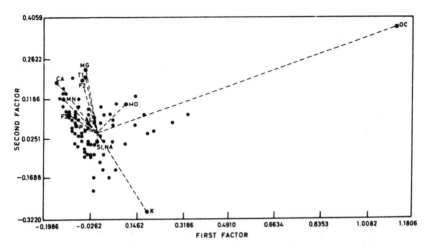

Figure 8.9. Plot of the first two axes of correspondence analysis for the Canadian metamorphic data. Both variables and rock specimens are plotted on the same graph.

Table 8.XII lists the loadings on the variables, and Fig. 8.9 displays a plot of the first two factor axes with both the variables and the rock samples plotted simultaneously. Factor 2 has elements arranged along it that closely follow the expected order in a differentiation series: Mg, Ti, Fe^{2+}, Ca, and Mn at one end, through Fe^{3+}, Al, P, Si, Na, and K at the other end. This axis, applied to the rock samples, groups them according to their positions in the mafic to silicic differentiation series. Factor 1, heavily loaded with respect to CO_2 and H_2O (marked OC and HO in the diagram), was interpreted as being the product of the metamorphism. Rock specimens plotting toward CO_2 and H_2O are thought to have suffered higher degrees of metamorphism than those lying along the trend between K and Mg.

This example illustrates well the power of correspondence analysis in portraying relationships between variables and the effects of these relationships on the objects in the one diagram.

8.11 TRUE FACTOR ANALYSIS IN MARINE ECOLOGY

Statement of the problem

Applications of random case factor analysis in the geological literature are rare. As an example, we offer an unpublished study by Reyment. In an actuopaleoecological study of the Niger Delta and the Ivory Coast, West Africa, Reyment (1969) was interested in comparing properties of the sedimentary interstitial environment of the two areas. Among the many problems tackled in this investigation, including the population dynamics of shelled organisms and the interplay of chemical factors of the environment with the abundance of organisms, it was thought important to establish the relationships between physical and chemical forces of the interstitial milieu. The variables considered in this *R*-mode study are pH, Eh, undissociated oxygen, temperature, and salinity of the interstitial water of the sedimentary samples and the depth from which they were taken.

Materials

Twenty stations were sampled off the village of Bandama. The data matrix thus consisted of 6 columns and 20 rows. The determinations of all variables were made on board ship, the Ivorian research vessel *Reine Pokou*, as soon as possible after the sampling device had been raised. Extra attention was paid to the redox determinations, which were made according to a procedure outlined by Professor Rolf Hallberg. Samples of the kind studied in the present connection are notoriously difficult to get into reliable shape for statistical analysis and there is usually a great deal of slack in the observations, leading to random fluctuations in the results of the computations. It is therefore somewhat surprising that many of the correlation coefficients turned out to be high.

Method

Jöreskog's method of factor analysis (Equation [4.19]) was deemed to provide the most logical vehicle for analyzing the data. It was felt that the statistical aspects of the study were such as to necessitate a reasonable level of control over the random fluctuations liable to exist in the material; in addition, the nature of the problem was more in line with psychometrically conceived factor analysis than the usual kind of "static" situation pertaining in the general run of factor analyses in the natural sciences in that a few simple relationships were expected to be able to account for the associations in the environment.

It was found that 90.07% of the total variance was localized to the first eigenvector of the correlation matrix and 4.26% thereof to the second eigenvalue. The third root accounted for about 2% of the total variance. Selecting $k = 3$ for the number of factors gave a residual correlation matrix with acceptably low values with zeros in almost all of the first decimal places (exception $r_{23} = 0.25$). With $k = 4$, all residuals were zero in the first decimal place. The unrotated factor loadings for

Table 8.XIII. *Unrotated factor loadings for the Ivorian data (k = 3)*

Variables	Communalities	Factor 1	Factor 2	Factor 3
pH	0.54	0.69	0.17	0.19
Eh	0.72	0.50	0.68	0.07
Free O_2	0.46	0.37	0.42	0.39
Temperature	0.85	−0.82	−0.34	0.26
Salinity	0.98	0.99	−0.04	−0.04
Depth	0.98	−0.98	0.05	−0.06

Table 8.XIV. *Varimax rotated factor loadings for the Ivorian data (k = 3)*

Variables	Factor 1	Factor 2	Factor 3
pH	0.57	0.17	0.43
Eh	0.21	0.56	0.60
Free O_2	0.17	0.12	0.64
Temperature	−0.64	−0.63	−0.23
Salinity	0.93	0.27	0.21
Depth	−0.93	−0.19	−0.28

three factors are listed in Table 8.XIII and the varimax rotated factor loadings in Table 8.XIV.

Results

The first rotated factor is largely concerned with an association between pH, temperature, salinity, and depth such that pH and salinity form one set of variables operating against temperature and depth in the other set, which relationship is clearly a bathymetrical factor.

The second factor weighs Eh against temperature, possibly a reflection of differing chemical rates of reaction in sediments at different temperatures.

The third factor relates pH, Eh, and free oxygen, which can be interpreted as being the outcome of the passage of free oxygen in the interstitial liquid to ionized form, capable of influencing the redox potential.

By admitting a fourth significant factor, the only appreciable difference in the foregoing interpretation is that pH is removed from the relationship expressed by Factor 3 and relocated in a fourth factor on its own; this is not an unrealistic solution, as it is to be expected that Eh and O_2 will be bonded closely. These results are tentative, for the sample is very small.

8.12 A PROBLEM IN SOIL SCIENCE

Statement of the problem

Campbell, Malcahy, and McArthur (1970) applied principal coordinate analysis and hierarchical sorting to the classification of some soils in Australia. This study was concerned with testing the usefulness of the so called *Northcote "factual key"* applied to seven readily observable variables. The results of the numerical analysis were found to agree closely with the original field grouping.

The soil profiles were mapped on the Swan River coastal plain, Western Australia at Coolup (eight sites), Fairbridge (four sites), Boyanup (nine sites), Blythewood (three sites), and Dordnup (four sites).

Materials and methods

For the purposes of the present illustration, we use the correlation matrix published by Campbell et al. (1970, p. 52, Table 2). Some of the variables are quite subjective and difficult to determine so that doubts could be entertained about the feasibility of a standard multivariate analysis. The traits are labeled as follows:

1. Hue
2. Value [i.e., decreasing darkness of soil from black ($= 0$) to white ($= 1$)]
3. Chroma (increasing purity of color)
4. Texture [increasing clay content from coarse sand ($= 1$) to heavy clay ($= 10$)]
5. Gravel (proportion of the soil volume made up by gravel)
6. Mottling [degree of intensity of color variations from nil ($= 1$) to $> 50\%$ ($= 9$)]
7. Reaction [decreasing acidity from pH $= 4.0$ ($= 1$) to pH $= 8$ ($= 9$)]

The colors were estimated by reference to the Munsell Soil Color Charts. These variables are of the category "qualitatives" (Gower, 1970) and the correlation matrix therefore cannot be regarded in the same light for accuracy as one obtained from directly measurable variables.

Principal component factor analysis

Campbell et al. concluded, on inspection of the correlation matrix (Table 8.XV), that Hue (variable 1) was poorly associated with all other six variables. Five principal components were noted to account for 90% of the trace of **R**.

Table 8.XV. *The correlation matrix for the western Australian soils*

	1	2	3	4	5	6	7
1	1.000	0.110	−0.186	−0.153	0.078	0.180	0.064
2		1.000	0.240	0.333	0.441	0.505	0.398
3			1.000	0.518	−0.070	0.026	0.312
4				1.000	0.234	0.483	0.532
5					1.000	0.458	0.359
6						1.000	0.509
7							1.000

1 = Hue of soil; 2 = Value; 3 = Chroma; 4 = Texture; 5 = Gravel; 6 = Mottling; 7 = Reaction.

Table 8.XVI. *Principal component decomposition for the correlation matrix of western Australian soils*

	1	2	3	4	5	6	7
Eigenvalues	2.83592	1.46795	0.85773	0.64855	0.50549	0.42853	0.25578
Percentage of Trace of **R**	40.51	20.97	12.25	9.27	7.22	6.12	3.65
Eigenvectors							
1	0.0397	−0.5417	0.7817	0.7819	0.1526	−0.2455	0.0978
2	0.4286	−0.1429	−0.0191	−0.7214	−0.3129	0.3077	0.2871
3	0.2494	0.6036	0.3504	−0.3549	0.2156	−0.2041	−0.4865
4	0.4456	0.3521	0.0478	0.3092	−0.1716	−0.4523	0.5878
5	0.3573	−0.3567	−0.4939	−0.1523	0.5519	−0.4068	−0.0888
6	0.4601	−0.2610	−0.0373	0.3145	−0.5483	−0.0608	−0.5617
7	0.4632	0.0495	0.1339	0.3681	0.4445	0.6554	0.0488

The analysis reported here supports the original work to a fair extent, but there seems to be a little more to the story. The first eigenvector (Table 8.XVI) indicates all variables, but Hue, to covary equally (this could be a sampling artifact). The second vector includes Hue together with Chroma, Texture, Gravel, and, perhaps, Mottling.

Varimax factor matrix

A varimax factor matrix with four factors (Chapter 7) was computed to see whether this could simplify interpretation (Table 8.XVII). The first factor is dominated by Texture, Mottling, and Reaction, with minor contributions from Gravel. There are only two larger loadings, to wit, Chroma (dominant) and Texture (subordinate). Hence the indication

Table 8.XVII. *Varimax factor matrix for the western Australian soil data*

Variable	Communality	Factors 1	2	3	4
1	0.9599	0.0438	−0.1177	**0.9700**	−0.0570
2	0.8888	0.1771	0.3286	0.1460	**−0.8533**
3	0.8981	0.1793	**0.9231**	−0.1158	−0.0218
4	0.8090	**0.7526**	**0.4455**	−0.1931	−0.0831
5	0.7731	0.3511	−0.2898	−0.1001	**−0.7455**
6	0.7657	**0.7441**	−0.1222	0.1644	**−0.4123**
7	0.7154	**0.7954**	0.2052	0.0785	−0.1858
Variance		27.74	18.76	15.09	21.42
Cumulative variance		27.4	46.50	61.59	83.00

given by the second principal component is not repeated in the varimax matrix. If we pass to the third eigenvectors, accounting for 12% of tr **R**, it seems as though Hue dominates with a minor contribution from Chroma and a somewhat larger contribution from Gravel. However, in the varimax matrix, the third factor is dominated by Hue alone – a one-component factor.

This example is one that shows how the rotation of axes can profoundly influence the message delivered by the linear combinations of variables. Here, it is the status of Hue that is under fire. The results of the principal component factor analysis support the original work, but new interpretations could be added. The type of problem represented by the Australian soils occurs quite commonly in the natural sciences. We offer this work as an illustration of how one can cope with data that at first sight may appear intractable.

8.13 THE METHOD OF PRINCIPAL WARPS

The subject broached in this section may seem to lie outside the scope of our book, but this is not so by any means. Consider the prominence given in biometrical and statistical literature to the analysis of size and shape by standard principal components (e.g., Flury, 1988; Flury and Riedwyl, 1988; Seber, 1984; Jackson, 1991). The ensuing presentation is of necessity more mathematical in parts than the level adopted in the rest of the book, the reason for this being that an essential element of the topic derives from differential geometry and mechanics, subjects that seldom confront statisticians, biologists, and geologists. The material presented in this section was extracted from Reyment and Bookstein (1993).

We begin the survey of shape analysis with measured dimensions of an organism, such as underlie the classic concepts of allometry. Differential growth between two lengths certainly supplies a descriptor of shape change in the ratio of rates. But if the problem is generalized to more than two distances, then as soon as those dimensions begin to share endpoints, any understanding of shape variability must take account of the relative locations of the endpoints that are shared. The analysis is perforce of the locations of those points as recorded by any sufficiently numerous set of distances, not really of the distances themselves. This implies that the usual distances of multivariate morphometrics are always arbitrary.

The appropriate archive of data is the set of Cartesian coordinates of labeled points, or *landmarks*. A landmark is at root a natural, easily recognizable feature of a very small spatial extent that is biologically homologous across all the forms of a data set. However, many "pseudo-landmarks" have good operational definitions; for example, site of maximum rounding of an ostracod margin. We offer the present exposé, extracted from Reyment and Bookstein (1993), as an example of techniques of factor-analytical type applied to a typical biological problem.

The multivariate analysis of distance-measures

Consider now a traditional multivariate morphometric analysis of the kind documented by Reyment, Blackith, and Campbell (1984). These types of analyses use conventional distance-measures, with which biologists are familiar. In crustaceans, for example, the natural point of departure for quantification is provided by the dimensions of the carapace: for example, length and height of the shell, its breadth, distances from the anterior and dorsal margins to the adductorial tubercle, and so on. These are *distances* from one endpoint of the character to the other. Please keep this statement in mind for what follows. Most obviously, variation in size is embodied in these characters, but there is also some shape information there that is jumbled to a greater or lesser extent, depending on the choice of variables. Clearly, length and height of a roughly rectangular shell will contain some information on the shape. So will other measures of the carapace, but there is not likely to be much specific information on regions of exaggerated growth rate.

The principal component decomposition of size and shape. In order to bring out the analytical significance of the methods reviewed in this section, we must refer to earlier work on size and shape. Following

Teissier (1938), Jolicoeur and Mosimann (1960) and Jolicoeur (1963) developed a model of size and shape decomposition by means of principal components, based on a posteriori considerations. Briefly, their argument for the principal component decomposition runs as follows.

The eigenvalues and eigenvectors of a covariance matrix represent successive extractions of variation. The vectors are mutually orthogonal and mutually uncorrelated. In the application of covariance matrices to the logarithms of distance measures, the first latent vector, bound to the greatest latent root, usually has all elements positive. The second latent vector, corresponding to the second-greatest latent root, contains both negative and positive elements, as do all subsequent latent vectors, down to the last of them. The relationships for the first two vectors was seized upon by Jolicoeur as being interpretable in terms of size and shape, the argument being that a vector with all its elements positive must describe variability in size. By the same token, the second vector expresses variation in shape.

The principal component model has captured the imagination of statisticians and biologists alike, owing to its simple appeal. However, the following weaknesses must be considered:

1. Growth and shape change encompass a tensorial component – they are at different rates in different directions through the general point of "tissue." This directionality, in general, is not accessible in a principal component decomposition, and spatial variability is likewise quite difficult to pin down.
2. Principal component analysis of size and shape presupposes that shape variation is subordinate to variability in size. If a group varies "more in shape than in size," we have no way of defining either term by means of principal component analysis.
3. Whenever it is not a multiple of a vector of 1s, the first eigenvector embodies an expression of differential growth.

It does sometimes happen that a sensible reification of eigenvalues and eigenvectors can be made in some cases. It is, however, good to remember that the principal component decomposition of size and shape is not the outcome of a particular model.

Principal component factor analysis of carapace dimensions. The five distance dimensions (labeled 1 through 5) illustrated schematically in Fig. 8.10 were analyzed by standard principal component analysis to fit Jolicoeur's (1963) model (N.B. logarithms of the observational vectors used). There are 18 specimens of right valves, 4 of which are juveniles (or paedomorphic females) and 14 adults. The results of the analysis of the raw data are summarized in Table 8.XVIII. The

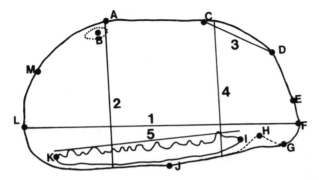

Figure 8.10. Landmarks and distance-measures on the shell. The distance-measures are denoted by arabic numerals, the landmarks and pseudolandmarks by uppercase Roman letters.

Table 8.XVIII. *Principal component analysis for five distance traits (N = 18)*

		\multicolumn{5}{c}{Latent vectors}				
	Latent roots	X_1	X_2	X_3	X_4	X_5
1	0.22193 (81.29%)	0.1612	0.2436	0.9306	0.1610	0.1510
2	0.04208 (15.41%)	0.4744	0.5671	−0.3616	0.4934	0.2812
3	0.00480 (1.76%)	0.0521	−0.7110	0.0109	0.4884	0.5031
4	0.00283 (1.04%)	0.0820	0.0918	−0.0481	−0.6473	0.7507
5	0.00138 (0.51%)	−0.8600	0.3242	−0.0289	0.2703	0.2855

values for the logarithmic data are given in Table 8.XIX.

1. The posterodorsal measure, X_3, is the most variable of the five traits. This turns out to be due to two of the adult specimens, which deviate from the rest of the material. These two specimens may represent a shape-morph in the population and, in effect, they do differ from the other specimens in being more pointed posteriorly. One of these atypical individuals was picked up in the geometric analysis.
2. The first vector of the decomposition, the "size vector," is dominated by X_3, the posterodorsal length measure.
3. Whatever the first vector is expressing, it can hardly be an indicator of pure size.

Table 8.XIX. *Principal component analysis of the logarithmic covariance matrix for 16 specimens*

		Latent vectors				
	Latent roots	X_1	X_2	X_3	X_4	X_5
1	0.13493 (83.52%)	0.3083	0.3880	0.7796	0.2983	0.2401
2	0.01902 (11.77%)	0.3260	0.6031	−0.6099	0.3809	0.1138
3	0.00369 (2.29%)	0.0043	−0.5950	−0.0972	0.6408	0.4753
4	0.00273 (1.69%)	0.0268	0.0818	−0.0947	−0.5505	0.8250
5	0.00119 (0.74%)	−0.8933	0.3536	0.0433	0.2286	0.1514

A simple varimax rotation of the first two axes was made (that is, a principal component factor analysis). This yielded a first factor entirely dominated by X_3, and a second factor in which every character except X_3 is represented. Both these factors combine aspects of size and shape.

Geometry and morphometry

Analysis of distances thus far "lifts" them off the form (just as is done in reality by the ruler or calipers): It treats them as freely translatable and rotatable. The remaining approaches we shall demonstrate here do not permit these maneuvers. The essential added element contributed by a geometrically based methodology are the factors of location and orientation. It is thus possible to view growth in relation to the relative motion of landmarks, which form trajectories traced out over the passage of time. These trajectories can yield important information concerning reorganization of shape. The usual methods of multivariate analysis contain no element permitting the recognition of location of points (viz., landmarks). In order to introduce the essential feature into the study of shape, the techniques of differential geometry become necessary.

Beginning with one form, conveniently to be thought of as the *reference*, a comparison is to be made with a second form, which can be designated as the *resultant*. The changes in morphology induced on growing (evolving) from the one to the other can be described mathematically in terms of deformation in that the first shape must be *warped* before its landmarks can be exactly superimposed on those of the second. (The word "warp" is used in its root sense, meaning the *crooked state brought about by uneven shrinkage or expansion*.) In general terms this can be referred to as a Procrustean comparison.

Principal axes of deformation

If we are to take advantage of a model of growth as deformation, we need some means of quantification. One method that has proved both descriptively helpful and statistically tractable is an adaptation of the old concept of strain analysis from continuum mechanics. Consider a small circle of tissue in one organism and the homologous piece of tissue after its deformation onto the figure of another. If the patch of tissue is small enough, the transformation of the circle is very nearly an ellipse. For many purposes, it is sufficient to describe the rate of strain (ratio of change of length less unity) as a function of direction through the point at the center of the circle. These rates are clearly proportional to the radii of the ellipse representing the form after deformation. Now, there is a longest and shortest radius of the ellipse, at 90°; these are called the major and minor *principal strains* of the transformation, and the logarithms of their ratio of lengths, or log-anisotropy, prove to be the most useful way of summarizing the "magnitude" of the transformation in this bit of tissue.

Shape coordinates. For variations in shape that are not too great, the analysis and interpretation of changes in the locations of a set of landmarks can be made in an informative manner by considering the landmarks, three at a time, over the form of interest. For the complete coverage of N landmarks, there must be at least $N - 2$ triangles in a *rigid configuration*. The classical T^2 statistic of multivariate analysis may be used for testing shape differences between two populations of triangles.

Landmarks on Neobuntonia. The landmarks on *Neobuntonia airella*, the ostracod species used to illustrate the method, are indicated schematically in Fig. 8.10:

A. the cardinal angle of the valve (landmark);
B. the location of the eye tubercle (a landmark);
C. the maximum dorsal convexity (a pseudolandmark);
D. the posterodorsal angle (a landmark);
E. the posterior marginal concavity (a pseudolandmark);
F. the posterior tip of the posterior process (a landmark);
G. the ventral tip of the posterior process (a landmark);
H. the posteroventral concavity (a pseudolandmark);

I. the posterior rib termination (a landmark);
J. the maximum ventral rounding (a pseudolandmark);
K. the anterior rib termination (a landmark);
L. the midpoint of the anterior marginal rounding (a pseudoland-
mark);
M. the midpoint of the anterodorsal margin (a pseudolandmark).

The material comprises 29 left and right valves (LV and RV), adults of both sexes and a few specimens referable to the final moulting.

Principal warps and the thin-plate spline.   The thin-plate spline is a general purpose *interpolation function* permitting us to visualize sets of landmark displacements as deformations. In principle, we could use as our visualization any mapping whatever that interpolated the landmark correspondence. The spline is a uniquely determined member of this class of interpolants, the function optimizing a very special measure of "bending." To explain what we mean by the "bending" of an interpolant, we need to think about our two-dimensional data as if they really existed in three dimensions.

Consider a completely *uniform transformation*, the kind that takes square graph paper to graph paper with all its cells shaped liked the same parallelogram. This is the uniform shape change you get if you distort all circles into ellipses of the same shape; such transformations are also called *affine* – they leave parallel lines parallel. Any change in shape of a set of landmarks has a uniform, or affine, part and a nonuniform, or nonaffine, part. The uniform part of shape change represents a subspace of shape coordinates of rank 2 and the nonuniform part a complementary subspace of rank $2K - 6$, where K is the number of landmarks. These two parts are incommensurate, so that there is no useful measure of "net shape distance." The measure of uniform shape distance is usefully taken as some variant of the bending energy described previously.

The complete description of a change of form involves separate descriptors for uniform shape change, nonuniform shape change, and change in centroid size. These must be combined by correlational methods, not as any "net change."

The theory of principal warps is based on the mathematical description of the deformation of a thin metal plate. Any single nonuniform transformation may be expressed as a finite sum of *principal warps*, which are the characteristic functions of the bending energy corresponding to Procrustean orthogonal displacements of the conceptual metal plate at landmarks. The sharper the degree of bending of the metal plate, the greater is the bending energy required to cause the deformation.

The thin-plate spline model and shape. The analysis of shape by the thin-plate spline is based on the properties of a function that describes the surface

$$z(x, y) = -U(r)$$
$$= -r^2 \log r^2, \qquad [8.5]$$

where r^2 is the distance $(x^2 + y^2)$ from the Cartesian origin. The function U satisfies the second-order partial differential equation

$$\Delta^2 U = \left[\frac{\partial^2}{\partial x^2} + \frac{\partial^2}{\partial y^2} \right]^2 U \propto \delta_{(0,0)}. \qquad [8.6]$$

The right-hand term denotes proportionality to the generalized function $\delta_{(0,0)}$. It is zero everywhere except at the origin. Its integral is 1.

The partial differential equation applied to a linear combination of delta functions of different centers and heights actually describes the form of an infinite, infinitely thin metal plate (the so-called thin-plate spline), originally flat, and now constrained to fixed distances above or below any set of landmarks in the plane of the "shadow." The formula of the plate is a linear combination of the kernels U relative to the landmark locations together with an affine term. In that version, the algebra represents the shape of a thin plate as a function $z(x, y)$ above the (x, y) plane. For the application to morphometrics, take the "shadow" after bending, add such a function $z(x, y)$, first to the x coordinate, then to the y coordinate of the original square grid. There results a very fast and flexible implementation of Thompson's original idea, a smooth transformation exactly matching the landmark locations.

The bending energy matrix. The bending energy matrix is defined as $\mathbf{L}_K^{-1}$ (Bookstein, 1989b) for the function $U(r)$ ([8.5]). Its calculation proceeds as follows.

Define a matrix

$$\mathbf{P}_K = \begin{bmatrix} 0 & U(r_{12}) & \cdots & U(r_{ik}) \\ U(r_{21}) & 0 & \cdots & U(r_2 k) \\ \vdots & \vdots & \cdots & \vdots \\ U(r_{k1}) & U(r_{k2}) & \cdots & 0 \end{bmatrix}, \qquad [8.7]$$

which is of order $K \times K$. $\mathbf{Z}_K = z_1(x_1, y_1), \ldots, z_k(x_K, y_K)$ are K points in the Euclidean plane in a Cartesian system. Here,

$$r_{ij} = |z_i - z_j|$$

is the distance between any two points. Also write

$$\mathbf{Q} = \begin{bmatrix} 1 & x_1 & y_1 \\ 1 & x_2 & y_2 \\ \vdots & \vdots & \vdots \\ 1 & x_K & y_K \end{bmatrix},$$

and

$$\mathbf{L} = \begin{bmatrix} \mathbf{P}_K & \mathbf{Q} \\ \mathbf{Q}' & \mathbf{0} \end{bmatrix},$$

where $\mathbf{Q}$ is a $K \times 3$ matrix and $\mathbf{0}$ is a 3×3 null matrix.

The bending energy matrix is $\mathbf{L}_K^{-1}$, the upper-left $K \times K$ submatrix of $\mathbf{L}^{-1}$. It is called the bending energy matrix because the bending energy of the plates we have been talking about can be expressed in terms of the values of $\mathbf{YL}_K^{-1}\mathbf{Y}'$, where $\mathbf{Y}$ is a vector of coordinates of the "heights" of the plate above the starting landmarks. The equation of the interpolating spline itself – its affine part together with the sum of the Us – is embedded in the vector $\mathbf{L}_K^{-1}\mathbf{Y}$ (Bookstein 1989b, 1992).

The latent vectors of $\mathbf{L}_K^{-1}$ can be interpreted as the coefficients of a thin-plate spline attached to the base-plane because they are coefficients of the function U located at K landmarks. These latent vectors were termed the *principal warps* by Bookstein (1989b). They are useful for depicting the nonzero latent roots (N.B., there are three zero latent roots because $\mathbf{0}$ in $\mathbf{L}$ is 3×3). The characteristic value of each eigenvector can itself be construed as a specific bending energy, namely, that of the (partial) warp corresponding to its eigenvector when each coefficient is treated as the height above the corresponding landmark. The nonuniform parts of any deformation can be decomposed as a sum of these principal warps as they apply first to the x coordinate of the deformation and then to the y coordinate. In effect, we have multiplied each warp by a 2-vector and used it to displace all the landmarks by the pattern of the heights of the principal warp in this single direction (or its reverse). These components of deformation are called the *partial warps* of the thin-plate spline.

Example. Principal Warps for *Neobuntonia*

   Presentation of the problem.   The material of larval and adult right and left valves will be examined for shape differences with respect to the chosen landmarks and pseudolandmarks as illustrated in Fig. 8.10. Some of the details of the material were presented in connection with the formal multivariate analysis on p. 266.

   The essential features of the principal warp calculations are summarized in Table 8.XX. Note first that the uniform part of the transformations is not always insignificant and for some comparisons, relatively

Table 8.XX. *Principal warp comparisons of left and right valves*

	Male–female			Male–larval			Female–larval		
	x	y	Energy	x	y	Energy	x	y	Energy
Left valves									
1	−0.036	−0.053	0.003	0.026	0.021	0.001	−0.065	−0.059	0.016
2	0.110	0.192	0.014	0.099	0.219	0.017	0.208	−0.036	0.014
3	−0.220	−0.007	0.011	0.002	−0.032	0.000	0.004	0.055	0.001
4	0.102	0.192	0.003	0.143	0.165	0.003	0.032	0.004	0.000
5	0.046	−0.004	0.000	−0.065	−0.043	0.000	0.079	0.043	0.001
6	−0.383	−0.152	0.006	−0.471	−0.130	0.009	−0.141	−0.050	0.001
7	−0.111	0.087	0.000	0.177	0.306	0.001	0.263	0.201	0.001
8	0.163	−0.106	0.000	−0.017	0.289	0.001	−0.118	0.392	0.002
Percent anisotropy									
	5.1			5.4			8.8		
Right valves									
1	−0.036	0.026	0.002	−0.017	−0.076	0.007	0.015	−0.093	0.012
2	0.096	−0.021	0.003	0.176	0.002	0.008	0.086	0.018	0.002
3	−0.346	−0.009	0.027	−0.258	−0.024	0.015	0.038	−0.015	0.000
4	0.330	0.127	0.010	−0.050	0.192	0.003	−0.187	0.054	0.005
5	0.080	0.018	0.000	0.199	−0.065	0.002	0.267	−0.118	0.005
6	−0.010	0.026	0.000	0.337	−0.098	0.006	0.384	−0.124	0.009
7	−0.049	0.120	0.000	−0.543	0.274	0.005	−0.336	0.224	0.002
8	−0.142	−0.181	0.001	0.286	0.059	0.001	0.534	0.144	0.004
Percent anisotropy									
	8.7			20.7			12.9		

large values occur. This is particularly so for right valves. The numbering of warps is from smallest scale (largest specific bending energy) to largest scale (smallest specific bending energy), following the convention of Rohlf's TPSPLINE.*

Significant warps, as expressed by geometric length (values in Table 8.XX) are as follows:

1. left males and females, warps 2, 3, 6, and 8;
2. left males and final moults, warps 6 and 7;
3. left females and final moults, warps 7 and 8;
4. right males and females, warps 3 and 4;
5. right males and final moults, warps 3, 6, 7, and 8;
6. right females and final moults, warps, 5, 6, 7, and 8.

---

* TPSRW and TPSPLINE are two programs for doing the calculations of this section that were distributed as part of the proceedings of the Michigan (Ann Arbor) Workshop on Morphometrics (Rohlf and Bookstein, 1990).

In the fit between males and females, the second warp is mainly concerned with a small-scale change, that involving negative allometry between the location of the eye tubercle and the cardinal angle. The third and sixth warps are likewise concerned with a local feature, to wit, the posteroventral region. The fourth warp describes corresponding anteroventral change. The shape relationships between left valve adults and moults are concentrated in warps 6 and 7, both of which encompass many landmarks moving in different directions. The sixth warp reflects deformation in the entire ventral region – the posterior process becoming narrower and the posteroventral concavity more acute. The seventh warp is likewise an indicator of change in ventral shape but on a larger scale than for the sixth warp. For females and the moulting class, warps 7 and 8 are likewise large-scale descriptors. The former describes shape-changes in the posterior half of the shell, the latter changes spread over the whole shell.

The uniform part of the transformations between the three categories being contrasted is relatively larger for right valves (Table 8.XX). The results for the nonaffine part of the transformation are, however, similar to those found for the left valves. There is for males and females a warp (third) describing the change in location of the eye tubercle, and the fourth warp also here depicts change in landmarks 3, 4, and 5. Males contrasted with the final moult resemble the left valve for the sixth and seventh warps. In addition, there is a descriptor of shape change in the ocular region (warp 3) and one that expresses a relationship between ventral inflation and maximum length of the shell (warp 8). Females contrasted with the penultimate growth-stage differ in several details of left valves. The fifth warp describes posterior growth changes, the sixth warp a relationship between ventral convexity and the posterior, the seventh warp is global, and the eighth warp is a descriptor of the relationship between total length and ventral curvature, just as was found for males and moults.

The analysis by principal warps shows that the shape changes have a significant affine component. This is not unexpected: Organisms that moult, such as crustaceans, tend to produce magnifications of the previous instar. In the case of ostracods, this is most marked for the shift from the final moult to the female, as is amply demonstrated in our material.

Relative Warps

The foregoing exposition has dealt with the decomposition of *single changes* of landmark configuration, representing means of two groups at a time. We have exhaustively re-expressed these configurations as the superposition of vector multiples of the principal warps, eigenfunctions of bending energy, together with one affine transformation, which can be thought of as a partial warp corresponding to the eigenspace of bending energy zero (the uniform transformations).

This methodology can be easily expanded into an analog of ordinary principal component analysis for the description of *within-sample varia-tion*. For an extended exposition, see Bookstein (1992, Section 7.6). We may interpred each form of an ostensibly homogeneous sample as a deformation of the sample mean form into that single form. This deformation may be described, like any other comparison of the same set of landmarks, by its affine component and its partial warps with respect to the principal warps of the mean configuration. This decom-position generates a set of derived shape variables over the list of specimens in the homogeneous sample. The reader well versed in "multivariate imagination" may construe this set of variables as a linear transformation of any other basis for the same shape space, such as the set of shape coordinates of the landmarks with respect to any two as baseline, or the set of residuals from an affine Procrustean fit (Rohlf and Slice, 1990) together with the specification of that affine fit in shape-coordinate form. In this representation, the uniform part may be estimated by the factorial approximation of Bookstein (Rohlf and Bookstein, 1990, p. 243) to any suitable baseline. [The best baseline (Bookstein, 1992, Section 7.2) will often be an interlandmark segment approximating the long axis of the within-specimen landmark scatter.]

The derived variables so obtained describe the variation of form first at the "largest possible" level and then at a series of decreasing geometric scales, from the largest nonlinearity, that which is most like a linear growth-gradient, through the reconfiguration of the nearest pair of landmarks with respect to their surround. A metric representing anisotropy corresponds to the uniform part, to wit, the ratio of axes of the ellipse into which a "typical" circle in the mean form is taken by the deformation into the individual specimen. The principal components of sample variation in this two-dimensional space may be extracted visu-ally. The analysis of the complementary space of shape variation – the space of changes that are not the same all across the form, and so may be thought of as "regionalized" – may be surveyed by ordinary princi-pal component analysis after each principal warp is deflated by its specific bending energy. The principal components will then represent *aspects of shape variation in order of variance per unit bending*. In this way, they exactly realize the purpose of principal component analysis in conventional multivariate morphometrics, which attempts to extract aspects of shape–size variation from measured distances in order of "variance explained." What is taken for granted in the conventional analysis, namely, that the "scale" of a pattern is its sum of squared loadings, is explicitly controlled in the geometric analysis by reference

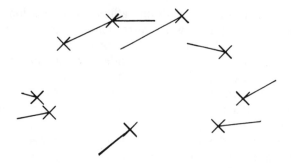

Relative warp 1,   "eigenvalue"   0.210519

Figure 8.11. First relative warp for one sample of 18 specimens of right valves.

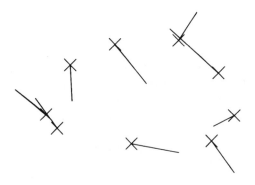

Relative warp 2,   "eigenvalue"   0.124042

Figure 8.12. Second relative warp for one sample of 18 specimens of right valves.

to the actual geometric scale of the patterns, as calibrated by inverse specific bending energy. (Remember that the *closer* the landmarks, the more energy it takes to "bend" a plate to a specific nonlinear pattern of "heights" above them.)

Example. Relative warps for *Neobuntonia*

We return now to our example. We shall now apply this technique of relative warps to a sample of 18 right valves of *Neobuntonia airella*, the

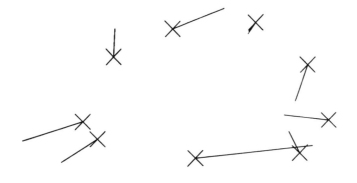

Relative warp 3, "eigenvalue" 0.054457

Figure 8.13. Third relative warp for one sample of 18 specimens of right valves.

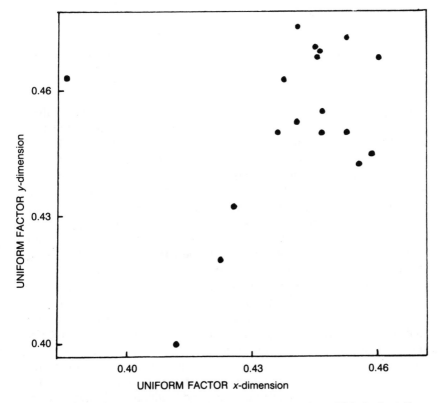

Figure 8.14. Uniform factor: *x*- and *y*-dimensions. This is the affine part in the relative warp analysis.

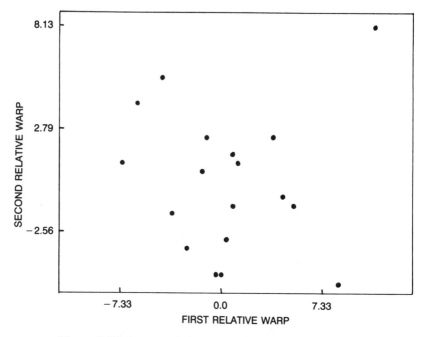

Figure 8.15. Scatter of the scores for the first and second relative warps.

same specimens as were used in the conventional multivariate analysis earlier on. Analysis of these data to an *F–L* baseline, along the long axis of the animal, results in interesting relative warps (Figs. 8.11–8.13) and corresponding scatters of scores. The uniform component of these 18 valves shows one clearly outlying individual (see Fig. 8.14). It is strongly sheared (about 17%) in relation to its fellows. A scatter of the first two relative warps shows that this outlier from the uniform distribution is also an outlier in the largest scale of nonlinearity (Fig. 8.15); in fact, it is wildly aberrant with respect to the distance between landmarks A and M. A plot of the first relative warp against the third (Fig. 8.16) shows a very suggestive structure of subsamples. There appear to be two extreme forms on relative warp 3, those located at the bottom of this plot (marked by a square and a circle). Of these, the one at right (circle) is the outlier we have already seen on the uniform component and the most prominent atypicality found in the analysis by cross-validation. The other, at lower left in the plot (square) of the first and third relative warps, is the extreme form at lower center in the uniform plot: end-member of an atypical dimension of within-sample form. Landmarks D and C are unusually far apart, and landmark D is unusually far

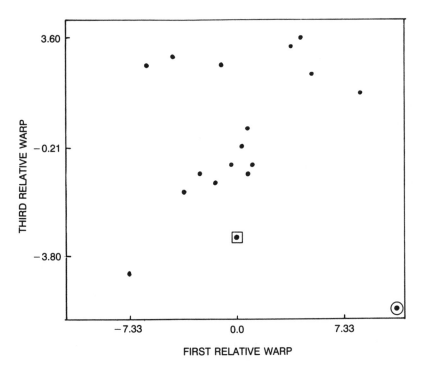

Figure 8.16. Scatter of the scores for the first and third relative warps.

posterior. Also the form is untypically squat (as encoded in the uniform *y*-component).

The third relative warp, on which both our outliers score quite low (Fig. 8.16), is essentially a relative displacement of both poles of the organism with respect to the landmarks lying between. The first relative warp, for which these outlying forms constitute the two extreme scores, is likewise a large-scale feature. The contrast between dorsal and ventral extends along this boundary. With these extreme tendencies removed from the sample, the plot of the affine component appears circular, consistent with a model of variation that is isotropic at this largest scale. Referring back to the principal coordinate analysis, we experience that one of the specimens indicated as being atypical by the ordination was also picked out by the relative warps.

Conclusions

Affine shape change is important. This is not unexpected, bearing in mind the manner in which crustaceans grow – whereby each successive stage tends to be an expanded image of the preceding stage. Changes in shape would then tend to be substantially linear. Affine change is more noticeable for right valves than for left.

The location of the eye tubercle is subject to a shift parallel to a baseline F–L corresponding to maximum length of the shell, whereby it tends to be displaced in a backward direction.

Right and left valves differ markedly in shape on all landmarks except the location of the eye tubercle, landmark B, which incorporates some overlap.

The analysis of the sample of right valves by the method of relative warps was successful in isolating and characterizing two atypical individuals, here interpreted as shape-morphs. When these extreme tendencies were deleted from the analysis, the affine component appears circular, which conforms with a model of isotropic variation. We could also identify several structural modifications in the material. One of these encompasses the posterior process, narrow or broad, and the other general shifts in the ventral configuration of the carapace.

Classical multivariate analysis was also successful in picking up atypicalities; in one case, the same specimen was identified by both methodologies.

It is now appropriate to compare and contrast the information yielded by the two classes of analysis invoked here. First, with regard to extracting information on variation in shape, there is little doubt that geometric morphometrics is inestimably more efficient than standard multivariate analysis. Not only does it highlight the geometry of sites at which changes occur, but also successfully describes, in analytical terms, uniform shape differentiation contrasted with nonuniform separation.

The shape content of principal components in the present example derives principally from the X_3/X_1 ratio and hence indirectly from posterior pointedness.

We have also used the relative warps to ordinate specimens. In this connection both approaches yield similar results, although with slight shifts in emphasis, presumably the outcome of size-confounding in principal components.

8.14 TWO SIMPLE CASE HISTORIES

It can be useful and instructive to see what can be done with the aid of a suitably constructed multivariate statistical analysis. The skills that have now been imparted to you place you in a favorable position for attacking any kind of multidimensional problem amenable to a factor-analytical treatment in the broad sense. The methods are quite general. In order to help you realize this, we have included two case histories, both of which deal with topics of considerable interest at the time of writing of these pages. The first deals with the mineral chemistry of *impacts* caused by a body from outer space striking the Earth. The second is related to the *hydrology* of inland water bodies. Neither of these examples is particularly exciting from the statistical point of view; their main interest derives from the opportunities they provide to illustrate how an analysis can be constructed. The data of the first example are easily identifiable as being compositional and therefore requiring special attention. The presence of constraints in the second example is far less obvious.

8.14.1 Impact glass at the K-T boundary

Introduction. There is a layer 50 cm thick at the Cretaceous–Tertiary (K-T) boundary at Beloc, Haiti, that contains spherules of silicic black glass. This layer has been interpreted as having been produced by the fusion of continental crust following on the energy released by the impact of an object from outer space (Sigurdsson et al., 1991). Yellow glass occurs as both a coating on the black spherules and as inclusions within them. This has been attributed to melting of carbonate-rich sediments or, alternatively, melting of limestone containing organic matter or pyrite. Sigurdsson et al. (1991) tested the likelihood of these suggestions by experiments involving the production of Ca-glass by melting sediments rich in evaporites and melting andesite with evaporites.

It was concluded that the silicic glass came from continental crust of andesitic composition, whereas the high-Ca glass was formed by melting of evaporite-rich sediment. Based on the variation of the sulfur content, it was inferred that the Haitian high-Ca glasses were formed at a temperature of about 1300°C (the range of trials was 1100, 1200, 1300, and 1400°C). It was moreover suggested that the Haitian impact resembles closely the K-T structure at Chicxlub, Mexico, which lies in evaporites of Cretaceous age. The study made by Sigurdsson et al. (1991) was linked to the problem of emissions of sulfur dioxide and short-term hemispheric cooling of the troposphere (amounting to about 4°C).

Statistical methods. A statistical comparison of the published analyses can point to possible interesting geochemical relationships that cannot be adequately appreciated from mere inspection of analyses. There are eight samples in the data matrix, which may seem very little, but each of the compositions represents averages obtained from several

Table 8.XXI. The correlations, simplex (upper triangle) and usual (lower triangle), for the impact glass from Hau:

	SiO_2	Al_2O_3	MgO	CaO	Na_2O	K_2O	SO_3
SiO_2	1.0000	0.9607	0.4904	-0.8848	0.7914	0.7122	-0.7899
Al_2O_3	0.8698	1.0000	0.5925	-0.9375	0.8346	0.7723	-0.8311
MgO	0.0437	0.3322	1.0000	-0.7010	0.6106	0.4593	-0.5365
CaO	-0.9627	-0.9652	-0.2111	1.0000	-0.8752	-0.8533	0.8523
Na_2O	0.8110	0.9224	0.3406	-0.8773	1.0000	0.9563	-0.9950
K_2O	0.7511	0.9061	0.2657	-0.8486	0.9648	1.0000	-0.9647
SO_3	-0.6591	-0.6685	-0.2596	0.6642	-0.8400	-0.7696	1.0000

Table 8.XXII. *Constrained eigenvalues and eigenvectors for the compositional correlation matrix for the impact glasses*

	1	2	3
Eigenvalues greater than 5% of tr **R**			
	5.74501	0.65920	0.47229
Eigenvectors			
SiO_2	0.3741	−0.0629	0.6108
Al_2O_3	0.3927	0.0329	0.4511
MgO	0.2808	0.8786	−0.2790
CaO	−0.4029	−0.1347	−0.1431
Na_2O	0.4020	−0.1436	−0.3130
K_2O	0.3814	−0.3472	−0.3799
SO_3	−0.3970	0.2523	0.2878

independent determinations. The oxides included in our analysis are SiO_2, Al_2O_3, MgO, CaO, Na_2O, K_2O, and SO_3.

These data are constrained in the usual manner of chemical tables, which precludes the use of standard multivariate methods. They are typically compositional in that the rows of the data matrix have a constant sum. The *R*-mode analysis must therefore be made by the procedure appropriate to simplex sample space (see Chapter 4, Section 4.7).

The analysis. Table 8.XXI lists the simplex correlations (upper triangle) in relation to the ordinary correlation coefficients (lower triangle) for the same data set. You will see that even if some of the correlations do not differ by very much, there are other comparisons that are very unlike. This is a common situation for compositional data, as we have had occasion to point out earlier in this book.

The principal components given in Table 8.XXII for the compositional data differ little from those yielded by the standard method with respect to the first eigenvector. There are significant differences in some elements for the second pair of principal components, particularly with respect to the entries corresponding to K_2O and SO_3. The first eigenvector can be expressed in the form of an equation:

$$z_1 = 0.37SiO_2 + 0.39Al_2O_3 + 0.28MgO - 0.40CaO$$
$$+ 0.40Na_2O + 0.38K_2O - 0.40SO_3,$$

associated with 82.07% of the information in the sample. The second principal component is

$$z_2 = 0.88MgO - 0.35K_2O + 0.25SO_3.$$

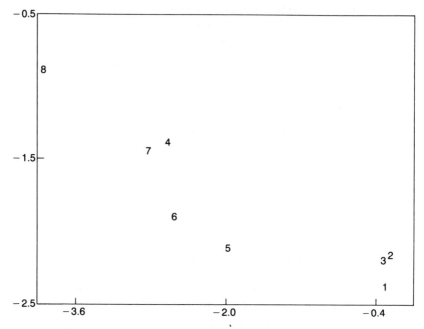

Figure 8.17. Ordination using the scores of the first two principal components for the impact glasses: (1) Average black impact glass, Haiti. (2) Synthetic high-silica glass from melt of andesite with gypsum. (3) Synthetic high-silica glass, melt of andesite with calcite. (4) Yellow high-Ca impact glass, Haiti. (5) Synthetic yellow glass from melt of andesite with evaporite (= gypsum) at 1402°C. (6) Synthetic yellow glass from evaporite melted with andesite at 1308°C. (7) Synthetic yellow glass from melting evaporite with andesite at 1308°C. (8) Synthetic yellow glass from melting evaporite with andesite at 1167°C.

This combination represents only 9.42% of the information in the data. The first principal component seems to point to the anhydrite (evaporitic) connection postulated by Sigurdsson et al. (1991).

Concluding analysis. The constrained component analysis puts us in a favorable position for providing further comments regarding the K-T impact episode and the chemical effects manifested at the site of the impact.

1. The first principal component for the impact glasses and the synthetic counterparts shows equal participation of all components, but with $CaSO_4$ balanced against the remaining components embodying the chemical properties of andesitic crust.
2. The ordination diagram obtained by plotting the first and second principal component scores (Fig. 8.17) supports the

hypothesis for the origin of the black impact glass from continental crust of andesitic composition. The results for the yellow glasses are not in such good agreement with the conclusions of the study analyzed here. The variety of synthetic glass most like the Haitian yellow high-Ca glass is the one that was made at a temperature of 1208°C, and not the one formed at 1308°C, as was originally concluded, at least on the ordinations of the data for the first three principal components. That product plots somewhat further from the actual specimen of high-Ca impact glass. Hence, the weight of evidence points to a slightly lower temperature of formation than was originally surmised.

8.14.2 Sensitivity parameters in a hydrological (ecometric) study

Introduction. Wallin (1991) collected data from a number of small enclosed coastal areas of archipelago type in southeastern Sweden. Part of his project was concerned with determining the *sensitivity* of these areas, as expressed by environmental state variables such as coastal morphology, water turnover, and bottom dynamic conditions, all of which can be linked to nutrient retention in coastal waters. The work resulted from a practical appraisal of the effects of fish farming. The following 11 variables were selected from among those used in the sensitivity study:

1. D_{max} = maximum depth in meters;
2. l = coastline length in kilometers;
3. I = average area of islands in square kilometers;
4. I_{ns} = island density expressed as a percentage;
5. L = strandline length in kilometers;
6. A = total area in square kilometers;
7. a = area of water surface in square kilometers;
8. A_t = average area in square kilometers;
9. A_b = bottom area in square kilometers;
10. V = water volume in cubic kilometers;
11. D_m = average depth in meters.

Wallin did not try more advanced methods than bivariate correlations and regression, nor were his data screened for statistical conformity. Our own analysis shows that several of his variables diverge markedly from the requirements of the Pearsonian correlation coefficient. This is clearly apparent in some of the published scatter diagrams (e.g., Wallin, 1991, pp. 131, 132, 145, 155, etc.).

Statistical methods. Rather uncertain data such as those analyzed in this example necessitate particular attention to statistical

Table 8.XXIII. *First four eigenvalues and eigenvectors for the correlation matrix of the environmental sensitivity data*

	1	2	3	4
Eigenvalues				
	5.4575	2.7121	1.6417	0.5956
Eigenvectors				
1	−0.2842	0.2153	−0.3645	0.5156
2	−0.0973	−0.4728	0.2009	0.5205
3	0.0220	−0.4244	−0.5115	−0.3240
4	0.0299	−0.4042	−0.5474	−0.0888
5	−0.3140	−0.3158	0.0700	0.3691
6	−0.4029	−0.1432	0.1469	−0.2105
7	−0.4044	−0.1073	0.1669	−0.2305
8	−0.3428	0.2070	−0.2045	−0.0076
9	−0.4044	−0.1068	0.1669	−0.2314
10	−0.4069	0.1145	−0.0435	−0.1950
11	−0.1933	0.4329	−0.3735	0.1582

detail. It is good policy to begin with standard univariate methods in order to prepare the ground for the coarser artillery. This preliminary analysis of Wallin's data showed that the distributions for his sensitivity variables 2, 3, 4, 8, and 10 are strongly skewed and would consequently have a detrimental effect on any multivariate analysis applied to them.

This condition could be "totally remedied" by transforming the data to their logarithms. The multivariate analysis was therefore made on the logarithmically transformed data matrix of order 22×11. Principal component factor analysis was used, supported by the methods of Section 4.7.

Results.   The cross-validation study indicates that there are three significant eigenvalues for the covariance matrix and four for the correlation matrix, so we shall proceed with four roots. Both matrices are of rank $p - 1 = 10$, with one zero eigenvalue. Eigenvalues 10 and 9 are so small as to make us suspect that the rank of these matrices are even less than $p - 1$. This is a condition provoking thought, the reason for which probably lies with the nature of the areas and volumes that seem to introduce a constraint.. The pertinent eigenvalues and eigenvectors are listed in Table 8.XXIII.

1. The cross-validation analysis shows that none of the variables is so vital that its deletion would markedly affect the results.
2. The deletion of cases for the analysis based on the covariance matrix gave the interesting outcome that none of the sites has

Table 8.XXIV. *Varimax factor matrix for the sensitivity data*

Variable	Communalities	Factors 1	2	3	4
1	1.00	0.28	**0.92**	0.00	0.04
2	0.95	0.24	−0.25	−0.20	**0.89**
3	1.00	0.00	0.00	−**0.99**	0.10
4	0.96	0.10	0.00	−**0.96**	0.11
5	0.94	**0.66**	0.11	−0.18	**0.57**
6	1.00	**0.98**	0.11	0.00	0.19
7	1.00	**0.98**	0.12	0.00	0.15
8	1.00	**0.52**	**0.57**	0.05	−0.07
9	1.00	**0.98**	0.12	0.00	0.15
10	0.99	**0.83**	**0.45**	0.07	−0.05
11	1.00	0.12	**0.92**	0.13	−0.28
Variance		40.24	21.21	18.02	11.98

Significant loadings on our ad hoc criterion are marked in bold type.

an influential effect on the variance relationships. On the other hand, there are several that influence the covariances. These are localities 2, 3, 8, 9, 16, 19, 20, and 22. This result is not unexpected if we inspect the bivariate plots for the data (Wallin, 1991). This promotes an attitude of healthy caution to the acceptance of Wallin's regressional and correlational studies.

3. If we go in via the correlation matrix, a somewhat different picture is painted. Localities 2, 4, 9, 11, and 17 show up as influential with respect to variances, and sites 2, 4, 9, 11, 13, and 17 as influencing the covariances. The presence of so many samples showing deviations from the main body of the material is a sure indication that any multivariate analysis applied to the data matrix is not going to be very reliable. Our advice would be to give careful thought to removing localities 2, 4, 9, 11, and 17, at least, from the study.

4. The factor analysis (Table 8.XXIV), using the varimax rotated vectors, indicates covariation in five properties for the first factor: L (strandline length), A (total area), a (area of water surface), A_t (average area), A_b (bottom area), and V (water volume). This factor could be termed the *area factor*. The second significant varimax factor represents relationships in D_{max} (maximum depth), A_t (average area), V (water volume), and D_m (average depth). This factor could be named the *volume factor*. The third factor can be interpreted as showing

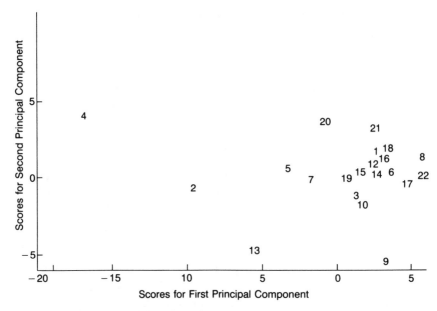

Figure 8.18. Plot of the first and second principal component scores.

covariation in I_m (mean island size), and I_{ns} (island density). This is a factor for *distribution*. The fourth factor represents covariation in l (total length of the coastline) and L (strandline length). This factor cannot be interpreted.

If you look again at the list of variables, you will see that several of them are redundant in that they can be produced by simple subtraction or addition from a lesser number of variables or they form part of another variable. This explains the fact that the rank of the correlation matrix is less than its order. The fifth and subsequent varimax factors show only one significant loading or no nonzero loadings at all and are therefore not interesting for the analysis, and the variances fall off very sharply after the fourth vector. This is further support for the conclusion that there are only four significant factors. The next stage in a complete analysis of Wallin's data would involve suppression of redundant variables followed by a new suite of computations.

Figure 8.18 shows the plot of the principal component scores for the first and second axes. Most of the sites form a cluster in the right part of the diagram. There are a few places that are obvious outliers dispersed out in the left part of the plot.

Program supplement to applied factor analysis in the natural sciences

Leslie F. Marcus

Department of Biology, Queens College of CUNY, Flushing, NY 11367; Department of Invertebrates, American Museum of Natural History, CPW at 79th, New York, NY 10024; E-mail: LAMQC@CUNYVM or LAMQC@CUNYVM.CUNY.EDU

INTRODUCTION

A set of programs is provided to support many of the methods discussed in the text by Reyment, Jöreskog, and Klovan. These programs have been written in a very special computer matrix language called MATLAB®, available from The MathWorks, Inc. Versions of MATLAB run on mainframes, many minicomputers, work stations, IBM PC computers and clones, and Macintosh computers. The programs are available on $3\frac{1}{2}$-in. disk from the author, or by anonymous FTP through the Internet at SBBIOVM.SUNYSB.EDU (User: Guest; Password: ANONYMOU; directory MORPHMET.192).

MATLAB is a statement-based language with functions, subroutines, and programs all called scripts. It is a very powerful interactive tool for handling data arrays and doing multivariate and multidimensional computations, as well as graphics. A similar set of programs was written for *Multidimensional Palaeobiology* by Richard Reyment (1991). Those programs were written for the PROCedure IML in SAS (Statistical Analysis System). The syntax of IML and MATLAB are very similar, so that programs can be translated from one to the other with minimum effort. Most of the Supplement for Reyment's earlier book is also now available in MATLAB. I intend to provide the more important programs included here in SAS IML as well, at a later date.

For this supplement, I chose MATLAB because it does not take so much disk space and is available on more kinds of computers than SAS, allowing larger data arrays. A very large number of matrix and graphics operations are built into MATLAB, and there is a supplied library of scripts, or programs, designated *.M files. These latter augment the built-in functions. One can easily add more .M files to those already available, and I have added a few for this project. A low-cost student

289

version of MATLAB is now available for members of educational institutions for both the IBM PC and clones and the Macintosh.

This Appendix compares a short list of MATLAB matrix operations to those from Chapter 2, together with an example. Code for all of my MATLAB .M files is printed here, with a short explanation and cross-references to the main text.

MATLAB must be available in order to use these procedures. The version of MATLAB required is that from The MathWorks. The earlier small MATLAB written in FORTRAN by Cleve Moler was freely and widely distributed in source and compiled form. It has a different and more limited syntax, poorer graphics, and cannot handle very large arrays. The MathWorks MATLAB is much richer in commands and can consider very large arrays. It was also developed by Cleve Moler. New versions and updates are frequently available from MathWorks. For further information on MATLAB and accompanying toolboxes, contact The MathWorks, Inc. at 24 Prime Park Way, Natick, MA 01760; (508) 653-1415.

The PC MATLAB version requires 640K RAM and a math co-processor. The PC version runs below the 640K boundary, whereas the AT and 386 versions for IBM micros is loaded at execution time into extended memory (the memory above 1 megabyte), where it resides with all arrays. The more extended memory, the better, because array size is only limited by the amount of this memory. Most of the memory below 640K is available for running other programs. For example, this text was written with MATLAB resident in extended memory. Although a math co-processor is required for the full MATLAB program, the low-cost student version does not need one, but will use it if available. The PC and AT versions have upper limits of 8088 elements for an array (1024 for the student versions), whereas the 386 version is only limited by memory.

Very few routines are slow (I used a 10-MHz AT for development of the programs, but have recently shifted to a 33-MHz 80486 machine). If a program is slow, this will be pointed out in its description. Newer algorithms are being explored to increase speed. The programs were also tested on a Macintosh, but not on a mainframe or other computers. MathWorks claims that programs and data files are compatible across all the computers for which the package is available. A built-in benchmark program checks the speed of your computer compared to other available platforms and is most instructive.

As stated in the supplement for *Multidimensional Palaeobiology* and reiterated here, the purpose of these programs is to make the matrix algebra and algorithms explicit for the methodology. This will reinforce the discussions and examples in the text. The more important matrix and other MATLAB statements are highlighted in boldface. It is

Table A.I. *Similarities of matrix algebra and* MATLAB *functions*

	Matrix algebra	MATLAB
Symbol for a matrix	**A**	A
Transpose	**A′**	A '
Symbol for a vector	***a***	a (by usage)
Transpose of vector	***a′***	a '
Addition of two vectors	***a* + *b***	a + b
Minor product of two vectors	***a′ · b***	a ' *b
Major product of two vectors	**M = *a* · *b′***	M = a*b '
Matrix multiplication	**C = AB**	C = A*B
Major matrix product	**C = XX′**	C = X*X '
Minor matrix product	**E = X′X**	E = X ' *X
Vector of 1s (length *n*)	***j*** or ***1***	ones(n,1)
Identity matrix	**I**	eye(n)
Individual *x* element	x_{ij}	x(i,j)
Matrix inverse	X^{-1}	inv(X) or X ^(-1)

recommended that the reader compare the code to the prose and formulas in the main text.

MATLAB, unlike SAS IML and the earlier smaller free version, is case-sensitive (command **casesen**), although this can be suppressed, and not being sensitive to case may be a better option when first learning the program. The programs included here are case-sensitive. Strings are defined as character variables when used as labels and stay in the case in which they were entered. All functions and program names are in lowercase, unless the default **c a s e s e n** is deactivated.

Help

Extensive help is available in MATLAB – when one types **he l p** the list of all of the available built-in functions appears. This list is given at the end of the Appendix. To obtain help for one function, for example the singular value decomposition (**svd**) just type

» help svd

where » is the MATLAB prompt. Help for any *.m file will list the comment (**%**) lines at the beginning of the file, so this is a way to see what any program may do or to find out any special instructions. I have put comments of this type in front of all of the programs included here.

A complete manual comes with MATLAB, but it is fairly terse.

A second screen of provided MATLAB scripts is listed after the built-in MATLAB commands and is followed by those in directories

Table A.II. *Some simple statistics formulas in matrix algebra shown in MATLAB notation*

Statistic	Matrix algebra	MATLAB
Mean vector	$mn = j'X/j'j$	`mn = j'*X / j'*j` [where `j = ones(n,1)`]
Deviate score	$y_{ij} = x_{ij} - mn$	`y(i,j) = x(i,j) - mn(1,j)`
Variance j	$s_j^2 = y_j'y_j/(N-1)$	`s(j,j) = y(:,j)'*y(:,j) / (N-1)`
Standard score	$z_{ij} = y_{ij}/s$	`z(i,j) = y(i,j) / sd(1,j)` [where `sd(1,j) = sqrt(diag(s(j,j)))`]
Covariance	$s_{ij} = y_i'y_j/(N-1)$	`s(i,j) = y(:,i)'*y(:,j) / (N-1)`
Deviation matrix	$\mathbf{Y} = \mathbf{X} - j \cdot mn$	`Y = X- ones(n,1)*mn`
Covariance matrix	$\mathbf{S} = \mathbf{Y'Y}/(N-1)$	`S = Y'*Y / (N-1)`
Standardized matrix	$\mathbf{Z} = \mathbf{Y} \cdot \text{isd}$	`z = Y*isd` [where `isd` is the inverse square root diagonal of `S = inv(diag(sqrt(diag(s))))`]
Correlation	$r_{ij} = s_{ij}/s_i s_j$	`r(i,j) = s(i,j) / (sd(1,i)*sd(1,j))`
Correlation matrix	$\mathbf{R} = \mathbf{Z'Z}/(N-1)$	`R = Z'*Z / (N-1)`

Note: There are script functions to find means, standard deviations, covariance matrices, and correlation matrices. These are sometimes used.

that are indicated in a PATH statement in the supplied MATLAB.BAT, the BAT file that first invokes MATLAB.

The following complete program will do a principal component analysis on a small data array imbedded in the code. As will be shown later, data may be accessed from external files in a variety of ways. This complete example will work directly and it is suggested that readers try it by typing it in line by line or putting it in a file and running it. Comments begin with a percent (**%**) sign. Statement continuation on the next line is indicated by ..., three periods, at the end of the line to be continued. Special punctuation is not needed to terminate statements, but a **,** (comma) or **;** (semicolon) at the end of a MATLAB statement allows several statements per line. A semicolon also suppresses printing of arrays or results to the screen. The default is to print any array mentioned on the left-hand side of an equation if no terminal punctuation is used. Any array may be printed to the screen by just naming it after the prompt. For example:

» X

This will print the data array X to the screen if it has been already defined. In the program that follows, all MATLAB code has been put in **boldface** type, thereafter only the MATLAB key words will be highlighted in the programs.

The command **echo on** will echo all the commands in a program to the screen, which is useful when first running a program; **echo off** turns this feature off. Echoing is a nuisance when a graph is being constructed.

Example of MATLAB code and results in **princomp.m**

```
% SOME SIMPLE MATLAB STATEMENTS AND A SIMPLE PRINCIPAL
% COMPONENTS PROGRAM
x = [1  4  5 % This is how comments are entered
     2  7  8 % Data is entered in a named array and
            % including the values between
     1  5  2 % [ ], i.e. square brackets.  Rows are
            % separated by ;s or are as entered here.
     4  2  7];     % The ; after the ] suppresses printing
                   % of X
[N,p] = size(X)  % Returns N as a row count; M as a
                 % column count.
j = ones(N,1)    % Defines and prints the n X 1 vector
                 % of 1 's
XBAR = j' *X / N   % Computes and prints means
Y = X - j*XBAR    % Computes deviations from means
varcov = Y'*Y / (N- 1)     % Computes variance-covariance
                           % matrix
```

```
[U,L,U] = svd(varcov);      % Singular value decomposition
                            % of varcov for
                            % eigenvalues L and
                            % eigenvectors E
pause                       % stops execution until a key is
                            % pressed
L = [diag(L)  100*cumsum(diag(L) / sum(diag(L)))]
                            % Eigenvalues and as percentages
U                           % prints eigenvectors E
Scores = Y*U                % PC scores
clg                         % clears any previous graphics
pause
plot(Scores(:,1),Scores(:,2),'+w') % : selects all rows
                                     % and plots
                                     % column 1 X 2
                                     % of PC Scores
xlabel('PC1')               % with plus sign for points, and
                            % axes labeled
ylabel('PC2')
title('Plot of PC2 against PC1')
```

You see here, in very few lines, a principal component (PC) analysis on a covariance matrix, using statements, many of which look like ordinary matrix algebra. A special function **svd** returns the eigenvalues and eigenvectors, and another function, **plot**, plots the resulting PC scores. The **xlabel**, **ylabel**, and **title** statements are not necessary but make more readable graphs.

If you are in the directory that contains the program, the command **dir** will list the contents of the current directory in abbreviated form. If you don't see the program, you may invoke the **chdir** command, or using DOS syntax or any other DOS command (including executing another program) by typing ! followed immediately by the DOS command followed by an [Enter].

chdir ..\FACTOR

This command will change directories to the subdirectory **factor** in the MATLAB directory.

In order to run this program, assume you already have installed MATLAB and know how to run **demo** and so on.

1. Create a subdirectory FACTOR in the MATLAB directory.
2. Copy the supplied programs and data to this directory.
3. Start MATLAB.
4. Type the following MATLAB command after the prompt:
   » **chdir ..\FACTOR**
   (assuming you had started the program within \MATLAB\BIN).
5. Type after the prompt »

» **princomp**

The program will execute after you press ⟦Enter⟧ and will only stop if there are **pause** statements in the program, or some kind of error. Program statements will also appear if the command **echo on** is part of the program or if you invoke it by typing the command **echo on** before running the program. **echo off** turns the echoing feature off.

» **diary** fn.ft

sends all code and results to a file designated by fn.ft. The diary is closed by **diary off**. Selected parts of the output can be put to the named file by typing **diary on** before the statements for which you want output and **diary off** after. Saying

» **diary princomp.out**

before running **princomp** produced the following when **princomp.out** was retrieved. Some blank lines have been removed to save space, and **echo on** was not used.

» **princomp**

```
N =
   4
p =
   3
j =
   1
   1
   1
   1
XBAR =
   2.0000   4.5000   5.5000
Y =
  -1.0000  -0.5000  -0.5000
        0   2.5000   2.5000
  -1.0000   0.5000  -3.5000
   2.0000  -2.5000   1.5000
varcov =
   2.0000  -1.6667   2.3333
  -1.6667   4.3333   0.3333
   2.3333   0.3333   7.0000
L =
   7.9480   59.6098
   5.0136   97.2116
   0.3718  100.0000
E =
   0.3862  -0.3277   0.8623
  -0.0934   0.9161   0.3900
   0.9177   0.2312  -0.3232
```

```
Scores =
 -0.7983  -0.2460  -0.8957
  2.0606   2.8681   0.1670
 -3.6448  -0.0234   0.4638
  2.3825  -2.5988   0.2649
```

Typing on the next line

» **diary off**

closes the diary file **princomp.out**.

Note that only five digits are displayed. Small numeric values less than 0.1 will be printed in exponential form. The only other choice – if five digits are not enough – is **format long**, which then prints all numbers with 15 digits. Issuing the command **format short** returns to the usual default 5 digits. Limited formatting capabilities is a weakness of MATLAB. This feature will be improved in the next release. Also, labeling of output with alphameric indicators is not very versatile except for plots. Note that the plot did not appear in the output. I was using a high-resolution graphics screen, and MATLAB recognizes that fact and plots to the screen accordingly in color. High-resolution plots can be produced on dot matrix printers, laser printers, or plotters quite easily from within a program, or even after seeing any plot you wish to print, you may direct it to a file to be printed later. The MacIntosh has both a program window and a plot window, as do work stations.

Many of the features just discussed are illustrated in the programs to follow; when new features are invoked they will be briefly discussed along with the program example.

CHAPTER 1

Most of the results in the mining exploration example are produced by the following program, which starts with the causal variables in an array **amount** and their mix in an array **proport**. The plot of the causal variables is invoked by the small program **mkfig1_2.m**. The published graph is a hypothetical continuous graph, while the program contours the available 20 points.

mkfig1_2.m

```
% This program will reproduce figure 1.2 in the book, as
% well as the data in the table support the drawing.
load amount.dat
X = amount;
index;
```

```
for j = 1:3,
  one = reshape(X(I,j),5,4);
  one = one';
  if j == 1,1 = '- ';elseif  j == 2,1 = '- - ';else  1 = ':';end
  contour(one,3,1)
  t = [' - paleotemp, - - permeability, .. deformation'];
  title(t)
  if j == 1,hold on,end
end
hold off
pause
```

It would be best to return to this plotting program after you are more familiar with MATLAB functions. I will explain a few of the statements: **index** invokes the program **index.m**, which contains the rows of the table so that they can be reordered for ease of plotting.

index.m

```
I = [[16:20] [15:-1:11] [6:10] [5:-1:1]]
```

The data were in an array with 20 rows and 3 columns, one for each variable. The **reshape** function puts each column in the proper form to produce the plot. The **'** or transpose function then rotates the data for each variable in the form of the plot. The **if, elseif,** and **else** are fairly common programming statements, but notice that an **==** is required with an **if**. The command **contour** is a very powerful function for contouring data.

The next program, **prospect.m**, takes the causal data matrix in **amount.dat** and the mix matrix in **proport.dat**, to produce the artificial data in Table 1.I. The data are then sorted in a new array of dimension 20 × 10 called **prospec** and saved in a file **prospec.dat** using the **save** command.

prospect.m

```
% This program will load and compute fictitious
% Lead- Zinc prospecting data and then save as
% prospec.dat for factor program to analyze
load amount.dat
amount
pause
load proport.dat
proport
pause
prospec = amount*proport
save prospec.dat prospec /ascii
```

The next program produces Table 1.II, the correlation matrix for this data, and then does a factor analysis reported in Table 1.III.

tabs1.m

```
% This program will do a principal factoring on the
% artificial data, compute the factor scores, and
% show that the data has rank three.
% Note you will have to type 'return' after the Keyboard
% prompt to complete the program.  It also pauses
% several times as it calls another general program
% varimax which requests input - type the word return
load prospec.dat
X = prospec;
[N,p] = size(X);
Y = X- ones(N,1)*mean(X);   % mean centered data
Z = Y*inv(diag(std(X)));    % standardized data with
                            % st. dev.= 1
['Inter-correlations among ten geological properties']
r = corrcoef(X)
pause
[U,D,U] = svd(r); % Singular value decomposition of
                  % correlation matrix
D = diag(D);
['Variance and Cumulative Variance']
Eigens = [D cumsum(D)  100*cumsum(D)]  % Eigenvalues
pause
U = U(:,1:3);  %     Eigenvectors - only three kept
D = D(1:3,1);
['Factors']
lding = U*sqrt(diag(D))
pause
['Check of how close factors reconstruct correlations']
check = sum(sum(r- U*diag(D)*U'))
varimax  % gets varimax loadings
% Next line prints factor scores
F = Z*lding*inv(lding'*lding)   % formula from Chapter 7
                                % since rank = 3
save pros_scr.dat F /ascii
```

Note there are a number of pauses in this and subsequent programs. When echo is not turned on, an [Enter] will do no harm when the program stops for no apparent reason. Also it is necessary to type **return[Enter]** after the last keyboard prompt.

The data are put in the array X by the = assignment. The built-in function **corrcoef** is used here to produce an array of correlation coefficients among the columns of X reproducing the results of Table 1.II. Matrix statements to produce a correlation matrix will be given later in the matrix algebra section. The factor loadings, and eigenvalues are found using the singular value decomposition (**svd**) function as this returns the eigenvalues in descending order and are as reported in Table 1.III. A check is done to show that the correlation matrix is fully recovered from the matrices produced by the **svd**. The remaining calculations of the varimax rotation and varimax factor scores are done by a call to the program **varimax.m** and use of a formula discussed in Chapter 7. **varimax.m** is a separate program, which is given in the section on Chapter 7. Since the varimax program provides a prompt to enter the loadings and they are already supplied by the program, type **return[Enter]** after the keyboard prompt **K»**.

Finally, a program plots the varimax rotated PC scores saved in the last program as a file **pros_scr.dat** to nearly recover the initial figure. The program is called **mkfig1_3.m**.

mkfig1_3,m

```
% This program will reproduce figure 1.3 in the book,
% based on the Varimax factor scores
load pros_scr.dat
index;
X = pros_scr;
for j = 1:3,
   one = reshape(X(I,j),5,4);
   one = one';
   if j == 1,1 = '-';elseif j == 2,1 = '--'; else 1 = ':';end
   contour(one,3,1)
   t1 = ['_Factor 1 Scores, - - Factor 2 Scores'];
   t2 = [',.... Factor 3 Scores'];
   title([t1 t2])
   if j == 1,hold on,end
end
hold off
pause
```

Note that **load pros_scr.dat** loads the ascii file containing the varimax factor scores that were computed in **tab1_iii.m**. Any two-part name will suffice for a data set, except for a file type **.mat**, which is reserved for a special binary format. The **.mat** format may be used by MATLAB on any platform (it also saves space and time). The **.m** file type is reserved for MATLAB programs and functions. The binary form of data in **.mat** will not be used much in this supplement. The program **mkfig1_3** is essentially the same as **mkfig1_1**, the first program in this section.

CHAPTER 2

Some of the more extensive matrix calculation examples included in this chapter are done in small programs with names given by the section, table, or figure where the example is found. There is no better way to try matrix procedures than doing small examples by hand using pencil and paper, and then using a program package such as MATLAB (or one of the other matrix language programs).

The mineral composition of the rock specimens in Table 2.I are stored in a file called **rock.dat** used later. **labrock.m** sets up the array **min** containing the mineral names for labeling the plots, and **numbers.m** contains the labels for the specimen numbers put in an array called **num**. This is a general array of 20 numbers, and the number required is selected in any specific program. Both **labrock.m** and **numbers.m** are shown here.

labrock.m

```
min = [ 'Quartz    '
        'Hornblende'
        'Biotite   '
        'Feldspar  ']
```

numbers.m

```
num = [ ' 1'
        ' 2'
        ' 3'
        ' 4'
        ' 5'
        ' 6'
        ' 7'
        ' 8'
        ' 9'
        '10'
        '11'
        '12'
        '13'
        '14'
        '15'
        '16'
        '17'
        '18'
        '19'
        '20'];
```

The program **mkfigs2.m** calls **mkfig2_1.m** and **mkfig2_3.m**, using the rock data given in Table 2.I, and produces Figs. 2.1 and 2.3. The separate programs can make each figure when called separately.

mkfigs2.ms

```
% This produces the first two figures in the Chapter by
% calling the programs that make them
clear
clg
mkfig2_1
pause
clg
mkfig2_3
```

mkfig2_1.m

```
% This reproduces figure 2.1
load rock.dat
[N,p] = size(rock);
labrock
numbers
num = num(1:N,:);
new = ones(2*N-1,2);
ind = [1:2:2*N-1];
new(ind,:) = rock(:,1:2)
ind2 = [2:2:2*(N-1)];
dum = zeros(N-1,2);
clg
new(ind2,:) = dum;
temp = [rock(:,1:2)
        new(:,1:2)];
v = [0 max(temp(:,1)) 0 max(temp(:,2))];
axis(v);
plot(rock(:,1),rock(:,2),'+ w',new(:,1),new(:,2),'- ')
text(rock(:,1),rock(:,2),num)
title('Figure 2.1 in Text')
xlabel(min(1,:))
ylabel(min(2,:))
```

mkfig2_3.m

```
% This reproduces figure 2.3
load rock.dat
rock = rock';
[p,N] = size(rock);
labrock
new = ones(2*p-1,2);
ind = [1:2:2*p-1];
```

```
new(ind,:) = rock(:,1:2)
ind2 = [2:2:2*(p-1)];
dum = zeros(p-1,2);
new(ind2,:) = dum;
temp = [rock(:,1:2)
        new(:,1:2)];
v = [0 max(temp(:,1)) 0 max(temp(:,2))];
axis(v);
plot(rock(:,1),rock):,2),' + ',new(:,1),new(:,2),'- ')
text(rock(:,1),rock(:,2),min)
title('Plot of Figure 2.3 in Text')
xlabel('ROCK SP. 1')
ylabel('ROCK SP. 2')
```

These programs just give examples of plotting routines, with additional functions to produce the vectors to the points and label them. You may wish to study this code later, but for now we will continue with more direct matrix algebra examples.

The program **tab2_i.m** loads and displays the data matrix **X** from Table 2.I and stored in the data file **rock.dat**. The program **prod _mom.m** takes the transpose of **X** and forms the major and minor products defined in the text in Section 2.3 and illustrated with the **rock.dat** in Section 2.6. Note the similarity in the matrix algebra and the MATLAB code. This is a case where one would want to use **echo on** and this is built into the program.

tab2_i.m

```
% Data in Table 2.I
clear
load rock.dat
X = rock;
['              Example of a Data Matrix          '
 '  Spec.    Quartz   Hornblende Biotite Feldspar']
row = [1:8];
disp([row ' X])
```

prod_mom.m

```
% Illustrations of major and minor product moments
clear
load rock.dat
echo on
X = rock
% Press space bar to continue
pause
X'    % transpose of X
```

```
pause
C = X*X'       % Major product moment
pause
E = X'*X       % Minor product moment
pause
echo off
```

Note the frequent use of **pause** statements. This stops the scrolling so you can examine each array and step through the results by pressing [Enter] after each pause.

The next example for the chapter computes the covariance matrix and correlation matrix, using matrix statements from Section 2.8 instead of the function **corrcoef** used in the first principal components example. Data standardization is also illustrated. These operations are illustrated using the **rock.dat**. Note that I have used $n = 8$, following the text, for the denominator of the variance rather than $n - 1 = 7$, which would generally be used.

z_stand.m

```
% Illustrates Descriptive Statistics in Section 2.8 in
% Matrix Notation
load rock.dat
X = rock;      % gets raw data into X
echo on
X
pause
[N,p] = size(X) % finds number of n = rows and m = columns
l = ones(N,1) % vector of ones for mean operation
pause
xbar = 1'*X./ (1'*1)    % means using formula in text
pause
Y = X- 1*xbar           % deviations from the mean -
                        % formula [2.16]
pause
S = Y'*Y./ N            % covariance matrix (note n in
                        % denominator)
D = inv(diag(sqrt(diag(S))))    % multiplier to
                                % standardize columns
                                % reciprocals of standard
                                % deviations
pause
Z = Y*D         % Z scores as standardized data
save Z.rck Z /ascii  % This will be used later in
                     %Chapter 7
```

```
pause
R = Z'*Z./N     % correlation matrix
echo off
```

The ι vector (sometimes called a *j*-vector) is the very useful vector of 1s, produced by the **ones** function. Note how the mean, deviations from the mean **Y**, and the **S** and **R** matrices are formed. Also the $D^{-1/2}$ is complicated to compute and requires four function calls. I have written a small routine to do this, which we will use later. The **./** is scalar division, not really required here, but it is wise to get into the habit of using it.

We will reproduce Fig. 2.12, containing bivariate graphs of *z* standardized data for different correlation coefficients in the rock data, showing the power of the MATLAB graphics, where a maximum of four graphs can be put on one screen. Note that both axes are scaled to go from −2 to 2 using the **v** array, and the graph is made square by the command **axis('square')**. The grid is superimposed using the **grid** command.

mkfg2_12.m

```
%Construction of figure 2.12 - calling up data saved
%in Z.rck
load Z.rck
echo off
clg
v = [-2  2  -2  2];
axis(v);
axis('square');
subplot(221)   % procedure for dividing screen into
               % 2 rows and 2 columns of
               % plots, where 1 is first plot
plot(Z(:,1),Z(:,2),'*')
grid
xlabel('Var 1')
ylabel('Var 2')
title('r12 = 0.00')
subplot(222)        % second subplot - 1st row 2nd column
plot(Z(:,1),Z(:,3),'*')
grid
xlabel('Var 1')
ylabel('Var 3')
title('r13 = 0.71')
subplot(223),  % third subplot - 2nd row and 1st column
plot(Z(:,2),Z(:,4),'*')
grid
xlabel('var 2')
```

```
ylabel('var 4')
title('r24 = 0.87')
subplot(224)    % fourth subplot - 2nd row and 2nd column
plot(Z(:,3),Z(:,4),'*')
grid
xlabel('var 3')
ylabel('var 4')
title('r34 = 0.30')
subplot(111)    % returns plotting to 1 plot on a screen
```

We now illustrate some of the points brought up in Sections 2.9–2.12 on rank, eigenvalues, eigenvectors, and the singular value decomposition. We anticipated the singular value decomposition as a way of computing eigenvalues and eigenvectors because it sorts the eigenvalues in descending order and is a more direct representation of the data matrix. The next few programs are named for the matrix operations described.

This program illustrates some features of rank of product matrices. The program will work with any other matrices put in the places of **R1** and **R2**.

findrank.m

```
% Example to illustrate The rank of product matrices
% R1 is of full rank from rank.m function and from the
% determinant of R1 we see it is substantial
clear
echo on
 R1 = [1.000 0.500 0.866
       0.500 1.000 0.707
       0.866 0.707 1.000]
rank(R1)
det(R1)
pause
% R2 is of full rank as well because of round off, but
% to 4 or 5 places. Is essentially in 2 space.
% By setting a tolerance epsilon we can show that
% the rank is essentially 2.
 R2 = [1.000 0.500 0.707
       0.500 1.000 0.966
       0.707 0.966 1.000]
rank(R2)
eps = .00001
rank(R2)
pause
% Use regression to find linear combination of columns
% or rows to define 3rd column as function of 1st two
X = R2(:,1:2)
```

```
Y = R2(:,3)
pause
b = inv(X'*X)*X'*Y
R2est = X*b
R2(:,3)
echo off
```

Eigenvalues, eigenvectors, and the singular value decomposition

The examples use the data in the text (also found in the array called **X.DAT**). The program may be run with any other data set by setting an array equal to **X** and removing the lines with

```
X = [- 8   -1
```

.   .   .   .   .   .

```
    2   6]
```

from the program. Yet another way to enter the data is mentioned in the comment lines.

eigens.m

```
% Note that instead of the eigen routine we have
% anticipated the use of the singular value
% decomposition discussed in sect 2.12. In this
% case it gives the same values as another
% routine eig which however does not sort the
% diagonal elements and corresponding vectors
% by size.
echo on
X = [- 8    -1
       6    10
      -2   -10
       8     1
       0     3
      -6    -6
       0    -3
       2     6]
C = X'*X
pause
[U,L,U] = svd(C)
pause
check = (C*U- U*L)
U'*U
```

The singular value decomposition is illustrated in the next program. The data this time are called from the file X.DAT. Any other data can be substituted by saying load fn.ft and on the next line typing X = fn. This program also checks that the results are the same for the major product moment X*X'.

eck_yng.m

```
echo on
% We start with the same data, but load it as X and use
% the singular value decomposition directly to produce
% the results in the text. Earlier in eigens.m we
% computed X'X etc. This is not required. We assume
% that L has been computed when ran eigens.m - so do
% that first if haven't already.
load X.dat
pause
[V,Gamma,U] = svd(X,0) % 0 takes advantage of less
                       % columns than rows for output
pause
Check = V*Gamma*U'
pause
V'*V
pause
% Then checking as well works for major product moment
X*X'*V
pause
[A,Lambda,A] = svd(X'*X);
V*Lambda
```

Canonical analysis of asymmetry considers the small data sets in the text in Section 2.1.4. The method is illustrated more thoroughly later with a real data example.

skew_sym.m

```
% Demonstration of some ideas of Skew Symmetric. See
% skewmc.m later for more complete analysis.
% This material is in section 2.14
A = [5  1  -4
     3  7   8
    -2  0   3]
A'
M = .5*(A + A')
N = .5*(A - A')
check = A - M - N
```

CHAPTER 3

The first factor analysis given is a maximum likelihood factor analysis of artificial data in the form of correlation coefficient matrix in Section 3.8 that is completely explained by two factors. An iterative maximum likelihood algorithm based on Morrison (1976, pp. 308–9) yields the same answer as in the book. It is slow for these data unless you have a very fast computer.

From this point on most of the programs are written so that they can use any data set in the right format. Instructions will be given in each program. Unless otherwise stated the data matrix will consist of n rows and p (or m) columns. Once the program has started one is prompted with

K»

This means the appropriate data set is to be entered using an *.m program, or with the **load** command. The next program does a maximum likelihood factor analysis of the data in Section 3.7, and detailed instructions for using it are given after the listing.

mlfact.m

```
% Factor analysis using artificial correlations from
% Section 3.7.  Maximum likelihood factor analysis and
% algorithm modified from Morrison (1976) second edition,
% pages 308-309.
% Put data into array called Rho. For the correlation
% matrix in Sect. 3.7. You need to type Rho(CR) after
% the K > ; then the word return(CR) after K >
keyboard
k = input('Give number of factors to determine')
[p,p] = size(Rho);
[A,L,A] = svd(Rho);    % preliminary singular value·
                       % decomposition of Rho
A1 = A(:,1:k)*sqrt(L(1:k,1:k)) % Prin. Comp. loadings
pause
Uni = diag(diag(Rho- A1*A1'))   % Uniqueness matrix
pause
Rh1 = sqrt(inv(Uni))*(Rho- Uni)*sqrt(inv(Uni))
                       % Matrix to iterate
pause
[A,L,A] = svd(Rh1);       % First estimate of Maximum
                          % Likelihood Loadings
A1 = sqrt(Uni)*A(:,1:k)*sqrt(L(1:k,1:k))
iter = input('Give number of iterations - ');
```

```
t = ['Tolerance for loadings - usually .001 or less - '];
check = input(t);
for i = 1:iter
   Uni = diag(diag(Rho- A1*A1'));
   Rh1 = sqrt(inv(Uni))*(Rho- Uni)*sqrt(inv(Uni));
   [A,L,A] = svd(Rh1);
   A2 = sqrt(Uni)*A(:,1:k)*sqrt(L(1:k,1:k));
   if max(max(abs(A1- A2))) < check;break;end
   A1 = A2;
end
i
A1
FacVar = diag(A1'*A1)
pause
Com = diag(A1*A1')
pause
Uniq = ones(p,1)- Com
pause
Resid = Rho- A1*A1'
```

If you type

```
help mlfact
```

the first five lines of the program will appear on the screen with instructions for running the program. In this case after the **K»** the user types the command to load the data. As the help lines indicate, one would type **Rho** after the prompt as

```
K» Rho
```

followed by an [Enter]. In this case a little program **Rho.m** is executed, which loads the correlation matrix into an array called **Rho**. Type the word **return** after the next **K»**

```
K» return
```

followed by an [Enter], and execution will continue. Once the program is run, the data need not be called again. A **return** after the prompt will continue execution of the program with the data previously defined.

An alternative way of loading the data is to type the command **load fact.dat** after the prompt, and then type **Rho = fact;**. The **;** suppresses printing; **return** works as before.

The user is also asked the number of factors, 2 in this example, and then the number of iterations and tolerance desired. Somewhat less than 400 iterations and a tolerance of 0.00001 will give a solution comparable to the one in the book to nearly 3 places. The algorithm is simplified over that in Morrison, in that it finds all loadings simultaneously. This program does not work well for some of the other examples later in the book.

The loadings for the factor solution are plotted with the program **mkfig3_2.m**, using the solutions and rotations given in the text. The residuals are given as well for each solution, but the oblique solution cannot be easily plotted using MATLAB.

mkfig3_2.m

```
% This program produces the plots in figure 3.2
% Since I cannot show non-orthogonal axes here,
% figure C is different. The numbers given in the
% text are in fact used for the results.
clg
echo on
A = [.889 -.138
     .791 -.122
     .501  .489
     .429  .419
     .358  .349
     .296 -.046];
pause
T = [.988 -.153
     .153  .988]
Astar = A*T'
pause
labs = ['x1'
        'x2'
        'x3'
        'x4'
        'x5'
        'x6'];
axis([-1 1 -1 1])
echo off
subplot(221)
  plot(A(:,1),A(:,2),'*w')
  text(A(:,1),A(:,2),labs)
  xlabel('I(f1)')
  ylabel('II(f2)')
  title('A Results')
  grid
subplot(222)
  plot(Astar(:,1),Astar(:,2),'*w')
  text(Astar(:,1),Astar(:,2),labs)
```

```
   xlabel('I(f1)')
   ylabel('II(f2)')
   title('B of Fig 3.2')
   grid
pause
echo on
commun = diag(A*A')
pause
comstar = diag(Astar*Astar')
pause
Rho;
Aresid = Rho- A*A'
pause
Astares = Rho- Astar*Astar'
pause
T2 = [1   0
      .6  .8]
A2star = Astar*inv(T2)
pause
Phi = [1   .6
       .6  1]
A2st_str = A2star*Phi
pause
Resid = Rho- A2star*Phi*A2star'
```

CHAPTER 4

Here are all of the results for the examples in Section 4.2. I would advise studying this code carefully. The program **pctab4_i.m** computes principal components for a covariance matrix and also produces the results in Table 4.II for standardized data and the correlation matrix.

pctab4_i.m

```
echo on
% Illustration of Principal Components using data in
% Table 4.1 Enter X_Tab4_I[Enter] after K > and then
% type return[Enter] after next K >
keyboard
% Solution 1 - Press Enter after each PAUSE
pause
[N,p] = size(X);
Mean = mean(X)
StDev = std(X)
```

```
pause
Y = X- ones(N,1)*Mean
S = Y'*Y./ (N- 1)
pause
[V,Gam,U] = svd(Y,0);
U2 = U(:,1:2)
G2 = Gam(1:2,1:2)
pause
L2 = G2*G2./ (N- 1)
A = U2*G2./ sqrt(N- 1)
pause
F = V(:,1:2)*sqrt(N- 1)
check = F'*F
pause
Ds = inv(diag(StDev));
C = Ds*A
% Solution 2
pause
A = U2
pause
F = V:,1:2)*G2
pause
check1 = F'*F
check2 = U2'*U2
pause
C = Ds*U2*sqrt(L2)
% Solution for Standardized Matrix and Correlations
% This makes Table 4.II
pause
Z = Y*Ds
pause
R = Z'*Z / (N- 1)
[V,G,U]svd(Z,0);
L = G*G./ (N- 1)
pause
U2 = U(:,1:2)
A = U2*sqrt(L(1:2,1:2))
pause
F = V(:,1:2)*sqrt(N- 1)
pause
check = F'*F
```

In the next program all of the parts of Table 4.III are reproduced for the methods discussed in Section 4.3.

tab4_iii.m

```
% Computations for Table 4.IIIa - merely follows
% formulae in book
```

```
S = [1.000   .466   .456   .441   .375   .312   .247   .207
      .466  1.000   .311   .296   .521   .286   .483   .314
      .456   .311  1.000   .185   .184   .300   .378   .378
      .441   .296   .185  1.000   .176   .244   .121   .341
      .375   .521   .184   .176  1.000   .389   .211   .153
      .312   .286   .300   .244   .389  1.000   .210   .290
      .247   .483   .378   .121   .211  0.210  1.000   .504
      .207   .314   .378   .341   .153   .289   .504  1.000];
[m,m] = size(S);
pause
sii = ones(m,5);
sii(:,1) = diag(inv(S));
sii(:,2) = sqrt(sii(:,1));
sii(:,3) = diag(inv(diag(sii(:,2))));
sii(:,4) = diag(inv(diag(sii(:,1))));
sii(:,5) = ones(m,1) - sii(:,4);
sii(:,6) = sqrt(sii(:,5));
sii
pause
Sstar = diag(sii(:,2))*S*diag(sii(:,2))
pause
[U,Lam,U] = svd(Sstar);
U3 = U(:,1:3)
ns = 0;
Lam2 = Lam - eye(Lam)*(2 - min(diag(Lam)));
%determines criterion for roots to keep
diag(Lam2)
pause
for i = 1:m,if Lam2(i,i) > 0,ns = ns + 1;end,end
ns
Theta = sum(diag(Lam(ns + 1:m,ns + 1:m))) / (m - ns)
A = diag(sii(:,ns))*U3*sqrt(Lam(1:ns,1:ns) - Theta*eye(ns))
% equation 4.20
```

I am not able to reproduce the results in Table 4.IV. The data array of correlations is given in **tab4_iva.dat**. You can experiment with this data using **mlfact.m** and **varimax.m**.

The Krzanowski cross-validation procedure in Section 4.5 is supported by three programs:

1. **validate.m** to determine the number of principal components to keep;
2. **var_info.m** to determine the number of significant variables to retain for the PC analysis;
3. **spe_infl.m** to see if there are any unduly influential specimens or rows in the data matrix.

The code for all three is presented next, with instructions for setting up data. The data is loaded using the **load** statement and then set equal to the appropriate array name. Note that the program pauses in response to **pause** command, but this is only obvious when you use **echo on**. In **var_info.m** and **spe_infl.m** the preliminary array calculations are done in a program called **modstore.m**. This program is listed after the third program and is called by each of these last two programs.

validate.m

```
% Data are expected to reside in an array y
% so load data and put in y
pause
[n,p] = size(y)
j = ones(n,1);
y = y- j*mean(y);
S = y'*y / (n- 1);
dSd = inv(diag(sqrt(diag(S))));
t1 = ['For standardized data type (1);'];
t2 = [' otherwise mean centered - '];
yans = input ([t1 t2])
if yans == 1;
    y = y*dSd;   % Standardizes data
end
R = y'*y / (n- 1)
pause
yest = zeros(n,p);
[u,d,v]svd(y,0);
 ye = u*d*v';
 d = diag(d);
press = zeros(p- 1,1);      % to store Krzanowski 's
                            % Press values
ujall = ones(p*n,p- 1);     % temporary storage svd results
djall = ones(p*(p- 1),1);   % loop creates some values
                            % needed later
 jn1 = ones(n- 1,1);
 for j = 1:p;
     uindl = (j- 1)*n + 1;
     uindu = j*n;
     vindl = (j- 1)*(p- 1) + 1;
     vindu = j*(p- 1);
      if j == 1,z = y(:,2:p);  % these 4 lines delete one
                               % column of y
      elseif j == p,z = y(:,1:p- 1);
      else L = [1:j- 1];L = [L j + 1:p];z = y(:,L);
      end
       [uj,dj,vj] = svd(z,0);
     ujall(uindl:uindu,:) = uj;
     djall(vindl:vindu,:) = diag(dj);
end
```

```
      for i = 1:n;
      iter = i
      if i == 1,z = y(2:n,:);     % these 4 lines delete one
                                  % row of y
       elseif i == n,z = y(1:n-1,:);
       else L = 1:i-1;L = [L i+1:n];z = y(L,:);
       end
      zmn = mean(z);
      z = z- jn1*zmn;
      [ui,di,vi] = svd(z,0);
      di = diag(di);
      for j = 1:p;

         uindl = (j-1)*n+1;
         uindu = j*n;
         vindl = (j-1)*(p-1)+1;
         vindu = j*(p-1);
         uj = ujall(uindl:uindu,:);
         dj = djall(vindl:vindu,:);
      for t = 1:p-1;   % actual computation of Press values
            ytemp = uj(i,t)*sqrt(dj(t,1))*sqrt(di(t,1))···
               *vi(j,t);
            ytemp0 = u(i,t)*d(t,1)*v(j,t);
            if ytemp*ytemp0 < 0, ytemp = -ytemp;end
            yest(i,j) = yest(i,j) + ytemp;
            press(t,1) = press(t,1) + (yest(i,j)-ye(i,j)).^2;
      end
      end
end
for t = 1:p-1
   press(t,1) = press(t,1) / (n-1);
end
% ydif = ye-yest;  if want differences between values and
% estimates print ydif
press
pause
ss0 = sum(sum(ye.*ye)) / (n-1)
               % sum of squares of deviations from svd
dr = (n-1)*p; % values required to compute W values
dm = n+p-2;
dr = dr-dm;
w = ones(p-1,1);    % to store W values - below
                    %computations of wi*/
w(1,1) = ((ss0-press(1,1))./dm) / (press(1,1)./dr);
for j = 2:p-1;
   dm = n+p-2*j;
   dr = dr-dm;
   w(j,1) = ((press(j-1,1)-press(j,1))./dm) /···
   (press(j,1)./dr);
```

```
end
w


var_info.m

% Krzanowski module to select variables
% Must load data into an array called Y
% If data already entered and equals Y then type return
keyboard
X = Y;
modstore
wselect = input('Maximum number of pcs - ')
DSD = invsqrdi(S);
t1 = ['For standardized data type (1);'];
t2 = [' otherwise mean centered - '];
yans = input([t1 t2])
if yans == 1;
    Y = Y*DSD;
end
R = Y'*Y / (N-1)
pause
[U,d,V] = svd(Y,0);
ZSC = U(:,1:wselect)*d(1:wselect,1:wselect);
clear U d V;
mn = mean(ZSC);
ZSC = ZSC- ones(N,1)*mn;
msq = zeros(wselect,p);
 for j = 1:p;
  if  j == 1,Z = Y(:,2:p);
  elseif j == p,Z = Y(:,1:p-1);
  else Z = [Y(:,1:j-1)Y(:,j+1:p)];end
   ztemp = Z(1,:);
   [Uj,dj,Vj] = svd(Z,0);
   for  k = 1:wselect
      ZSCj = Uj(:,1:k)*dj(1:k,1:k);
      mn = mean(ZSCj);
      ZSCj = ZSCj- ones(N,1)*mn;
      Temp = ZSCj'*ZSC(:,1:k);
      [Us,ds,Vs] = svd(Temp);
      Q = Vs*Us';
      msq(k,j) = sum(sum(ZSC(:,1:k). ^2)) + ...
      sum(sum(ZSCj. ^2))- 2*sum(diag((ZSCj*Q'*...
      ZSC(:,1:k)')));
   end
end;
var = [[1:p]
         msq];
var = [[0 [1:wselect]]' var]
```

```
spe_infl.m
% Krzanowski module for angles
% Must load data into an array called X
% If data already entered and equals X then
% type return (CR)
keyboard
wselect = input('Give max number of prin. components - ')
modstore
DSD = istd(X);
t1 = ['Type 1 if data standardized,'];
t2 = [' otherwise mean centered -'];
yans = input([t1 t2])
if yans == 1
      Y = Y*DSD;
end
R = Y'*Y / (N-1); % if use Correlation coefficients
[U,d,V] = svd(Y,0);
angle = ones(N,wselect);
for i = 1:N
  spec(i,1) = i;
  if  i == 1,Z = Y(2:N,:);
    elseif i == N,Z = Y(1:N-1,:);
    else 1 = [[1:i-1] [i+1:N]];Z = Y(1,:);
  end
  zmn = mean(Z);
  Z = Z-ones(N-1,1)*zmn;
  [Ui,di,Vi] = svd(Z,0);
  for j = 1:wselect
      Temp = V(:,1:j)'*Vi(:,1:j);
      [Ut,dt,Vt] = svd(Temp,0);
      angle(i,j) = acos(min(diag(dt)))*180 / pi;
  end
end
lab = ['  Specimen   Minangle']
angle = [spec angle]

modstore.m
% this program creates a module for the
% computation of covariance matrix and Mahalanobis Dşq
% for each observation to be used in other modules
% The module is stored in the current directory
[N,p] = size(X);
mn = mean(X);
Y = X-ones(N,1)*mn;
S = Y'*Y / (N-1);
SI = inv(S);
Dsq = ones(N,1);
  for i = 1:N
```

```
    Dsq(i,1) = Y(i,:)*SI*Y(i,:)';
% Mahalanobis Dsq to each observation
    end
```

The example in the book is run on the data in Hubberton et al. (1991) by loading the file HUBBERT.DAT in an array **y** after the keyboard prompt. One has a choice of using mean centered or standardized data.

We now consider the Aitchison procedure for compositional data discussed in Section 4.6. This is essentially the program from *Multidimensional Palaeobiology*, now generalized for any data set and rewritten in MATLAB. It also plots the simplex before and after adjustment. (The data set in **echino.dat** contains a column with numerical indicators for the rows. Remove this column and put the appropriate part in **X**.) After the K» prompt, type **load rhesus.dat**, then on the next line set the data equal to **X** and look at it by typing **X = rhesus**. A semicolon at the end of the statement would suppress printing. This program follows the notation in Aitchison's book and gives several other ways of depicting the simplex matrix as discussed there.

aitch.m

```
%   Aitchison analysis of constrained frequency data use
%   data by putting in array X. Make sure you remove
%   any identification columns or rows, for example
%   the first column of the echino.dat file contains
%   identification data. The rhesus data is found in
%   RHESUS.DAT, so load rhesus.dat, and set X = rhesus.
%   This is essentially the IML program from Multi-
%   dimensional Paleobiology translated into MATLAB.
clg
keyboard
[N,D] = size(X)
% Original Data
X
pause
delta = .005; % Round off value for 0 replacement
for i = 1:N; % 12 lines do Aitchison 0 replacement
    cnt = 0;
    for j = 1:D;
        if X(i,j) == 0,cnt = cnt + 1;end
    end;
    zero_rpl = delta*(cnt + 1)*(D - cnt) / (D*D);
    correc = delta*cnt*(cnt + 1) / (D*D);
    if cnt > 0
        for j = 1:D;
            if X(i,j) == 0,X(i,j) = zero_rpl;
            else X(i,j) = X(i,j) - correc;
            end;
        end
```

```
end
% Adjusted Data
X
pause
LogX = log(X);
dsmall = D-1;                 % Dimension of simplex
X = inv(diag(sum(X')))*X
% Column Sum = 1 if not already done
pause
mnX = mean(X);
Xdev = X- ones(N,1)*mnX;
Var = Xdev'*Xdev / (N-1);    % CoVariance Matrix of
                             % Proportions
Dig = invsqrdi(Var); % Note this is a function I have
                      % supplied
RX = Dig*Var*Dig       % Correlation Matrix of Proportions
pause
[E,a,E]svd(Var);
  a = a(1:D,1:D)
%   Reduced Roots and Vectors since Simplex
  E = E(:,1:D)
pause
Xscore = Xdev*E;
rmnLogX = mean(LogX')';
% Eigenvalues a
% Eigenvectors E
Z = LogX- rmnLogX*ones(1,D); % produces logratio matrix
sumz = sum(Z')
pause
% check that rows sum to 0
mnz = mean(Z);
Zdev = Z- ones(N,1)*mnz;
Gamma = Zdev'*Zdev / (N-1) % SimpleX covariance matrix
Dg = invsqrdi(Gamma);
Rg = Dg*Gamma*Dg            % SimpleX correlation matrix
pause
% Gamma Correlation
Colgam = diag(Gamma)*ones(1,D);
Tau = Colgam + Colgam'- 2*Gamma % Tau form of simplex
pause
DcolTau = Tau(:,D)*ones(1,D);
Sigma = .5*(DcolTau + DcolTau'- Tau);
Sigma = Sigma(1:dsmall,1:dsmall)  % Sigma form of Simplex
pause
[E,a,E] = svd(Gamma);
%Eigenvectors of Gamma E
E = E(:,1:dsmall)
a = a(1:dsmall,1:dsmall)
pause
```

```
Zscore = Zdev*E;
label = setstr([97:96+N]');
axis('normal');
subplot(211)
  v = [min(Zscore(:,1)) max(Zscore(:,1))];
  v = [v min(Zscore(:,2)) max(Zscore(:,2))]
  axis(V)
  plot(Zscore(:,1),Zscore(:,2),'*w')
  title('Simplex Scores');
  text(Zscore(:,1),Zscore(:,2),label),pause
subplot(212)
  v = [min(Xscore(:,1)) max(Xscore(:,1))];
  v = [v min(Xscore(:,2)) max(Xscore(:,2))]
  axis(V)
  plot(Xscore(:,1),Xscore(:,2),'*w')
  title('Raw Data Scores');
  text(Xscore(:,1),Xscore(:,2),label)
```

Path analysis and Wrightian factor analysis (Section 4.7)

A program called **wrightfa.m** does iterative principal factoring and produces the single-factor solution, shown in Table 4.X, for the fowl correlation matrix stored in the file **fowl.cor** from Table 4.X. This program was constructed following the discussion in Bookstein et al. (1985). The data are loaded as in the previous programs from this file and are put into an array called **r**. The program has several options, as indicated in the comment (%) statements in the program. Three options will be illustrated, including sequestering of the head, front limb, and hind limb factors. An iterative subroutine **wfacsub.m** is called as needed for the various options.

wrightfa.m

```
% Wright factor analysis developed from discussion in
% Bookstein et al. (1985).  Includes:
%   1. Principal components - for which set sel < 0
%   2. Single factor    -      for which set sel > 1
%   3. Sequestered pairs of variables - for example
%      for fowl data       for which set sel = [1 2
%                                               3 4
%                                               5 6]
%   4. A sub-array for a second factor - for which sel
%      is a column array containing the variables in
%      the sub-array
%   After the first Keyboard prompt specify the
%   correlation matrix as r. Type return for size r;
```

```
%   and then after second Keyboard prompt give the
%   scalar sel for 1. or 2. above, or the array for
%   3. or 4. A second return will complete the run
tol = .0001;   % for accuracy of factor determination
keyboard       % enter or load correlation matrix as r
[p,p] = size(r)
keyboard       % enter sel as scalar or array
[nrs,ncs] = size(sel);
corr = r
pause
[e,a,e] = svd(corr);   % Corresponds to principal
                       % components - 1. above
a
pause
e
pause
lding = e*sqrt(a)
pause
if sel(1,1) > 0;       % Finds the first factor by
                       % iteration - 2. above
   swi = 1;
   lding = e(:,1)*sqrt(a(1,1));
   lding0 = lding;
   ntemp = p;
   wfacsub             % subroutine for iteration
   corr
   pause
   lding
   pause
   resr = r- lding*lding';
end;
   if nrs > 1;if ncs == 1;   % solution 4 for list of
                             % variables   4. above
      swi = 2;
      ntemp = p;
      wfacsub
      lding
      pause
      resr = r- lding*lding';
      pause
      corr = resr(sel,sel)
      lding0 = lding0(sel,:);
      lding = lding(sel,:)
      ntemp = nrs;
      swi = 1;
      ntemp = nrs;
      wfacsub
      sel
```

```
        [sel lding]
        pause
        sec = ones(p,1);
        sec(sel,:) = lding;
        rresfin = resr- sec*sec';
        rresfin
        pause
  end;end;
if nrs > 1;if ncs == 2;  % solution for 3. above
  swi = 2;
  ntemp = p;
  wfacsub
  corr;
  lding;
  res = r- lding*lding';
  anew = ones(nrs,1);
  for i = 1:nrs;
        anew(i,1) = sqrt(res(sel(i,1),sel(i,2)));
  end;
  % [sel anew]
  for i = 1:nrs;
   res(sel(i,1),sel(i,2)) = res(sel(i,1),sel(i,2))...
                              -anew(i,1) ^2;
   res(sel(i,2),sel(i,1)) = res(sel(i,2),sel(i,1))...
                              -anew(i,1) ^2;
  end;
  nrs1 = nrs + 1;
  Factors = ones(p,nrs1);
  Factors(:,1) = lding;
  for j = 2:nrs1;
        Factors(:,j) = zeros(p,1);
        Factors(sel(j- 1,:),j) = ones(2,1)*anew(j- 1,1);
        if j == nrs1;
                    corr
                    pause
                    Resid = r- Factors*Factors'
                    pause
                    Factors
        end
  end;
end;end;
```

wfacsub.m

```
% Subroutine called by wrightfa.m
% Algorithm is basically iterated principal components
for i = 1:20; % arbitrary - seems to converge quite fast
      lding0 = lding;
```

```
corr = r- eye(ntemp) + diag(diag(lding*lding'));
if swi == 1;lding = e(:,1)*sqrt(a(1,1));end
if n == nrs;lding = e(:,1)*sqrt(a(1,1));end
if nrs > 1; if ncs == 2;for i = 1:nrs;
corr(sel(i,:),sel(i,:)) = lding(sel(i,:),1)...
                          *lding(sel(i,:),1)';
    end;end;end;
if nrs > 1;if ncs == 1;
    eest = lding*lding';
    corr(sel,sel) = eest(sel,sel);
    end;end;
[e,a,e] = svd(corr);
lding = e(:,1)*sqrt(a(1,1));
if max(abs(lding0- lding)) < tol;break;end;
end;
```

In order to load the **fowl.cor** data and do several options, complete step-by-step instructions are given next.

   1. Principal components: After the first keyboard prompt type

```
K» load fowl.cor

K» r = fowl

K» return
```

You will then be told the size of the correlation matrix. After the second keyboard prompt type

```
K» sel = -1

K» return
```

Any value less than 0 for **sel** will give a PCA.

   2. In order to get a single factor solution for the **fowl.cor** data load it as before; if it is already loaded, then **return** will take you to the next step after the prompt. Then after the second keyboard prompt type

```
K» sel = 1

K» return
```

Any value of **sel** greater than 1 will produce this solution.

   3. In order to get four factors and the Wright solution you will need to designate the number of additional factors and the correlations to be considered. Enter the data as before. Then after the second

keyboard prompt enter the **sel** matrix as indicated:

```
K» sel = [1  2

     3  4

     5  6]

K» return
```

This indicates a skull factor from r_{12}, a front limb factor from r_{34}, and a hind limb factor from r_{56}. The **sel** matrix will be printed unless you put a **;** after the right square bracket. Since **sel** has three rows and two columns, three factors will be found, one for each of the indexed correlation coefficients in each row.

Here is an application of principal components to a larger data set where we can see that the clustering of the data supports an hypothesis of more than one species. It is based on the **echinocy.dat**, which represents means of ostracods from 103 samples from a borehole. The program will work in general for any data set and will plot scores for the first two principal components. See **biplot4** program in the section on Chapter 6.

pca.m

```
%  Program to do general principal component analysis of
%  mean centered or standardized data. Use echinocy.dat
%  to produce graph in book. Data goes into an array
%  called x. Note the first two columns of the
%  Echinocytheris data contain sample sizes and level,
%  so must not be included in the analysis. Therefore
%  x = echinocy(:,3:7). Where 7 is the number of columns
%  of x. Obtained as second value of size (echinocy).
clg
keyboard
[n,m] = size(x)
r = corrcoef(x)
[u,d,v] = svd(r)
mn = mean(x)
stdev = std(x)
z = (x- ones(n,1)*mn)*inv(diag(stdev));
scres = z*v;
v = [min(scres(:,1)) max(scres(:,1))];
v = [v min(scres(:,2)) max(scres(:,2))];
axis(v);
plot(scres(:,2),scres(:,1),'w+')
```

```
xlabel('Principal Component 1')
ylabel('Principal Component 2')
title('Plot of Principal Component Scores')
```

CHAPTER 5 – Q-MODE METHODS

The first two programs do *Q*-mode analyses using the method of principal coordinates. The first uses Gower's association matrix. When this is used with continuous data, it frequently produces Kendall's horseshoe. For continuous data, Gower's coefficient is an association measure based on Manhattan distances of data standardized by range.

The small program **mkimbdat.m** constructs the data from two arrays as indicated in the program.

mkimbdat.m

```
% This program will combine the end members Section 5.1
% called endmem.dat in the proportion indicated
% by propend.dat to produce the data X
load endmem.dat
load propend.dat
X = propend*endmem / 100
```

The two principal coordinate programs for Gower's association and for Euclidean distances follow, making use of the data previously defined in X. One should therefore run **mkimbdat.m** at the keyboard prompt to use these data to produce the results in the text.

princrd1.m

```
echo on
%    Illustration of Principal Coordinates - adapted from
%    Gowcrab in Multidimensional Paleobiology where used
%    for first 10 female and first 10 male orange crabs.
%    Uses Gower 's Association Measure. (See and run
%    similar program based on distance squared).
%    Data goes into array called X here - for example
%    use MKIMBDAT.M
clg
keyboard
[N,p] = size(X)
Range = abs(max(X)-min(X))    % Required for Gower 's
                              % association measure
% DRange = diag(inv(diag(Range)))
H = eye(N); % set aside space for association matrix a
for i = 1:N-1;  % these next 10 steps computes Gower 's
                % measure.between all crabs
```

```
  for J = i + 1:N;
    H(i,j) = (p- abs(X(i,:)-X(j,:))
              *diag(inv(diag(Range)))).//p;
    H(j,i) = H(i,j);
  end
end
H
pause
rmean = mean(H);    %  these three steps centers matrix -
                    %  see Gower
gmean = mean(rmean);
j = ones(N,1);
Q = H- j*rmean- rmean'*j' + ones(N,N)*gmean;
[E,a,E] = svd(Q);
EigVal = a(1:3,1:3)
EigVec = E(:,1:3)
pause
labl = ['    PCoord1        PCoord2        PCoord3']
Coord = EigVec*sqrt(EigVal) % find principal Coordinates
% Note the form of graphs and contrast to
% Euclidean Distance
numbers % loads number 1-20 as strings in array num
num = num(1:N,:)
echo off
subplot(111)
v1 = [min(Coord(:,1)) max(Coord(:,1))];
axis([v1 min(Coord(:,2)) max(Coord(:,2))]);
plot(Coord(:,1),Coord(:,2),'*')
title('Plot of PCoord1 and PCoord2')
text(Coord(:,1),Coord(:,2),num)
pause
subplot(221)
axis(v1[min(Coord(:,2)) max(Coord(:,2))])
plot(Coord(:,1),Coord(:,2),'*')
title('Plot of PCoord1 and PCoord2')
text(Coord(:,1),Coord(:,2),num)
% pause
subplot(222)
axis([v1 min(Coord(:,3)) max(Coord(:,3))])
plot(Coord(:,1),Coord(:,3),'*')
title('Plot of PCoord1 and PCoord3')
text(Coord(:,1),Coord(:,3),num)
%pause
subplot(223)
v = [min(Coord(:,2)) max(Coord(:,2))];
axis ([v min(Coord(:,3)) max(Coord(:,3))]);
plot(Coord(:,2),Coord(:,3),'*')
title('Plot of PCoord2 and PCoord3')
```

```
text(Coord(:,2),Coord(:,3),num)
```

This second program uses Euclidean distances between observations. In this case the result is the same as if one did a principal component analysis on the covariance matrix among the variables and plotted the principal component scores.

princrd2.m

```
echo on
% Illustration of Principal Coordinates - adapted from
% crabcord in Multidimensional Paleobiology where
% first 10 female and first 10 male orange crabs used.
% (This data is also available here)
% Uses Euclidean distance between specimens to contrast
% Gower 's Association. See and run similar program
% that is based on Gower 's measure.
% Data goes into array called X here - for example use
% MKIMBDAT.M
keyboard
[N,p] = size(X)
XSS = X*X'; % major product moment of data for crabs
ss = diag(XSS); % vector elements are sums of squares for
                % each crab
j = ones(N,1); % form j vector of ones
Dist = ss*j' + j*ss' - 2*XSS % computes distances
rmean = mean(Dist); % form mean centered association
                    % matrix
gmean = mean(rmean); % grand mean
Q = -.5*(Dist- j*rmean- rmean'*j' + ones(N,N)*gmean);
[E,a,E] = svd(Q);
EigVal = a(1:3,1:3)
EigVec = E(:,1:3)
pause
% Results below are very different from Tables in Text
% since the Association Measure is based on Euclidean
% Distance Squared. Also plot is very different -
% as does not produce horseshoe.
labl = [' PCoord1     PCoord2      PCoord3']
Coord = EigVec*sqrt(EigVal) % find principal Coordinates
clg
numbers;
num = num(1:N,:);
echo off
v1 = [min(Coord(:,1)) max(Coord(:,1))];
if rank(a)<3;subplot(111);
    axis([v1 min(Coord(:,2)) max(Coord(:,2))])
    plot(Coord(:,1),Coord(:,2),'i')
```

```
                          title('Plot of PCoord1 and PCoord2')
                          text(Coord(:,1),Coord(:,2),num)
                          xlabel('Coord1')
                          ylabel('Coord2')
else;subplot(221);
                          axis([v1 min(Coord(:,2)) max(Coord(:,2))])
max(Coord(:,2))])

                          plot(Coord(:,1),Coord(:,2),'i')
                          title('Plot of PCoord1 and PCoord2')
                          text(Coord(:,1),Coord(:,2),num)
                          xlabel('Coord1')
                          ylabel('Coord2')
subplot(222)
axis([v1 min(Coord(:,3)) max(Coord(:,3))])
plot(Coord(:,1),Coord(:,3),'i')
title('Plot of PCoord1 and PCoord3')
text(Coord(:,1),Coord(:,3),num)
xlabel('Coord1')
ylabel('Coord3')
subplot(223)
v1 = [min(Coord(:,2)) max(Coord(:,2))];
axis([v1 min(Coord(:,3)) max(Coord(:,3))])
plot(Coord(:,2),Coord(:,3),'i')
title('Plot of PCoord2 and PCoord3')
text(Coord(:,2),Coord(:,3),num)
xlabel('Coord2')
ylabel('Coord3')
end
```

Q-mode factor analysis following Imbrie's procedure is given for the data in Table 5.II; it produces all of the results given in Tables 5.III–XIII and displayed in Figures 5.6 and 5.7.

The program **qfact3.m** is tailored to produce the tables and graphs in the text, and one might want to reduce the amount of output for practical runs with other data. I have tried to follow the notation in the text as much as possible. Since **varimax.m** is called at least twice, type **return[Enter]** after each keyboard prompt.

qfact3.m

```
echo on
% Some developments of Q-mode factor analysis with data
% in Section 5.4. For this example load the Raw Data
% X which is in tab5_ii.m
clg
keyboard
[N,p] = size(X)
Centroid = mean(X);
Rowss = diag(X*X');
```

```
[X                  Rowss
  Centroid sum(Centroid. ^2)]     % produces Table 5.II for
                                  % example
pause
W = inv(sqrt(diag(Rowss)))*X;
Wmn = Centroid / sqrt(Centroid*Centroid');
[W          % produces Table 5.III
Wmn]
pause
H = W*W'    % produces Tble 5.IV
pause
[u,Lam,u] = svd(H);
rnk = rank(W)
u = u(:,1:rnk);
Lam = Lam(1:rnk,1:rnk);
Lout = diag(Lam);
[Lout 100*Lout / sum(Lout)   100*cumsum(Lout) / sum(Lout)]
% Table 5.V
pause
A = u*sqrt(Lam);
[A % Table 5.VI
  100*cumsum(Lout') / sum(Lout)]
pause
[diag(A*A') A(:,1:2)] % Table 5.VII for 2 factors
pause
F = W'*A*inv(Lam) % Table 5.VIII Factor scores for data
pause
F(:,1:2) % Table 5.IX - Two Unrotated factor scores
check = W - A*F'
pause
%Alternative computations by Rmode
WpW = W'*W
pause
[V,L,V] = svd(WpW);
L
pause
V
pause
%  Plotting part of program to plot objects and make
%  Figure 5.6
A2 = W*V
axis('square')
val = [-1 1 -1 1];
axis(val)
% compute factor scores for Centroid and put on graph
astar = Wmn*F
% Note is different from text which says w'*Fstar?
echo off
```

```
plot(A(:,1),A(:,2),'+w',astar(1,1),astar(1,2),'ow')
numbers;
num = num(1:N,:)
text(A(:,1),A(:,2),num)
text(astar(1,1),astar(1,2),'Ce')
title('Plot of objects on two, unrotated factors')
     % Fig. 5.5
grid
xlabel('Factor 1')
ylabel('Factor 2')
pause
echo on
% Next part does varimax rotation of results above
lding = A(:,1:2)
A0 = lding
pause
echo off
varimax        % Remember to type return to continue
echo on
A2 = lding     % This is result in Table 5.X
T = inv(A0'*A0)*A0'*A2   %   transform matrix
pause
lding = F(:,1:2)    % To rotate factor scores
echo off
varimax        % Remember to type return to continue
echo on
F = lding      % This is the result in Table 5.XI
echo off
% New plot to produce Figure 5.6
astar2 = astar(:,1:2)*T
% Note is different from text which says w'*Fstar?
echo off
plot(A2(:,1),A2(:,2),'+w',astar2(1,1),astar2(1,2),'ow')
numbers;
num = num(1:N,:);
text(A2(:,),A2(:,2),num)
text(astar2(1,1),astar2(1,2),'Ce')
title('Plot of objects on two, varimax factors')
     % Fig. 5.5
grid
xlabel('Factor 1')
ylabel('Factor 2')
pause
echo on
% Remaining part of the program does oblique rotation
% and produces the remaining tables in the section
B = A2
[Y,I] = max(B)
```

```
V = B(I,:)
pause
C = B*inv(V)   % Table 5.XI
pause
C = inv(sqrt(diag(diag(X*X'))))*C;    % divide rows by
                           % original lengths of objects
C = C*inv(C(1:2,:)); % divide through by reference
                     % object values
Cde = 100*inv(diag(sum(C')))*C
% "De-normalized" projection matrix of Table 5.XIII
```

Section 5.5 discusses Gower's procedure for analyzing skew-symmetric data and gives two data examples. The program **skewmc.m** works with the data example in Table 5.XIV in the text. It will also work with other appropriate data sets. Load the data from Table 5.XIV by executing the program **tab5_xiv.m** either before you start the program, or after the **K»** prompt. A **return** after the **K»** prompt will continue execution of the program in either case. You will be asked if you want labels for the plots; they should be stored in an array called **labels**. If you supply labels, say so and put in the array. For the data in Table 5.XIV **symlab.m** will load the appropriate labels into **labels**.

skewmc.m

```
% Analysis of Asymmetrical comparison matrix - see
% Chapters 5 and 7. The data go into an array X and
% the labels for plotting in an array called labels.
clg
keyboard
t = ['Will you supply labels (1) or use defaults - '];
Y = input(t);
[N,p] = size(X)
if Y == 1;keyboard;else numbers;labels = num(1,N);end;
pt = fix(p / 2);
% NeXt three rows turn McCammon data into Proportions
% would usually leave these rows out
Z = diag(sum(X'));
Prop = inv(Z)*X + eye(7);
pause
% Next steps compute symmetrical part and do a Principal
% Coordinates Analysis on the M matrix
M = (Prop + Prop') / 2
pause
mr = mean(M);
mg = mean(mr);
Mcent = M - ones(p,1)*mr - (ones(p,1)*mr)' + mg*ones(p,p);
[E,a,E] = svd(Mcent);
```

```
% Principal Coordinate Scores
Es = E(:,1:p-1)*sqrt(a(1:p-1,1:p-1))
pause
subplot(111)
v = [min(Es(:,1)) max(Es(:,1)) min(Es(:,2)) max(Es(:,2))];
axis(v);
plot(Es(:,1),Es(:,2),'+')  % could do 3:4 also
   xlabel('PCoord1')
   ylabel('PCoord2')
   text(Es(:,1),Es(:,2),labels)
   title('Principal Coordinates of Symmetrical Part')
% Rest of Program does Skew Symmetric Part
% Skew Symmetric Partition
N = (Prop-Prop')/2
pause
[U,sing,V] = svd(N);
% The U matrix is treated as pairs of columns, and a PCA
% analysis is done on each pair to rotate the plots
U
  for i = 1:pt;
     ind = (i-1)*2+1;
     Ut = U(:,ind:ind+1);
     umn = sum(Ut);
     Devu = Ut-ones(p,1)*umn;
     Su = Devu'*Devu/(p-1);
     [E,a,E] = svd(Su);
     Sct = Ut*E*sqrt(sing(ind,ind));
     v = [min(Sct(:,1)) max(Sct(:,1))];
     v = [v min(Sct(:,2)) max(Sct(:,2))];
     axis(v);
     plot(Sct(:,1),Sct(:,2),'+');
        title('Plots for Pair of Axes for Plane')
        text(Sct(:,1),Sct(:,2),labels),pause
     if i == 1
        Scu = Sct;
     else
        Scu = [Scu Sct];
     end
  end;
% Scores for SVD of Skew Symmetric Matrix
Scu
```

CHAPTER 6 – Q-R METHODS

Correspondence analysis is discussed in Section 6.1, and a program is given here using the same data constructed in **mkimbdat.m**. Note that these data are not frequency data, but as the text points out, if they were multiplied by 10, then we would have integer values – which would be the kind of data we would get if we were using frequencies. This scale change would not in any way change the analysis.

corresp.m

```
% Correspondence analysis program - to use artificial
% data in text load into an array called K after the
% prompt. mkimbdat.m will do that, but the
% data must be renamed K.
keyboard
tot = sum(sum(K)) % finds sum of elements in matrix K
K = K / tot;    % standardizes all elements by dividing
                % through by total
[N,p] = size(K)
t = ['Type 1 if give labels in varlab,'];
t = [t ' otherwise defaults used - '];
Y = input(t);
if Y == 1;keyboard;else varlab;alph = alph(1:p,:);end
CD = inv(sqrt(diag(sum(K)))); %  These two produce
                % diagonal matrices
RD = inv(sqrt(diag(sum(K')'))); % of reciprocal sq. root
                % of row and col.totals
X = RD*K*CD;  % X is rescaled matrix
Sim = X'*X;   % Association matrix for rescaled K
pause
[U,q,V] = svd(X); % Singular Value Decomposition
lat_root = q(2:p,2:p) ^2 % Rescales singular values
                % to roots
total = trace(lat_root)
inertia = diag(100*lat_root / total);
percent = cumsum(inertia);
Out2 = [inertia  percent]
XSC = RD*U(:,2:3)*q(2:3,2:3)  % Correspondence axial
YSC = CD*V(:,2:3)*q(2:3,2:3)  % loadings
minxy = min([XSC
          YSC]);
maxy = max([XSC
          YSC]);
v = [minxy(1,1) maxy(1,1) minxy(1,2) maxy(1,2)];
axis(v);
echo off
plot(XSC(:,1),XSC(:,2),'wo')
  title('Q-Mode o and R-Mode +')
  hold on
        new = ones(2*p-1,2); % makes vectors borrowed from
                % mkfig1_2.m
```

```
ind = [1:2:2*p-1;
new(ind,:) = YSC(:,1:2);
ind2 = [2:2:2*(p-1)];
dum = zeros(p-1,2);
new(ind2,:) = dum;
plot(YSC(:,1),YSC(:,2),'w + ',new(:,1),new(:,2),'- ')
numbers;
num = num(1:N,:);
xlabel('FIRST AXIS')
ylabel('SECOND AXIS')
text(XSC(:,1),XSC(:,2),num)
text(YSC(:,1),YSC(:,2),alph)
grid
pause
hold off
```

The biplot method is given in Section 6.3. A general program
biplot4.m, which is discussed in Marcus (1992), will produce the
biplots of Gabriel, as well as several other forms. It is illustrated with
the data on diabetes given in the file **diabetes.dat** as in the text.
The variable or column labels are given in a program **diabcol.m**, and
the row or specimen labels are given in a program **diabrow.m** for this
data set. Differences in the plots produced here and in the text are due
to the reflection that occurs because eigenvectors (and therefore scores)
change their signs but not their properties when they are multiplied by
-1. Detailed instructions for running this program are given after the
listing.

biplot4.m

```
% This biplot program does four different forms of the
% biplot, as described on each plot. Also it allows
% choice of logging the data, non-centered data,
% centered data, or standardized data.
% Enter the data by saying after the prompt.
% load fn.ft --- where fn.ft is the two part name of
%                  your data file
% X = fn        --- this sets X = to the data in fn.ft.
% later you can assign custom variable labels
% interactively or in an array called labl for the
% vectors (variables); and ptlab for the rows.
% Note that new scales for the axes can produce more
% acceptable plots. Also a "fudge" factor for length-
% ening or shortening the variable vectors
% is provided. fudge = 0 advances to the next form
% of the biplot.
clg
a = ['Input the data matrix.'];
```

```
a = [a ' Type RETURN after data loaded in X'];disp(a)
keyboard
yans = input('Type 1 if you want logs, otherwise 0 - ')
if yans == 1,X = log(X);end
t-['Type 0 if mean centered, 1 if standardized'];
yans = input([t ', 2 if non-centered - '])
mn = mean(X)
[N,p] = size(X)
labl[97:96 + p]';
labsp = ' ';
lab = ones(1:2*p- 1,1)*labsp;
lab(1:2:2*p-1,1) = setstr(labl);
t = ['Type 0 if default variable labels OK,'];
yans2 = input([t '1 to supply an array labl - '])
if yans2 == 1,
    keyboard
    [nl,pl] = size(labl)
end
t = ['Type 0 if +label ok for points,'];
yans3 = input([t 'or 1 to give an array ptlabl - '])
if yans3 == 1
    keyboard
end
t = ['If you will want to adjust axes later'];
yv = input([t 'enter 1, 0 otherwise - '])
dev = X- ones(N,1)*mn;
if yans == 2,dev = X;end
s = dev'* / (N-1);
isd = inv(diag(std(X)));
z = dev*isd;
if yans ~= 1,
    [V,D,U] = svd(dev,0);
 else [V,D,U] = svd(z,0);
end
nv = sum(any(D));
D = D(1:nv,1:nv);
eig = diag(D. ^2)./ (N- 1);
st = std(X)
eig = [eig 100.*cumsum(eig)./ sum(eig)]
pause
r = corrcoef(X);
V = V(:,1:2);
D = D(1:2,1:2);
U = U(:,1:2);
if yans == 0,
    s
    pause
    res = s- U*D. ^2./ (N- 1)*U'
```

```
      pause
      r
      pause
      resr = isd*res*isd
pause
 else r
      pause
      res = r- U*D. ^2./ (N- 1)*U'
      pause
end
for plots = 1:4
   if plots == 1;
       clg
       fudge = 1;
       G = V*D;
       H = U*fudge;
       t = ['G = VD and H = U - pts. at Euc. Dist. m'];
       top = [t 'vectors orthogonal'];
       biplot;
   end;
   if plots == 2;
       clg
       fudge = 1;
       G = V*sqrt(N- 1);
       H = U*D*fudge / sqrt(N- 1);
       t = ['G = sqrt(N- 1)V and H = UD / sqrt(N- 1) - pts.'];
       top = [t 'Mah. Dist., vectors cos r'];
       biplot;
   end
   if plots == 3;
       clg
       fudge = 1;
       G = V*sqrt(D);
       H = U*sqrt(D)*fudge;
       t = ['G = V*sqrt(D) and H = U*sqrt(D)'];
       top = [t '- compromise of previous two'];
       biplot;
   end
   if plots == 4;
       clg
       fudge = 1;
       G = V*D;
       H = U*D*fudge./ sqrt(N- 1);
       t = ['G = V*D and H = U*D / sqrt(N- 1)-not'];
       top = [t'biplot, pts. Euclidean; vectors r'];
       biplot;
   end
end
```

biplot.m

```
% Subroutine to do all of the plots for biplot4.m
% Made from biplot3.m - dissecting to main routine and
% general plotter
while fudge>0;
  H = fudge*H;
  clg
  Haug = zeros(2*p-1,2);
  aug = [1:2:2*p-1];
  Haug(aug',:) = H(:,1:2);
  Temp = [G
          Haug];
  v = [min(Temp(:,1)) max(Temp(:,1))];
  axis([v min(Temp(:,2)) max(Temp(:,2))]);
  if yv == 1,
      v
      ['Give new minx, miny, maxx,maxy as array v']
      keyboard
      axis(v)
  end
  plot(G(:,1),G(:,2),'+w',Haug(:,1),Haug(:,2),'- ')
  if yans3 == 1, tex(G(:,1),G(:,2),ptlab),end
  text(H(:,1),H(:,2),labl)
  xlabel('Principal Component I')
  ylabel('Principal Component II')
  title(top)
pause
yg = input('if you want to print last graph type 1 - ')
if yg == 1,meta,end
  t = ['Give a multiplier for the graph, '];
  fudge = input([t ' 0 for next form - ']);
end
```

In order to run this program with the diabetes data given in the file **diabetes.dat** the following steps are required:

1. Type **biplot4** after the prompt
   » **biplot4**
2. After the first keyboard prompt type
   load diabetes.dat
   return
3. In order to reproduce the plot in the text (except for scaling and rotation) respond with a **0** to the questions for logs.
4. Request mean centering by typing **0** after that question.
5. Type **1** for request for variable labels and then after the keyboard prompt type
   diabcol
   return

6. Type **1** for request for point labels and then type, after the keyboard prompt,
   diabrow
   return
7. If you want to adjust axes later, type **1** after next query, otherwise type **0**. For now we use **0** and explain the option later.
8. You will then be asked two questions after each of the four different "biplot" graphs are presented. The first asks if you want to print a graph – the program **meta.m** is invoked. Type **0** if no print is desired. This is recommended until you are used to the program. The second asks for a multiplier, called **fudge** factor, which allows the vectors to be rescaled so the graph will be more interpretable. If the vectors are short relative to the point scatter, then a fudge value greater than 1 is desirable. A value of about 30 is required for the first graph. If the points are crowded, then a fudge factor less than 1 (sometimes much less than 1) will produce a more pleasing plot. A factor of 0.07 is about right for the second graph. You can adjust the plot as long as you do not set **fudge = 0**. A value of 0 will take you to the next form of the biplot.

If you had requested the ability to adjust the axes in item 7 by typing 1, you would be asked before each graph if you want to set new limits. If you type return after the K prompt, they will stay the same. Otherwise you will have to supply the axes limits as follows:

v = [minx maxx miny maxy]

For plotting or publishing purposes you would want to scales to have the same proportions, and you can adjust them using this option. I have only used this option for published graphs.

Another example is given in the section on Chapter 8.

CHAPTER 7

Sections 7.8 and 7.9 have extensive discussions of rotation and estimation of factor scores, respectively.

The example in Tables 7.I–7.III comes from an analysis of the **rock.dat** introduced in Chapter 2. In order to produce the results of these tables, a small program is given (**tab7_iii.m**) that calls **mktab7_i.m** (which does a principal component analysis and creates Table 7.I) and puts the loadings into the array **lding**. Then the program **varimax.m** is called. The data used are the Z transformed data stored in **Z.rck**. Type **load Z.rck** after the first keyboard prompt, and **return[Enter]** to continue execution of the program. When **varimax** is called you will have to type **return[Enter]** after each keyboard prompt.

tab7_iii.m

```
% This little program will do a principal components
% calling mktab7_i.m and then find the varimax factor
% loadings in Table 7.iii for 3 factors by running
% varimax.m. The varimax.m program shows each cycle
% and iteration, but is rapid. Finally it computes the
% rotation matrix of Table 7.iː using the procrustes
% formula (2) from Rohlf (1990).
% The mktab7_i.m program will issue a keyboard prompt K»
% after which you load Z.rck.
% After the next K» type the word return, as loadings
% already are in lding
mktab7_i
A2 = A(:,1:3) % to set aside three factors for the example
lding = A2;
varimax % invokes varimax.m
T = inv = (lding'*lding)*lding'*A2;
T'       % this is the result found in Table 6_ii
check- lding- A2*T'
```

The two programs called in **tab7_iii.m** are given next.

mktab7_i.m

```
% General pca analysis which starts with Z standardized
% data and produces pca loadings. Enter array and call Z
% after first keyboard prompt. Note that this is set up
% for data which had standard deviations computed with N
% rather than N-1 in the denominator.
keyboard
[V,D,U] = svd(Z,0);
U
[N,p] = size(Z)
A = U*D / sqrt(N) % loadings for all p pc 's are in A
```

varimax.m

```
% This procedure follows algorithm as spelled out in
% Harman (1960) in Chapter 14, section 4. To run the
% program - the loadings are put in an array called
% lding. Type return to continue processing.
% The notation follows Harman. The routine vfunct.m is
% called to compute the Variance of the loadings
```

```
% squared.
keyboard
b = lding;
[n,nf] = size(lding)
pause
hjsq = diag(lding*lding')    % communalities
hj = sqrt(hjsq);
pause
vfunct                       % function to compute
                             % variances of loadings ^2
v0 = Vtemp
pause
for it = 1:10;  % Never seems to need very many iterations
for i = 1:nf- 1  % Program cycles through 2 factors at a
  jl = i + 1;        % time
  for j = jl:nf
      xj- lding(:,i)./ hj;  % notation here closely
      yj = lding(:,j)./ hj;  % follows harman
      uj = xj.$4xj- yj.*yj;
      vj = 2*xj.*yj;
      A = sum(uj);
      B = sum(vj);
      C = uj'*uj- vj'*vj;
      D = 2*uj'*vj;
      num = D- 2*A*B / n;
      den = C- (A ^2- B ^2) / n;
      tan4p = num / den;
      phi- atan2(num,den) / 4;% This function finds the
                              % right quadrant
      angle = phi*180 / pi;
      [i j it angle]  %     This statement can be deleted
      if abs(phi)>.00001;
          Xj = cos(phi)*xj + sin(phi)*yj;   % actual
          Yj = -sin(phi)*xj + cos(phi)*yj;  % rotation
          bj1 = Xj.*hj;
          bj2 = Yj.*hj;
          b(:,i) = bj1;
          b(:,j) = bj2;
          lding(:,1) = b(:,i);
          lding(:,j) = b(:,j);
      end
  end
end;
lding = b;
vfunct;
V = Vtemp;
if abs(V- V0)>.0001;break;else V0 = V;end;
end;
```

```
lding
V
pause
```

vfunct.m

```
% This little program computes the value of V (Harman's
% notation) given also in formula 7.4 in the text.
bh = lding./ (hj*ones(1,nf) );
Vtemp = n*sum(sum(bh. ^4))- sum(sum(bh. ^2). ^2);
```

The next example **scores.m** computes factor scores by both the direct method and the regression formulas given in Section 7.9. Again the **rock.dat** is called in Z standardized from the file **z.rck**. One needs to type return after each prompt as earlier programs **mktab7_i.m** and **varimax.m** are invoked.

scores.m

```
%   This program loads the Z scores for the rock data,
%   and then computes pca factor scores, Fpca - Table
%   7.VIII, and varimax factor scores using both the
%   direct formula for Fdir, and the regression
%   formula 7.11. Several of the programs used earlier
%   are called here. Notation from the text is used as
%   much as possible.
load Z.rck
mktab7_i     % produces loadings from pca
Fdir = Z*A*inv(A'*A); % formula for direct Factor
                      % scores
lding = A(:,1:3)    % sets up lding matrix for
                    % varimax rotation
varimax
A = lding;        % puts loadings back in A
R = Z'*Z / N;
['Results for Table 7.VIII']
Fdir
pause
['Results for Table 7.IX']
Fvm = Z*inv(R)*A      % this is formula 7.11 in text
pause
```

The next program **mkfig7_9.m**, repeats some analysis of earlier programs and calls **plot_scr.m** several times to plot factors on principal component axes, varimax axes, and promax axes. Unfortunately I am unable to plot on nonorthogonal axes using MATLAB. A program **promax.m** computes oblique factors.

mkfig7_9.m

```
% This loads the save results of earlier programs
% and runs promax.m to produce various factor axes
% Need to load Z.rck at first keyboard prompt
% Type return at second keyboard prompt
% Type B = lding at third keyboard prompt
tab7_iii
scr = A2;titl = ['Principal Component Axes'];
clg
plot_scr   % Makes A part
pause
clg
scr = lding;titl = ['Varimax axes'];
plot_scr   % Makes B part
pause
clg
promax
scr = Pp;titl = ['Promax axes - orthogonal'];
plot_scr   % Makes C part
pause
```

This is a routine called from the foregoing program. It will plot any three vectors in three subplots, when the data is put into **scr**; and the title in **titl** is plotted in place of the fourth subplot.

plot_scr.m

```
% General program to plot factor scores for any sets of
% of scores for first three axes
% function [] = plot_scr(scr,strng)
[nv,nf] = size(scr)
numbers;
num = num(1:nv,1:2);
subplot(221)
v = [min(scr(:,1)) max(scr(:,1)) min(scr(:,2))];
axis([v; max(scr(:,2))]);
plot(scr(:,1),scr(:,2),' + w')
text(scr(:,1),scr(:,2),num)
xlabel('Factor I')
ylabel('Factor II')
grid
subplot(222)
v = [min(scr(:,1)) max(scr(:,1)) min(scr(:,3))];
axis(v);max(scr(:,3))]);
plot(scr(:,1),scr(:,3),' + w')
text(scr(:,1),scr(:,3),num)
xlabel('Factor I')
ylabel('Factor III')
grid
```

```
subplot(223)
v = [min(scr(:,2)) max(scr(:,2)) min(scr(:,3))];
axis([v max(scr(:,3))]);
plot(scr(:,2),scr(:,3),'+w')
text(scr(:,2),scr(:,3),num)
xlabel('Factor II')
ylabel('Factor III')
grid
subplot(224)
title(titl)
```

promax.m

```
% Program to illustrate Promax rotations using data in
% Table 6.i tab6_i.dat Load data into varimax
% loading matrix B
keyboard
Bstar = B. ~4
pause
Tr = inv(B'*B)*B'*Bstar
Tr = Tr*sqrt(inv(diag(diag(Tr'*Tr))))  % Normalizes column
                                       % of Tr
pause
Tpp = inv(Tr)    % Tp' from definition
Tpp = inv(sqrt(diag(diag(Tpp*Tpp'))))*Tpp % Normalizes
                                          % rows of Tp'
pause
Sr = B*Tr            % Oblique reference structure
Phip = Tpp*Tpp'      % Correlation between primary factors
                     % Table 6.VI
pause
Sp = B*Tpp'  % Primary Factor structure matrix Table 6.V
pause
Pp = B*inv(Tpp) % Primary Factor pattern matrix Table 6.IV
pause
Pr = B*inv(Tr') % Reference pattern matrix
Phir = Tr'*Tr   % Correlations between reference axes
pause
```

Finally, the next program calls several of the foregoing programs and plots the factor scores in three ways using the **plot_scr.m** program listed earlier.

fact_scr.m

```
% This program loads the Z scores for the rock data,
% and then computes pca factor scores, Fpca - Table
```

```
%  7.VIII, and varimax factor scores using both the
%  direct formula for Fdir, and the regression
%  formula 7.11. Several of the programs used earlier
%  are called here.
load Z.rck
mktab7_i    % produces loadings from pca
Fdir- Z*A*inv(A'*A); % formula for direct Factor scores
lding = A(:,1:3)    % sets up lding matrix for
                    % varimax rotation
varimax
A = lding;
R = Z'*Z / N;
['Results for Table 7.VIII']
Fdir
pause
clg
scr = Fdir;titl = ['Principal factor scores'];
plot_scr    % calls plotting routine
pause
['Results for Table 7.IX']
Fvm = Z*inv(R)*A
pause
clg
titl = ['Varimax factor scores'];
scr = Fvm;plot_scr    % calls plotting routine
pause
B = lding
promax   % probably puts out more results than interested
         % in  here
pause
['Results for Table 7.X']
Fob = Z*inv(R)*Sp
pause
clg
scr = Fob;titl = ['Promax factor scores'];
plot_scr    % calls plotting routine
pause
```

CHAPTER 8

Several of the data sets mentioned in Chapter 8 are available in the text, and the examples can be run either from raw data or, in some cases, from the correlation matrices. It is left for the reader to invoke the appropriate programs for each example. A couple more examples are supplied by the author of the supplement and are instructive.

The example in Section 8.9 for crude oils uses the data in the file `kvalheim.dat`. A biplot is to be run on this data. As no row or column labels are supplied, defaults should be used.

The data are supplied for two examples in Section 8.15. The Haiti glass oxide data are supplied in a file called `haiti.dat`. Labels for the oxides are in a file called `haiticol.m`, and for the specimens, in `haitirow.m`. A simplex analysis of this data would be provided by running the program `aitch.m`.

Finally, the data for Section 8.15, Example 2, are supplied in a file called `wallin.dat`. A principal component analysis can be run using one of the programs supplied or using a new one–easily written by the reader. Cross-validation analysis to determine the number of pcas is given by running the programs `validate.m` and examining the W values; the variable importance evaluated, by running `var_info.m`; and a search for the influential data, by running `spec_infl.m`.

Logarithms can be invoked at almost any time by just inserting a line after the data are loaded. For example, for an array X they are converted to logarithms by the statement $X = \log(X)$.

The data `lhippo.dat` represents 46 skulls of four species of hippopotamus, measured for 10 variables. The species are well separated in a biplot if the appropriate variables are chosen. The specimen labels contain the species designators and are in `lhiprow.m`. The variables names are in `lhipcol.m`. The data are already in logarithmic form and should be mean centered. If one restricts the data only to variables `lei` and `llo`, better species separation obtains. This illustrates the point that sometimes a plot of the first two principal component scores may give poorer separation of groups than a bivariate scattergram. The characters `lei` and `llo` are diagnostic for these species. Running `validate.m` followed by `var_info.m` finds these two variables and one other.

PROGRAMS AND DATA USED IN THE APPENDIX

The following is a complete alphabetic list of all the programs as well as the data sets according to the chapters where they are described.

Programs

`aitch.m`	Chapter 4	Compositional data analysis
`biplot4.m`	Chapter 6	Biplot analysis
`biplotm.`	Chapter 6	Plotting routine of `biplot4`

corlab.m	General	Produces labels 'x1'-'x10'
corresp.m	Chapter 6	Correspondence analysis
crab20.m	Chapter 5	Other data for principal coordinates
diabcol.m	Chapter 6	Puts diabetes column labels in labl
diabrow.m	Chapter 6	Puts diabetes row labels in ptlab
eck_yng.m	Chapter 2	Eckardt–Young theorem and **svd**
eigens.m	Chapter 2	Eigenvalues and vectors using **svd**
fact_scr.m	Chapter 7	Computes and plots factor scores
findrank.m	Chapter 2	Illustrates rank principle
index.m	Chapter 1	Indices to make Figs. 1.2 and 1.3
invsqrdi.m	Function	Computes reciprocal of standard deviation
iqrnge.m	Function	Interquartile range
istd.m	Function	Reciprocal of standard deviation
labrock.m	Chapter 2	Puts rock names in min array
lhipcol.m	Chapter 8	Sets up species labels for lhippo.dat
lhiprow.m	Chapter 8	Sets up variable labels for lhippo.dat
mkfg^2_12.m	Chapter 2	Illustrates correlations
mkfig1_2.m	Chapter 1	Contours of prospect data
mkfig1_3.m	Chapter 1	Contours of prospect factors
mkfig2_1.m	Chapter 2	Plots of vectors for rock.dat
mkfig2_3.m	Chapter 2	Plots of vectors for rock.dat
mkfig3_2.m	Chapter 3	Plots of factors
mkfig4_1.m	Chapter 4	PCA plot of echinocy.dat
mkfigs2.m	Chapter 2	Calls mkfig2_1.m and mkfig2_3.m
mkimbdat.m	Chapter 5	Constructs artificial data
mktab7_i.m	Chapter 7	Does PCA on Z rock.dat in Z.rck
mlfact.m	Chapter 3	Maximum likelihood factoring
modstore.m	Chapter 4	Called in var_info.m and spec_infl.m
numbers.m	Chapter 2	Puts numbers 1–40 in num array
pctab4_i.m	Chapter 3	Plot of PCA scores for echinocy.dat
plot_scr.m	Chapter 7	Routine for making three subplots
princomp.m	Introduction	Simple example of PCA
princrd1.m	Chapter 5	Principal coordinates using Gower measure
princrd2.m	Chapter 5	Principal coordinates using Euclidean distance
prod_mom.m	Chapter 2	Illustrates $\mathbf{XX'}$ and $\mathbf{X'X}$
promax.m	Chapter 7	Promax factor rotation
prospect.m	Chapter 1	Constructs prospecting data
qfact3.m	Chatper 5	Q-mode factor analysis
rho.m	Chatper 3	Loads correlations used in **mlfact.m**

scores.m	Chapter 7	Computes factor scores
skew_sym.m	Chapter 2	Simple skews symmetric definition
skewmc.m	Chapter 5	Gower's skew symmetric analysis
spe_infl.m	Chapter 4	Specimen influence in PCA
ssq.m	Function	Sum-of-squares matrix
symlab.m	Chapter 5	Alpha labels A–P for Table 5.XIV
tab2_i.m	Chapter 2	Produces Table 2.I with rock.dat
tab4_iii.m	Chapter 4	Factor results in Table 4.III
tab5_xiv.m	Chapter 5	Data from Table 5.XIV for skewmc.m
tab7_iii.m	Chapter 7	PCA and varimax example for Z.rck
tabs1.m	Chapter 1	Principal factoring of prospec.dat
validate	Chapter 4	Krzanowski cross-validation procedure
var_info.m	Chapter 4	Krzanowski PCA variable evaluation
varimax.m	Chapter 7	Varimax rotation program
varlab.m	Chapter 5	Variable labels for Table 5.XIV
vfunct.m	Chapter 7	Function called by **varimax.m**
wfacsub.m	Chapter 4	Routine called in wrightfa.m
wrightfa.m	Chapter 4	Wright factor analysis
x_tab4_1.m	Chapter 4	Data to load for **pctab4_i.m**
z_stand.m	Chapter 2	Illustrates *Z* standardization

Data sets

causes.dat	Chapter 1	Causal data for prospecting example
diabetes.dat	Chapter 4	Used to illustrate **biplot4.m**
echinocy.dat	Chapter 4	Used to illustrate PCA scores
endmem.dat	Chapter 1	For first prospecting example
fact.dat	Chapter 3	Another version of data in **Rho.m**
fowl.cor	Chapter 4	Fowl correlations for **wrightfa.m**
geoprop.dat	Chapter 1	Geological proportions for prospecting example
haiti.dat	Chapter 8	Glass data for simplex in **aitch.m**
hubbert.dat	Chapter 4	Used for **validate.m** Krzanowski procedure
kvalheim.dat	Chapter 8	Data used for another biplot example
lhippo.dat	Chapter 8	Data for logarithms of skull measurements
propend.dat	Chapter 1	Proportions of endmem.dat
pros_scr.dat	Chapter 1	Prospecting factor scores
prospec.dat	Chapter 1	Stored prospecting data

rhesus.dat	Chapter 4	Used for simplex illustration
rho3_7.dat	Chapter 3	Same as fact.dat
rock.dat	Chapter 2	Used for matrix algebra examples
tab1_iii.dat	Chapter 1	PCA loadings from Table 1.III
tab1_iii.vma	Chapter 1	Varimax loadings from Table 1.III
tab4_iva.dat	Chapter 4	Data used in Table 4.IVA
tab7_i.dat	Chapter 7	Text PCA loadings
tab7_iii.dat	Chapter 7	Text varimaxloadings
wallin.dat	Chapter 8	Used in cross-validation, etc.
x.dat	Chapter 2	Used in eck_yng.m
z.rck	Chapter 7	Used as input in qfact.m

Keywords, commands, and symbols

The first list presents the MATLAB keywords, commands, and symbols used in this supplement. The second is a complete listing, with those actually used set in **boldface** type.

abs	**end**	**pi**
any	**eps**	**plot**
acos	**eye**	**reshape**
ascii	**for**	**return**
atan2	**function**	**save**
axis	**grid**	**setstr**
break	**help**	**size**
casesen	**hold off**	**sqrt**
clg	**hold on**	**subplot**
contour	**if**	**sum**
corrcoef	**input**	**svd**
cos	**inv**	**text**
cumsum	**keyboard**	**title**
det	**load**	**while**
diag	**log**	**xlabel**
diary on	**max**	**ylabel**
diary off	**meta**	**zeros**
echo off	**min**	**%**
echo on	**nan**	**=**
else	**ones**	**==**
elseif	**pause**	**!**

/	+	./
^	,	.*
;	-	...
:	[ ]	>
*	( )	<
		~

The following is a complete list of MATLAB keywords, commands, and symbols (those actually used in this supplement are in boldface type):

help	atan	**elseif**
[	**atan2**	**end**
]	**axis**	**eps**
(	balance	error
)	**break**	eval
.	**casesen**	exist
,	ceil	**exit**
;	**chdir**	exp
%	chol	expm
!	clc	**eye**
:	**clear**	feval
,	**clg**	fft
+	clock	**filter**
-	computer	find
*	conj	finite
\	**contour**	fix
/	**cos**	floor
^	cumprod	flops
<	**cumsum**	**for**
>	dc2sc	format
=	delete	fprintf
&	**det**	**function**
|	**diag**	getenv
~	**diary**	ginput
abs	**dir**	global
all	disp	**grid**
ans	**echo**	hess
any	**eig**	**hold**
acos	**else**	home
asin		

ident	min	script
i	nan	semilogx
ieee	nargin	semilogy
if	norm	setstr
imag	ones	shg
inf	pack	sign
input	pause	sin
inquire	pi	size
inv	plot	sort
isnan	polar	sprintf
isstr	polyline	sqrt
j	polymark	startup
keyboard	prod	string
length	prtsc	subplot
load	qr	sum
log	quit	svd
loglog	qz	tan
logop	rand	text
ltifr	rcond	title
ltitr	real	type
lu	relop	what
magic	rem	while
matlabpa	return	who
max	round	xlabel
memory	save	ylabel
mesh	sc2dc	zeros
meta	schur	

Special MATLAB functions applied as .m files:

corrcoef
cov
mean
std

Some functions I have written:

invsqrdi
istd
modstore

Bibliography

Aitchison, J. 1986. *The Statistical Analysis of Compositional Data*. Monographs on Statistics and Applied Probability. Chapman and Hall, London.

Aitken, T. W. 1951. *Determinants and Matrices*. University Mathematical Texts, Oliver and Boyd, Edinburgh.

Anderson, T. W. 1963. Asymptotic theory for principal components analysis. *Ann. Math. Statist.*, 34:122–48.

———— 1984. *An Introduction to Multivariate Statistical Analysis*, Second Edition (revised). Wiley, New York.

Andrews, D. F. 1972. Plots of high-dimensional data. *Biometrics*, 28:125–36.

Andrews, D. F., Gnanadesikan, R., and Warner, J. L. 1971. Transformation of multivariate data. *Biometrics*, 27:825–40.

Armands, G. 1972. Geochemical studies of uranium, molybdenum and vanadium in a Swedish alum shale. *Stockholm Contr. Geol.*, 27:1–148.

Atchley, W. R., Rutledge, J. J., and Cowley, D. E. 1981. Genetic components of size and shape. II. Multivariate covariance pattern in the rat and mouse skull. *Evolution*, 35:1037–55.

Autonne, L. 1913. Sur les matrices hypohermitiennes et sur les matrices unitaires. *Comptes Rendus Acad. Sci. Paris*, 156:858–60.

———— 1915. Sur les matrices hypohermitiennes et sur les matrices unitaires. *Ann. Univ. Lyon*, 38:1–77.

Barnett, V. 1976. The ordering of multivariate data. *J. Roy. Statist. Soc.*, 139–44.

Bartlett, M. S. 1953. Factor analysis in psychology as a statistician sees it. *Uppsala Symposium on Psychological Factor Analysis: Nord. Psykol. Monogr. Ser.* 3:23–4.

———— 1954. A note on the multiplying factor for various chi-square approximations. *J. Roy. Statist. Soc. Ser. B*, 16:296–8.

Bellman, R. 1960. *Introduction to Matrix Analysis*. McGraw-Hill, New York.

Bénzécri, J.-P. 1973. *L'Analyse des Données*, Vol. 2: *L'Analyse des Correspondances*. Dunod, Paris.

Birks, H. J. B. 1974. Numerical zonations of Flandrian pollen data. *New Phytologist*, 73:351–8.

Birks, H. J. B. and Gordon, A. D. 1985. *Numerical Methods in Quaternary Pollen Analysis*. Academic Press, London.

Bookstein, F. L. 1986. Size and shape spaces for landmark data in two dimensions. *Statistical Science*, 1:181–242.

1989a. "Size and shape": A comment on semantics. *Systematic Zoology*, 38:173–80.

1989b. Principal warps: Thin-Plane splines and the decomposition of deformations. *IEEE Trans. Pattern Analysis and Machine Intelligence*, 11:567–85.

1992. *Morphometric Tools for Landmark Data*. Cambridge University Press.

Bookstein, F. L., Chernoff, B., Elder, R., Humphries, J., Smith, G., and Strauss, R. 1985. *Morphometrics in Evolutionary Biology*, Special Publication No. 15, Academy of Natural Sciences, Philadelphia.

Butler, J. C. 1976. Principal component analysis using the hypothetical closed array. *Mathematical Geology*, 8:25–36.

Cameron, E. M. 1968. A geochemical profile of the Swan Hills Reef. *Canadian J. Earth Sciences*, 5:287–309.

Campbell, N. A. 1979. Canonical variate analysis: Some practical aspects. Unpublished doctoral thesis, Imperial College, University of London.

1980. Shrunken estimators in discriminant and canonical variate analysis. *Applied Statistics*, 29:5–14.

Campbell, N. A. and Reyment, R. A. 1980. Robust multivariate procedures applied to the interpretation of atypical individuals of a Cretaceous foraminifer. *Cretaceous Research*, 1:207–21.

Campbell, N. A., Mulcahy, M. J., and McArthur, W. M. 1970. Numerical classification of soil profiles on the basis of field morphological properties. *Australian J. Soil Research*, 8:43–58.

Carroll, J. B. 1961. The nature of the data – or, how to choose a correlation coefficient. *Psychometrika*, 26:347–72.

Cassie, R. M. 1962. Multivariate analysis in the interpretation of numerical plankton data. *N.Z. J. Science*, 6:36–59.

1967. Principal component analysis of the zoo-plankton of Lake Maggiore. *Mem. Ist. Ital. Idrobiol.*, 21:129–244.

Cassie, R. M. and Michael, A. D. 1968. Fauna and sediments of an intertidal mud flat: A multivariate analysis. *J. Experimental Marine Biol. Ecol.*, 2:1–23.

Cattell, R. B. 1952. *Factor Analysis*. Harper, New York.

Chang, W. C. 1983. On using principal components before separating a mixture of two multivariate normal distributions. *Applied Statistics*, 32:267–75.

Cheetham, A. H. 1971. Functional morphology and biofacies distribution of cheilostome Bryozoa in the Danian stage (Paleocene) of southern Scandinavia. *Smithsonian Contr. Paleobiol.*, 6:1–52.

Constantine, A. G. and Gower, J. C. 1978. Graphical representation of asymmetric matrices. *Applied Statistics*, 27:297–304.

Cooley, J. M. and Lohnes, P. R. 1971. *Multivariate Data Analysis*. Wiley, New York.

Darroch, J. N. and Mosimann, J. E. 1983. Canonical and principal components of shape. *Biometrika*, 72:242–52.

David, M., Campiglio, C., and Darling, R. 1974. Progress in *R*- and *Q*-mode analysis. Correspondence analysis and its application to the study of geological processes. *Canadian J. Earth Sciences*, 11:131–46.

Davis, J. C. 1970. Information contained in sediment-size analyses. *Mathematical Geology*, 2:105–12.

1986. *Statistics and Data Analysis in Geology*, Second Edition, Wiley, New York.

Davis, P. J. 1965. *The Mathematics of Matrices*. Blaisdell, New York.

Dempster, A. P. 1969. *Elements of Continuous Multivariate Analysis*. Addison-Wesley, Reading, Massachusetts.

Digby, P. G. N. and Kempton, R. A. 1987. *Multivariate Analysis of Ecological Communities*. Chapman and Hall, London.

Eastment, H. T. and Krzanowski, W. J. 1982. Cross-validatory choice of the number of components from a principal component analysis. *Technometrics*, 24:73–7.

Eckart, C. and Young, G. 1936. The approximation of one matrix by another of lower rank. *Psychometrika*, 1:211–18.

1939. A principal axis transformation for non-hermitian matrices. *Bull. Amer. Math. Soc.*, 45:118–21.

Efron, B. 1979. Bootstrap methods: Another look at the jacknife. *Ann. Statist.*, 7:1–26.

Erickson, D. B., Ewing, M., and Wollin, G. 1964. The Pleistocene Epoch in deep-sea sediments. *Science*, 146:723–32.

Everitt, B. 1978. *Graphical Techniques for Multivariate Data*. Heinemann, London.

Fenninger, A. 1970. Faktoranalyse nordalpiner Malmkalke. *Verh. Geol. Bundesanst.*, 4:618–36.

Fisher, R. A. 1940. The precision of discriminant functions. *Ann. Eugenics*, London, 10:422–9.

Flury, B. 1988. *Common Principal Components and Related Multivariate Models*. Wiley, New York.

Flury, B. and Riedwyl, H. 1988. *Multivariate Statistics*. Chapman and Hall, London.

Folk, R. L. and Ward, W. C. 1957. Brazos river bar: A study in the significance of grain-size parameters. *J. Sediment. Petrol.*, 27:3–26.

Fornell, C. 1982. *A Second Generation of Multivariate Analysis*, Vol. 1: *Methods*. Praeger, New York.

Friedman, G. M. 1961. Distinction between dune, beach and river sands from their textural characteristics. *J. Sediment. Petrol.*, 31:514–29.

Fries, J., and Matérn, B. 1966. On the use of multivariate methods for the construction of tree-taper curves. *Res. Notes, Dept. Forest Biometry, Skogshögskolan, Stockholm*, 9:85–117.

Gabriel, K. R. 1968. Simultaneous test procedures in multivariate analysis of variance. *Biometrika*, 55:489–504.

1971. The biplot graphic display of matrices with application to principal components analysis. *Biometrika*, 58:453–67.

Gnanadesikan, R. 1974. Graphical methods for informal interference in multivariate data analysis. *Proc. ISI Meetings Section*, 13:1–11.

1977. *Methods for Statistical Data Analysis of Multivariate Observations*. Wiley, New York.

Gordon, A. D. 1981. *Classification*. Monographs on Applied Probability and Statistics. Chapman and Hall, London.

Gould, S. J. 1967. Evolutionary patterns in Pelycosaurian reptiles: A factor analytic study. *Evolution*, 21:385–401.

Gower, J. C. 1966a. Some distance properties of latent root and vector methods used in multivariate analysis. *Biometrika*, 53:325–38.

1966b. A *Q*-technique for the calculation of canonical variates. *Biometrika*, 53:588–9.

1967. Multivariate analysis and multidimensional geometry. *The Statistician*, 17:13–28.

1968. Adding a point to vector diagrams in multivariate analysis. *Biometrika*, 55:582–5.

1971. A general coefficient of similarity and some of its properties. *Biometrics*, 27:857–74.

1977. The analysis of asymmetry and orthogonality. In *Recent Developments in Statistics*, J. Barra, ed., North Holland, Amsterdam.

Greenacre, M. J. 1984. *Theory and Applications of Correspondence Analysis*. Academic Press, London.

Greenacre, M. J. and Haste, T. 1987. The geometric interpretation of correspondence analysis. *J. Amer. Statist. Assoc.*, 82:437–47.

Guttman, L. 1953. Image theory for the structure of quantitative variates. *Psychometrika*, 18:277–96.

1954. Some necessary conditions for common factor analysis. *Psychometrika*, 19:149–61.

1956. "Best possible" systematic estimates of communalities. *Psychometrika*, 21:273–85.

Hampel, F. R., Ronchetti, E. W., Rousseeuw, P. J., and Statel, W. A. 1986. *Robust Statistics*, Wiley, New York.

Harman, H. H. 1960. *Modern Factor Analysis*. University of Chicago Press.

1967. *Modern Factor Analysis*, Second Edition, University of Chicago Press.

Harris, C. W. 1962. Some Rao–Guttman relationships. *Psychometrika*, 27:247–63.

Harris, R. J. 1975. *A Primer of Multivariate Statistics*. Academic Press, New York.

Hawkins, D. M. 1973. On the investigation of alternative regressions by principal component analysis. *Applied Statistics*, 22:275–86.

Healy, M. J. R. 1968. Multivariate normal plotting. *Applied Statistics*, 17:157–61.

Heijden, P. G. M. v. d., Falguerolles, A. de, and Leeuw, J. de. 1989. A combined approach to contingency table analysis using correspondence analysis and log-linear analysis. *Applied Statistics*, 38:249–92.

Hendrickson, A. E. and White, P. O. 1964. Promax: A quick method for rotation to oblique simple structure. *British J. Statist. Psychol.*, 17:65–70.

Hill, M. O. 1974. Correspondence analysis: A neglected multi-variate method. *J. Roy. Statist. Soc., Ser. C*, 23:340–54.

Hirschfeld, H. O. 1935. A connection between correlation and contingency. *Cambridge Philos. Soc.* (*Math. Proc.*), 31:520–4.

Hitchon, B., Billings, K. G., and Klovan, J. E. 1971. Geochemistry and origin of formation waters in the western Canada sedimentary basin. III. Factors controlling chemical composition. *Geochim. Cosmochim. Acta*, 35:567–98.

Hope, K. 1968. *Methods of Multivariate Analysis*. Unibooks, London.

Hopkins, J. W. 1966. Some considerations in multivariate allometry. *Biometrics*, 22:747–60.

Horst, P. 1963. *Matrix Algebra for Social Scientists*. Holt, Rinehart and Winston, New York.

1965. *Factor Analysis of Data Matrices*. Holt, Rinehart and Winston, New York.

Hotelling, H. 1933. Analysis of a complex of statistical variables into principal components. *J. Educ. Psychol.*, 24:417–41 and 498–520.

1957. The relations of the newer multivariate statistical methods to factor analysis. *British. J. Statist. Psychol.*, 10:69–79.

Hubberton, H. W., Morche, W., Westoll, F., Fütterer, D. K., and Keller, J. 1991. Geochemical investigations of volcanic ash layers from southern Atlantic legs 113 and 114. *Proc. Ocean Drilling Program* (*Scientific Results*), 114:733–49.

Hurley, J. R. and Cattell, R. B. 1962. The Procrustes program producing direct rotation to test a hypothesized factor structure. *Behavioral Science*, 7:258–62.

Imbrie, J. 1963. Factor and vector analysis programs for analyzing geologic data. Office of Naval Research Geographic Branch; Technical Report 6:1–83.

Imbrie, J. and Kipp, N. G. 1971. A new micropaleontological method for quantitative micropaleontology: Applications to a Late Pleistocene Caribbean core. In *The Late Cenozoic Glacial Ages*, K. Turekian, ed., pp. 71–181. Yale University Press, New Haven.

Imbrie, J. and Purdy, E. 1962. Classification of modern Bahamian carbonate sediments. *Mem. Amer. Assoc. Petrol. Geol.*, 7:253–72.

Imbrie, J. and Van Andel, T. H. 1964. Vector analysis of heavy mineral data. *Bull. Geol. Soc. Amer.*, 75:1131–56.

Ivimey-Cook, R. B. and Proctor, M. C. F. 1967. Factor analysis of data from an East Devon heath: A comparison of principal component and rotated solutions. *J. Ecology*, 55:405–13.

Jackson, J. E. 1991. *A User's Guide to Principal Components.* Wiley, New York.

Jeffers, J. N. R. 1962. Principal component analysis of designed experiment. *Statistician,* 12:230–42.

1967. Two case studies of the application of principal component analysis. *Applied Statistics,* 16:225–36.

Johnson, R. M. 1963. On a theorem stated by Eckart and Young. *Psychometrika,* 28:259–63.

Jolicoeur, P. 1963. The degree of robustness in *Martes americana. Growth,* 27:1–27.

Jolicoeur, P. and Mosimann, J. E. 1960. Size and shape variation in the painted turtle. *Growth,* 24:399–54.

Jolliffe, I. T. 1973. Discarding variables in a principal component analysis: Real data. *Applied Statistics,* 22:21–31.

1986. *Principal Component Analysis.* Springer-Verlag, New York.

Jones, K. G. 1964. *The Multivariate Statistical Analyzer.* Harvard University Bookstore.

Jongman, R. H. G., ter Braak, C. J. F., and Van Tongeren, O. F. R. (eds.) 1987. *Data Analysis in Community and Landscape Ecology.* Pudoc, Waageningen.

Jöreskog, K. G. 1963. *Statistical Estimation in Factor Analysis.* Almqvist & Wiksell, Uppsala.

1966. Testing a simple structure hypothesis in factor analysis. *Psychometrika,* 31:165–78.

1967. Some contributions to maximum-likelihood factor analysis. *Psychometrika,* 32:443–82.

1969. A general approach to confirmatory maximum likelihood factor analysis. *Psychometrika,* 34:183–202.

1971. Simultaneous factor analysis in several populations. *Psychometrika,* 57:409–26.

1973. Analysis of covariance structures. In *Multivariate Analysis – III,* P. R. Krishnaiah, ed., pp. 263–85. Academic Press, New York.

1976. Factor analysis by least squares and maximum likelihood. In Enslein, Ralston and Wilf (Editors). Statistical Methods for Digital Computers, Wiley, New York.

1981. Analysis of covariance structures. *Scandinavian J. Statist.,* 8:65–92.

Jöreskog, K. G. and Goldberger, A. S. 1972. Factor analysis by generalized least squares. *Psychometrika,* 37:243–60.

Jöreskog, K. G., Klovan, J. E., and Reyment, R. A. 1976. *Geological Factor Analysis,* First Edition. Elsevier, Amsterdam.

Jöreskog, K. G. and Lawley, D. N. 1968. New methods in maximum likelihood factor analysis. *British J. Math. Statist. Psychol.,* 21:85–96.

Jöreskog, K. G. and Sörbom, D. 1989. *LISREL 7 – A Guide to the Program and Applications,* Second Edition, SPSS Publications, Chicago.

Jöreskog, K. G. and Wold, H. 1982. *Systems under Indirect Observation. Part II.* North-Holland Publ., Amsterdam.

Kaiser, H. F. 1958. The varimax criterion for analytic rotation in factor analysis. *Psychometrika*, 23:187–200.

1959. Computer program for varimax rotation in factor analysis. *Educ. Psych. Meas.*, 19:413–420.

Kempthorne, O. 1957. *An Introduction to Genetic Statistics.* Wiley, New York.

Kendall, M. G. and Stuart, A. 1958. *The Advanced Theory of Statistics* Vol. 1: *Distribution Theory.* Griffin, London.

Klovan, J. E. 1966. The use of factor analysis in determining depositional environments from grain-size distributions. *J. Sediment. Petrol.*, 36:115–25.

1968. Selection of target areas by factor analysis. In *Proc. Symp. Decision-Making Exploration*, Vancouver (9 pp.).

Klovan, J. E. and Imbrie, J. 1971. An algorithm and FORTRAN IV program for large scale Q-mode factor analysis and calculation of factor scores. *Mathematical Geology*, 3:61–7.

Koch, G. S. and Link, R. F. 1971. *Statistical Analysis of Geological Data.* Wiley, New York.

Krumbein, W. C. and Aberdeen, E. 1937. The sediments of Barataria Bay. *J. Sediment. Petrol.*, 7:3–17.

Kruskal, J. B. 1964. Multidimensional scaling by optimizing goodness of fit to a non-metric hypothesis. *Psychometrika*, 29:1–27.

Krzanowski, W. J. 1983. Cross-validatory choice in principal component analysis: Some sampling results. *J. Statist. Comput.*, 18:294–314.

1984. Principal component analysis in the presence of group structure. *Applied Statistics*, 33:164–8.

1987a. Cross-validation in principal component analyses. *Biometrics*, 43:575–84.

1987b. Selection of variables to preserve multivariate data structure, using principal components. *Applied Statistics*, 36:22–33.

1988. *Principles of Multivariate Analysis.* Oxford Science Publications, Oxford University Press.

Kullback, S. 1959. *Information Theory and Statistics.* Wiley, New York.

Kvalheim, O. M. 1987. Methods for the interpretation of multivariate data: Examples from petroleum geochemistry. Thesis, Department of Chemistry, University of Bergen.

Lamont, B. B. and Grant, K. J. 1979. A comparison of twenty-one measures of site dissimilarity. In *Multivariate Methods in Ecological Work*, L. Orloci, C. R. Rao, and W. M. Stiteler, eds., pp. 101–26. International Coop. Publ. House, Fairland, Maryland.

Lawley, D. N. 1940. The estimation of factor loadings by the method of maximum likelihood. *Proc. Roy. Soc. Edinburgh, Ser. A*, 60:64–82.

1956. Tests of significance for the latent roots of covariance and correlation matrices. *Biometrika*, 43:128–36.

Lawley, D. N. and Maxwell, A. E. 1971. *Factor Analysis as a Statistical Method*, Second Edition. American Elsevier, New York.

Lebart, L. and Fénelon, J.-P. 1971. *Statistique et Information Appliquées*. Dunod, Paris.

Lefebvre, J. 1976. *Introduction aux Analyses Statistiques Multidimensionelles*. Masson et Cie, Paris.

Lefebvre, J., Desbois, N., Louis, J., and Laurent, P. 1973. Recherche de structures et essai de représentation par l'analyse des correspondances et l'analyse canonique de certaines fluctuations physicochimiques annuelles observées de 1967 à 1972 au Lac d'Annecy. *39éme Session de l'Institut International de Statistique*, Vienna, 1973 (16 pp. + annexe).

Li, C. C. 1955. *Population Genetics*. University of Chicago Press.

Mahé, J. 1974. L'analyse factorielle des correspondances et son usage en paléontologie et dans l'étude de l'évolution. *Bull. Soc. Géol. France, 7éme Série*, 16:336–40.

Mardia, K. V. 1970. Measures of multivariate skewness and kurtosis with applications. *Biometrika*, 57:519–30.

Maronna, R. A. 1976. Robust *M*-estimators of multivariate location and scatter. *Ann. Statist.*, 4:51–67.

Marquardt, D. W. and Snee, R. D. 1975. Ridge regression in practice. *Amer. Statist.*, 29:3–19.

Mather, P. M. 1972. Studies of factors influencing variations in size characteristics of fluvioglacial sediments. *Mathematical Geology*, 4:219–34.

1975. *Computational Methods of Multivariate Analysis in Physical Geography*. Wiley, London.

McCammon, R. B. 1968. Multiple component analysis and its application in classification of environments. *Amer. Assoc. Petrol. Geol.*, 52:2178–96.

1972. Map pattern reconstruction from sample data: Mississippi Delta region of south eastern Louisiana. *J. Sediment. Petrol.*, 42:422–4.

McElroy, M. N. and Kaesler, R. L. 1965. Application of factor analysis to the Upper Cambrian Reagan Sandstone of central and northwest Kansas. *Compass*, 42:188–201.

Melguen, M. 1971. Etude de sédiments pleistocène–holocène au nord-ouest du Golfe Persique. Thèse, Université de Rennes.

Middleton, G. V. 1964. Statistical studies on scapolites. *Canadian J. Earth Sci.*, 1:23–34.

Miesch, A. T. 1975. *Q*-mode factor analysis of compositional data. *Computers & Geosciences*, 1:147–59.

Miesch, A. T., Chao, E. C. T., and Cuttitta, F. 1966. Multivariate analysis of geochemical data on tektites. *J. Geol.* 74:673–91.

Miller, R. L. and Kahn, J. S. 1967. *Multivariate Statistical Methods in the Geological Sciences*. Wiley, New York.

Morrison, D. F. 1976. *Multivariate Statistical Methods*, Second Edition. McGraw-Hill, New York.

Mosier, C. 1939. Determining a simple structure when loadings for certain tests are known. *Psychometrika*, 4:149–62.

Mosimann, J. E. 1962. On a compound multinomial distribution, the multivariate β-distribution and correlations among proportions. *Biometrika*, 49:65–82.

1970. Size allometry: Size and shape variables with characterizations of the log-normal and generalized gamma distributions. *J. Amer. Statist. Assoc.*, 65:930–45.

Mulaik, S. A. 1972. *The Foundations of Factor Analysis*. McGraw-Hill, New York.

Murdoch, D. C. 1957. *Linear Algebra for Undergraduates*. Wiley, New York.

Noy-Meir, I. and Austin, M. P. 1970. Principal coordinate analysis and simulated vegetational data. *Ecology*, 51:551–2.

Orloci, L. 1967. Data centring – a review and evaluation with reference to component analysis. *Systematic Zoology*, 16:208–12.

Osborne, R. H. 1967. The American Upper Ordovician standard. *R*-mode factor analysis of Cincinnatian limestones. *J. Sediment. Petrol.*, 37:649–57.

1969. The American Upper Ordovician standard. Multivariate classification of typical Cincinnatian calcarenites. *J. Sediment. Petrol.*, 39:769–76.

Pearce, S. C. 1965. *Biological Statistics: An Introduction*. McGraw-Hill Series in Probability Statistics. McGraw-Hill, New York.

Pearce, S. C. and Holland, D. A. 1961. Analyse des composantes, outil en recherche biométrique. *Biométrie-Praximétrie*, 2:159–77.

Pearson, K. 1897. Mathematical contribution to the theory of evolution. On a form of spurious correlation which may arise when indices are used in the measurement of organs. *Proc. Roy. Soc., London*, 60:489–98.

1901. On lines and planes of closest fit to a system of points in space. *Philos. Mag., Ser. 6*, 557–72.

Phillips, B. F., Campbell, N. A., and Wilson, B. R. 1973. A multivariate study of geographic variation in the whelk *Dicathais*. *J. Exp. Mar. Biol. Ecol.*, 11:27–69.

Pielou, E. C. 1977. *Mathematical Ecology*, Wiley-Interscience, New York.

Preisendorfer, R. W. 1988. *Principal Component Analysis in Meteorology and Oceanography*. Elsevier, Amsterdam.

Press, W. H., Flannery, B. P., Teukolsky, S. A., and Vetterling, W. T. 1986. *Numerical Recipes*. Cambridge University Press.

Rao, C. R. 1952. *Advanced Statistical Methods in Biometric Research*. Wiley, New York.

1955. Estimation and tests of significance in factor analysis. *Psychometrika*, 20:93–111.

1964. The use and interpretation of principal components analysis in applied research. *Sankhyā*, 26:329–58.

Rayner, J. H. 1966. Classification of soils by numerical methods. *J. Soil Sci.*, 17:79–92.

Reyment, R. A. 1963. Multivariate analytical treatment of quantitative species associations: An example from palaeoecology. *J. Anim. Ecol.*, 32:535–47.

1971. Multivariate normality in morphometric analysis. *Mathematical Geology*, 3:357–76.

1989. Compositional data analysis. *Terra Nova*, 1:29–34.

1990. Reification of classical multivariate analysis in morphometry. In *Proceedings of the Michigan Morphometrics Workshop*, F. J. Rohlf and F. L. Bookstein, eds., pp. 122–44. Special Publication 2, University of Michigan Museum of Zoology.

1991. *Multidimensional Palaeobiology*. Pergamon Press, Oxford (with an appendix by L. F. Marcus).

Reyment, R. A. and Banfield, C. F. 1981. Analysis of asymmetric relationships in geological data. In *Modern Advances in Geomathematics*. R. Craig and M. Labovitz, eds., Pion Ltd., London.

Reyment, R. A. and Bookstein, F. L. 1993. Infraspecific variability in shape in *Neobuntonia airella*: An exposition of geometric morphometry. In *Proc. 11th Internat. Symp. on Ostracoda*, K. G. McKenzie, ed., Balkema, Rotterdam.

Reyment, R. A., Blackith, R. E., and Campbell, N. A. 1984. *Multivariate Morphometry*, Second Edition, Academic Press, London.

Richards, O. W. and Kavanagh, J. 1943. The analysis of the relative growths and changing form of growing organisms. *American Naturalist*, 77:385–99.

Rohlf, F. J. and Bookstein, F. L. (eds). 1990. *Proceedings of the Michigan Morphometrics Workshop*. Special Publication No. 2, University of Michigan Museum of Zoology.

Rohlf, F. J. and Slice, D. 1990. Extensions of the Procrustes method for the optimal superposition of landmarks. *Systematic Zoology*, 39:40–59.

Rouvier, R. R. 1966. L'analyse en composantes principales: Son utilisation en génétique et ses rapports avec l'analyse discriminatoire. *Biometrics*, 22:343–57.

Rummel, E. J. 1970. *Applied Factor Analysis*. Northwestern University Press, Evanston, Illinois.

Sampson, R. J. 1968. R-mode factor analysis program in FORTRAN II for the IBM 1620 computer. Geol. Surv. Kansas Computer Contrib. 20.

Saxena, S. K. 1969. Silicate solid solutions and geothermometry. 4. Statistical study of chemical data on garnets and clinopyroxenes. *Contr. Mineral. Petrol.*, 23:140–56.

1970. A statistical approach to the study of phase equilibria in multicomponent systems. *Lithos*, 3:25–36.

Saxena, S. K. and Ekström, T. K. 1970. Statistical chemistry of calcic amphiboles. *Contrib. Mineral. Petrol.*, 26:276–84.

Schöneman, P. H. and Carroll, R. M. 1970. Fitting one matrix to another under choice of a central dilation and a rigid motion. *Psychometrika*, 35:245–56.

Seber, G. A. F. 1984. *Multivariate Observations.* Wiley, New York.

Siegel, S. 1956. *Non-Parametric Statistics for the Behavioral Sciences.* McGraw-Hill, New York.

Siegel, A. F. and Benson, R. H. 1982. A robust comparison of biological shapes. *Biometrics*, 38:341–50.

Sigurdsson, H., Bonté, P., Turpin, L., Chaussidon, M., Metrich, N., Pradel, P., and D'Hondt, S. 1991. Geochemical constraints on source region of Cretaceous/Tertiary impact glasses. *Nature*, 353:839–42.

Smith, L. 1978. *Linear Algebra*, Springer-Verlag, New York.

Sneath, P. and Sokal, R. R. 1973. *Principals of Numerical Taxonomy.* Freeman and Cooper, New York.

Sokal, R. R. 1961. Distance as a measure of taxonomic similarity. *Systematic Zoology*, 10:70–9.

Solohub, J. T. and Klovan, J. E. 1970. Evaluation of grain-size parameters in lacustrine environments. *J. Sediment. Petrol.*, 40:81–101.

Sörbom, D. 1974. A general method for studying differences in factor means and factor structures between groups. *Brit. J. Math. Statist. Psych.*, 27:229–39.

Spurnell, D. J. 1963. Some metallurgical applications of principal components. *Applied Statistics*, 12:180–8.

Stone, M. and Brooks, R. J. 1990. Continuum regression-cross-validated sequentially constructed prediction embracing ordinary least squares, partial least squares and principal components regression. *J. Roy. Statist. Soc., Ser. B*, 52:237–69.

Sylvester, J. J. 1889. On the reduction of a bilinear quantic of the n-th order to the form of a sum of n products by a double orthogonal substitution. *Messenger of Mathematics*, 19:42–6.

Teil, H. and Cheminée, J. L. 1975. Application of correspondence factor analysis to the study of major and trace elements in the Erta Ale Chain (Afar, Ethiopia). *Mathematical Geology*, 7:13–30.

Teissier, G. 1938. Un essai d'analyse factorielle. Les variants sexuels de *Maia squinada. Biotypologie*, 7:73–96.

Telnaes, N., Björseth, A., Christy, A. A., and Kvalheim, O. M. 1987. Interpretation of multivariate data: Relationship between phenanthrenes in crude oils. *Chemometrics and Intelligent Laboratory Systems*, 2:149–53.

ter Braak, C. F. J. 1986. Canonical correspondence analysis: A new eigenvector technique for multivariate direct gradient analysis. *Ecology*, 67:1167–79.

1987. The analysis of vegetation–environment relationships by canonical correspondence analysis. *Vegetation*, 69:69–77.

Thompson, D. W. 1917. *On Growth and Form.* Cambridge University Press.

Thurstone, L. L. 1935. *The Vectors of the Mind*. University of Chicago Press.
    1947. *Multiple Factor Analysis*, University of Chicago Press.

Torgerson, W. S. 1958. *Theory and Methods of Scaling*. Wiley, New York.

Trowell, H. 1975. Diabetes mellitus and obesity. In *Refined Carbohydrate Foods and Disease*, D. P. Burkitt and H. C. Trowell, eds., Chapter 16 (pp. 227–49). Academic Press, London.

Tucker, L. R. 1966. Some mathematical notes on three-mode factor analysis. *Psychometrika* 31:279–311.

Wallin, M. 1991. Ecometric analysis of factors regulating eutrophication effects in coastal waters. Doctoral Dissertation, University of Uppsala, Faculty of Science, No. 353.

Wilkinson, C. 1970. Adding a point to a principal coordinates analysis. *Systematic Zoology*, 19:258–63.

Wilkinson, J. H. 1965. *The Algebraic Eigenvalue Problem*. Clarendon Press, Oxford.

Wold, S. 1976. Pattern recognition by means of disjoint principal component models. *Pattern Recognition*, 8:127–39.

Wright, S. W. 1932. The roles of mutation, inbreeding, crossbreeding and selection in evolution. In *Proceedings of Sixth International Congress on Genetics*, 1:356–66.

    1968. *Evolution and the Genetics of Populations*, Vol. 1, *Genetic and Biometric Foundations*. University of Chicago Press.

Zurmühl, R. 1964. *Matrizen und ihre technischen Anwendungen*. Springer-Verlag, Berlin.

Index

(Note abbreviations: PCA = principal component analysis; FA = factor analysis, CANOCO = canonical correspondence analysis.) Several entries for supplementary authors, not specifically referred to in the main text, are included for completeness.